Foundations of Engineering Mechanics

Series editors

V.I. Babitsky, Loughborough, Leicestershire, UK
Jens Wittenburg, Karlsruhe, Germany

More information about this series at http://www.springer.com/series/3582

Zinoviy Nazarchuk · Valentyn Skalskyi
Oleh Serhiyenko

Acoustic Emission

Methodology and Application

Springer

Zinoviy Nazarchuk
Karpenko Physico-Mechanical Institute
Lviv
Ukraine

Oleh Serhiyenko
Karpenko Physico-Mechanical Institute
Lviv
Ukraine

Valentyn Skalskyi
Karpenko Physico-Mechanical Institute
Lviv
Ukraine

ISSN 1612-1384 ISSN 1860-6237 (electronic)
Foundations of Engineering Mechanics
ISBN 978-3-319-84141-0 ISBN 978-3-319-49350-3 (eBook)
DOI 10.1007/978-3-319-49350-3

Printed on acid-free paper

This Springer imprint is published by Springer Nature
The registered company is Springer International Publishing AG
The registered company address is: Gewerbestrasse 11, 6330 Cham, Switzerland

Preface

A great amount of the industrial equipment in Ukraine long ago exhausted the lifespan for which it was designed. Its replacement today is either economically unprofitable or, for various other reasons, impossible. This causes us to use such objects of durable operation in close-to-critical conditions. To extend the safe operating life of the equipment, it is necessary to use new, modern methods and facilities of strength and durability diagnostics. Besides, in some fields of production, new materials and technologies are introduced, experience of their operation under conditions of intensive loading and the aggressive effect on the environment is insufficient and therefore these objects need to be monitored.

An analysis of the causes of failure of structural materials, which are widely applied in engineering, the power-generating industry, pipeline transport, aircraft construction, chemical and oil industry, etc., has showed that in most cases failure occurs due to the initiation and propagation of crack-like defects. Therefore, investigation of these processes becomes an extraordinarily important task. It is also conditioned by the fact that on the one hand, it is necessary to study the mechanisms of defect initiation, and propagation, and to develop on this basis the effective methods of improvement of the durability, reliability, and other service characteristics of materials—and on the other to investigate the phenomena that accompany deformation and fracture of materials and eventually provide bases of new physical methods of non-destructive testing.

The initiation and development of defects cause a number of accompanying physical phenomena. The most interesting are those related to the transfer of energy or matter; they are what allows us to effectively detect the defects and determine their parameters at certain (sometimes considerably large) distances from their locations. The radiation of heat, electrons, electromagnetic and elastic waves are among them.

A phenomenon of elastic wave radiation during the deformation or fracture of materials, phase transformations and precipitation of second-phase particles, magnetic or surface transformations in materials is called "acoustic emission." Experience over the last few decades testifies to the great potential possibilities for the acoustic emission method. It is especially appropriate in situations where it is impossible to visually check the cracks' initiation and propagation because of their tunneling in material, or when there is no access to the inspected object. The remoteness of the examination, high sensitivity, and possibility to detect the distant defects, which exceed their sizes by many orders of magnitude—independent of the shapes and sizes of the inspected object—the realtime recording of fracture development and so on, are the advantages that place the acoustic emission method in the lead among all the known prospective methods of non-destructive testing.

The successful application of the acoustic emission method for the observation of the state of materials and products is possible only if scientifically grounded methods and proper facilities for their realization are available. Today, despite a great number of publications and undoubted progress in the development of equipment, there are certain difficulties in correctly selecting and applying the methodological bases of this method.

In this monograph, the physical aspects of the problem are analyzed in detail; the methodological bases of the practical use of acoustic emission devices, known both in Ukraine and outside it, are described; the results of theoretical and experimental research of peculiarities of evaluation of the crack growth resistance of materials are discussed; and the selection of the useful AE signals, etc., are presented. The methods' efficiency is shown in the conditions of diagnostics of multi-purpose industrial objects. The results of experimental research have been obtained by the authors with the help of the new methods and facilities, and some of them are thanks to the support of the Science and Technology Center in Ukraine (grants Nos. 1689, 1628 and 3905).

The authors have made an attempt to synthesize and classify not only their own, but also other information culled from the literature, in order to create clear approaches to the practical use of acoustic emission. They hope that the results presented in this monograph will be of great interest to the specialists who deal with the problem of non-destructive testing and the technical diagnostics of products and structures operating in various fields of human activity.

Acknowledgements The authors are grateful to Orest Ya. Tsurkovskyi for translating and assisting in the preparation of this book.

Lviv, Ukraine

Zinoviy Nazarchuk
Valentyn Skalskyi
Oleh Serhiyenko

Contents

Abbreviations

ADC	Analog-to-digital converter
AE	Acoustic emission
AES	Acoustic emission signal
AET	Acoustic emission transducer
AFC	Amplitude-frequency characteristics
AG	Acoustic emission diagram
BM	Base metal
COD criterion	Crack opening displacement criterion
DAC	Digital-to-analog converter
DDAE	Directional diagram of acoustic emission
EAM	Electromagnetic-acoustic method
FZ	Fusion zone
GPUDBN	Generator of the pseudo-random uniformly distributed binary numbers
HPF	High-pass filter
IMS	Information-measuring systems
IO	Inspected object
KE	Kaiser effect
LCF	Low-cut filter
MC	Macro-crack
MN	Mechanical noise
NDT	Non-destructive testing
PD	Plastic deformation
PDS	Plain deformation state
PDV	Plastically deformed volume
PSS	Plain stress state
PZ	Plastic zone
SIF	Stress intensity factor
SSS	Stress-strain state
TD	Technical diagnostics

TH wave	Horizontally polarized wave
TV wave	Vertically polarized wave
WM	Welded metal

Nomenclature

α	Material constant, dimensionless multiplier, coefficient (factor), magnitude of a plane angle
β	Coefficient (factor), multiplier, magnitude of a plane angle
Γ	Integration contour
γ	Coefficient (factor), multiplier, density of surface energy, magnitude of a plane angle
Δ	Absolute value of the increment
Δl	Crack increment
δ_c	Crack opening displacement, relative elongation, relative error
ε	Conditional strain
$\dot{\varepsilon}$	Strain rate
ε_{ij}	Components of strain tensor
η	Amplification factor, variable of integration
θ	Angle of the polar coordinate system
Λ, λ, μ	Lame constants
λ	Wave length
v	Poisson's ratio, frequency
ρ	Material density, radius of a notch tip
σ	Conditional mechanical stress
σ_{ij}	Components of stress tensor
σ_r, σ_θ	Stresses in the polar coordinate system
τ	Tangential stresses, time
φ	Potential
Φ	Function
ϕ	Magnitude of a plane angle
χ	Potential
ω	Circular frequency
A_d	Amplitude of acoustic emission signal, multiplier
a	Coefficient (factor), geometrical size, material constant
a_j	Antisymmetric modes of the Lamb waves

a.u.	Arbitrary unit
B	Specimen thickness, multiplier, variable
b	Specimen height, coefficient (factor)
c	Velocity of elastic wave
D	External diameter
d	Internal diameter, distance
E	Young's modulus
F	Force, function of distribution, source activity
f	Frequency, deflection
G	Shear modulus, energy of acoustic emission signal
G_1	Elastic energy
$H(\ldots)$	Heaviside function
h	Geometrical size, notch length
i	Discrete variable
J	Energy flow, intensity of propagating wave
j	Discrete variable
K	Stress intensity factor, modulus of triaxial compression
k	Discrete variable
L	Distance
l	Length
$2l$	Crack length
M	Moment of force
m	Random value
N	Cumulative count of acoustic emission signals
$\dot{N}$	Rate of acoustic emission count
n	Constant
P	Loading force, work of elastic force, probability
p	Loading force, pressure
R	Radius, parameter, stress ratio, ultimate strength of concrete
$R(\ldots)$	Raleigh function
$r_{\rm p}$	Radius of plastic zone at the crack tip
r_0	Radius of a penny-shaped crack
S	Area of newly formed surface
T	Temperature
t	Time
U	Amplitude of electric signal, displacement of a surface half-space
$U_{\rm g}$	Threshold value
$u_{\rm r}$, $u_{\rm z}$	Components of displacement vector in cylindrical coordinate system
V	Volume
$V_{\rm p}$	Volume of the plastic deformed material
$v_{\rm t}$	Rate of a macro-crack tip propagation
W	Energy of elastic strain
Z	Mechanical impedance, number of tests

Chapter 1
The Generation of Elastic Acoustic Emission Waves Due to the Fracture of Solids

A real solid always contains defects that reveal themselves as sharp stress concentrators, i.e., cracks. In such cases, the approaches of classic continuum mechanics cannot be used in calculating the strength of materials. In order to solve the problem, new strength criteria have been formulated, taking into account the material characteristics that are invariant, both in the models of continuum mechanics and in those considering the structural peculiarities of the material. These criteria formed the basis of the theory of strength and fracture mechanics of materials.

It is known that under the effect of an external loading, stress concentration at the crack tip is formed and the multi-axial stress state arises. As a result, especially at low temperatures, structural materials can be the subject of brittle fracture. Critical values of force, strain, or energy criteria were introduced into fracture mechanics as the parameters that describe the effect of stresses and crack sizes on determining the characteristics of a material in order to estimate the state of the inspected object (IO). This enables one to determine the conditions of propagation of the available cracks that grow in two ways: a stable growth that takes place at increased stresses in the net-section of the specimen, and an unstable growth that occurs without such an increase. Thus, fracture mechanics should predict the crack propagation and determine the capability of machine elements to operate when such crack-like defects occur. The acoustic emission (AE) phenomenon is a modern, high-performance tool for the investigation of these processes as well as the basis for the creation of new methods of non-destructive testing (NDT).

1.1 Some Fracture Mechanics Criteria Under Quasi-Static Loading of Materials

Fracture is one of the types of strength breaking that can occur in a structure or in its element due to surplus (elastic or plastic) deformation, or due to the loss of their stability or division into separate parts. Fracture can be partial or complete. In the

Z. Nazarchuk et al., *Acoustic Emission*, Foundations of Engineering Mechanics,
DOI 10.1007/978-3-319-49350-3_1

first case, damages can appear (such as cracks or other structural defects) distributed over the body volume that cause the mechanical properties of a material to deteriorate. At a complete fracture, a solid is divided into pieces.

In fracture mechanics, a material is often considered as a homogeneous isotropic medium. Cracks in the initial state are assumed to be the ruptures of the material, the lateral surfaces (edges) of which are separated by an infinitely small distance with tips of an infinitely sharp geometry. The simplest concept, i.e., linear fracture mechanics, predicts the linear elastic properties of the material as well. This statement in the case of metals is satisfied only approximately, since during their loading, the plastic deformation always takes place at the crack tips, i.e., the plastic deformed zones appear [1–5]. The concept of elastic-plastic nonlinear fracture mechanics assumes not only the presence of large plastic deformed zones at the crack tip, but also a complete material yielding. Depending on the loading orientation with respect to the direction of crack propagation, the following three basic mechanisms of crack growth are distinguished in fracture mechanics (Fig. 1.1): Mode I, Mode II and Mode III cracks.

Fracture in a general case is divided into plastic, brittle, and fatigue. *Plastic fracture* is caused by a substantial plastic deformation (PD), which occurs over a whole (or almost whole) volume of a body. One of the types of plastic fracture is the specimen rupture after the 100% narrowing of the neck during tension that arises due to the loss of the material capability to resist the PD.

Brittle fracture appears due to the main crack propagation after macroscopically insignificant PD that concentrates in the crack subsurface region. For a perfect brittle fracture, PD is absent.

Under quasi-brittle (quasi-elastic) fracture, there is a plastic zone (PZ) ahead of the crack front and a plastically deformed (work-hardened) material near the crack surface. The remaining body volume (considerably greater) is in the elastic state. The term “quasi-brittle” also pertains to a fracture when the stresses in a net-section are higher than the yield strength, but lower than the ultimate strength value.

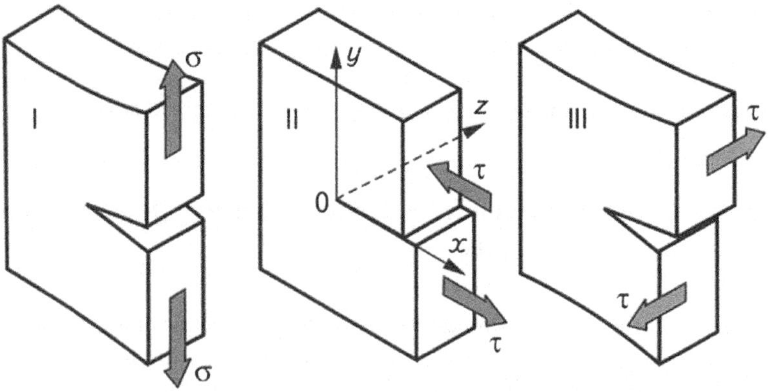

Fig. 1.1 Loading modes of a body with a crack

Fatigue fracture occurs under cyclic (repeated) loading of a body due to the accumulation of irreversible damages that cause crack initiation and propagation. The fatigue fracture surface of the material is macroscopically brittle, although the material is always cold-worked near the fracture surface.

Multi- and low-cycle fatigue is distinguished in the literature. The former, which is simply referred to as "fatigue," is characterized by nominal stresses lower than the yield strength; their repeated appearance in a deformed solid causes a macro-deformation in the elastic region. Under a multi-cycle fatigue, the number of loading cycles to a fracture is rather large.

Low-cycle fatigue (or repeated static loading) is characterized by nominal stresses that are higher than the yield strength, and during each loading cycle with such a level of loading, the PD rises in the body. Here, the number of loading cycles that fail is comparatively small.

The most important idea in the fracture mechanics of solids is that a fracture is interpreted as a process of crack initiation and propagation. Such an interpretation differs from the classical approaches because it foresees the mechanism of fracture itself, and a crack becomes the instrument of failure. Thus, the problem of a limiting-equilibrium state of the deformed solids with cracks, crack propagation under short-term or long-term action of a specified loading and service environment on the body, as well as evaluation of the characteristics of material resistance to a crack propagation (crack growth resistance characteristics), etc., turn out to be of prime importance.

At the beginning of the twentieth century, the criteria for and approaches to the assessment of the strength of materials and structural elements were formulated. They were based on the fact that the continuum with a set of rheological properties (e.g., elastic continuum) served as a calculation model of the real body, while the deformed solid element is in one of the following states (Fig. 1.2): continuous (C-state) or fractured (F-state) [4]. Transition of the element from the C-state to the F-state (fracture process) occurs instantly, only if the stress-strain state calculated within the framework of the accepted rheological model attains some critical value (e.g., tensile stresses at a given point of a deformable solid attain the ultimate strength σ_b).

Such a classical approach to the materials in the brittle state, whose structure contains sharp, crack-like defects, does not make it possible to solve the problem of their strength since it does not consider the special stress-strain state of the material at the tip of the sharp, crack-like defect during deformation of the body. This is conditioned by the fact that the rounded radius of the stress concentrator is commeasurable with the structural parameters of the material.

The basic idea of a non-classical approach (fracture mechanics of materials) is as follows: It is assumed (Fig. 1.2) that transition of the deformed body element from the C-state to the F-state takes place through an intermediate state (I-state), which is to be taken into account when solving the problem of the strength of a body containing crack-like defects. The basic feature of the deformed solid regions, where the F-states (process zones) arise, is that their material is always deformed beyond the elastic limit and just there the plastic yielding, interaction with service

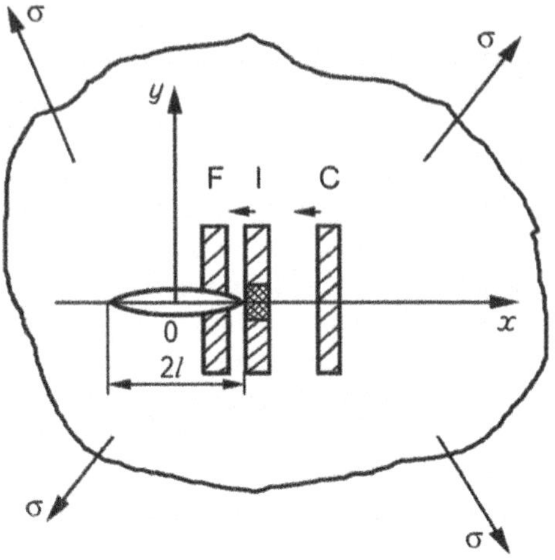

Fig. 1.2 Non-classical fracture (schematically)

environment, diffusion processes, material damaging, and other phenomena are most intensive, eventually determining the local fracture of the material, i.e., the C → I → F transition.

Thus, the non-classical fracture scheme foresees the account of F-states at the sharp defects in the deformed body and, first of all, the crack-like defects—stress concentrators whose curvature radius is commeasurable with the typical size of the material element structure. Thus, when assessing the strength of a solid, it is necessary to take into account its local physico-mechanical properties—for example, its ability to resist crack propagation, i.e., its crack growth resistance. For the account of the material F-states within the framework of continuum mechanics, it is necessary to introduce new (in comparison with classic ones) calculation models, concepts, and criteria. Since stresses, deformations and energy are the main characteristics of material behavior control at the crack tip, all criteria of fracture mechanics, similar to the classic strength theories, are subdivided into *energy*, *force* and *deformation* criteria.

1.1.1 Energy Criteria

Griffith started the development of brittle fracture criteria by introducing the energy criterion [6]. This criterion uses the principle according to which a crack initiates in a brittle body only on the condition that the rate of elastic energy release during crack propagation exceeds the increment of the crack surface energy.

$$\frac{\partial}{\partial l}[U(p^*, l_1) - W(p^*, l_1)] = 0, \tag{1.1}$$

where $U(l_1)$ is the surface energy of a crack; $W(p^*, l_1)$ is the elastic strain energy, conditioned by the opening of a crack of length $2l_0$ under the effect of an externally applied load p; p^* is the maximum value of p.

If, according to Griffith, the surface energy density γ is assumed to be the material constant, then using the energy balance (1.1), we get the critical value of stresses σ_c, at which unstable growth of the crack of length $2l_0$ occurs:

$$\sigma_c = \sqrt{2\gamma E/\pi l_0} \tag{1.2}$$

for the plane stress state (PSS) and

$$\sigma_c = \sqrt{2\gamma E/\pi l_0(1-v^2)} \tag{1.3}$$

for the plane deformation state (PDS); where E is the Young's modus; v is the Poisson's ratio.

Irwin [7], Felbek and Orowan [8] modernized the Griffith approach by replacing the surface energy density γ by a specific study of PD γ_{ef}, which are concentrated in the small zone of a material near the crack tip. Thus, the local PD at the crack tip are accounted for by the introduction of the effective surface energy for the formation of a new surface area unit.

In the papers by Morozov [9, 10], in the case of a perfectly elastic body, the energy criterion of limiting equilibrium of the bodies with cracks takes into account the law of energy conservation during a real or virtual crack increment:

$$\delta A = \delta W + \delta \Gamma, \tag{1.4}$$

where δA is the mechanical work of external forces; δW is the volume potential energy of elastic deformation of the body; and $\delta \Gamma$ is the fracture energy.

Cherepanov [11], using the idea of a "fine structure" of the crack tip, proposed a generalized energy approach to a description of the crack propagation in materials with arbitrary rheological properties:

$$R\int_0^{2\pi} [(Q+K-B)\cos\theta - P]\mathrm{d}\theta = 2\gamma. \tag{1.5}$$

Here R is the radius of a circle with a center at the crack tip, the size of which is small in comparison with typical linear sizes of the crack and the body; Q is the intrinsic work of internal forces; K is the kinetic energy; B is the work of volume forces; P is the work of surface forces (Q, K, B, and P are calculated directly from the singular solution of the problem); and θ is the angle of the polar coordinate system with a center at the crack tip.

For large plastic zones near the crack tip, the Griffith-Orowan criterion is incorrect. In this case, the criterion of *R-curves*, introduced by Irwin and Kies, is used [3] (Fig. 1.3). It is based on the balance of the rate of release of elastic strain

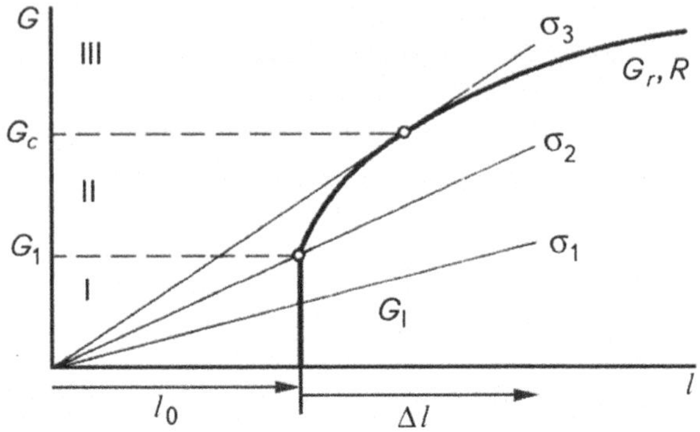

Fig. 1.3 Schematics of *R*-curve: G_{I} is the beginning of a stable crack growth and G_C is the unstable one

energy and the energy of crack propagation. Material resistance to the crack propagation in this case is determined by a specific work G_R during crack growth due to a release of the elastic strain energy, which is known in the literature as the intensity of elastic energy release G_1. Then, the criterion can be written as follows:

$$G_1 = R_c \tag{1.6}$$

where R_c is the specific material resistance to the crack length increase when a spontaneous fracture begins.

The drawback of *R-curves* is that the *R* parameter depends not only on the crack increment but also on the loading conditions and specimen geometry. For an infinite plate with an internal crack of a length of $2l_0$, the following equality is satisfied:

$$G_1 = \sigma^2 \pi l_0 / E'. \tag{1.7}$$

where $E' = E$ for PSS, and $E' = E/(1 - \nu^2)$ for PDS.

The limiting-equilibrium state of plastic materials can be also described by the *J*-integral, which is based on the Rice-Cherpanov energy approach [11, 12]. The idea of the approach is as follows: A large PD near the crack tip significantly affects the intensity of the elastic energy G_1 release. To evaluate this effect, it is essential to find the exact solution of the elasto-plastic problem on distribution of stresses at the crack tip. However, this proved to be a very complicated problem, and the indirect method incorporating the following expression is proposed for this purpose:

$$J = \int_{\Gamma} \left(W dy - T_i \frac{\mathrm{d}u}{\mathrm{d}x} \mathrm{d}s \right). \tag{1.8}$$

where Γ is the closed contour that bounds some region around the crack in the stress field; $W = W(x, y) = W(\varepsilon) = \int_0^{\varepsilon} \sigma_{ij} d\varepsilon_{ij}$ is the strain energy of a unit volume; $T_i = \sigma_{ij} n_j$ is the vector of stresses perpendicular to the contour Γ and directed outwards; u

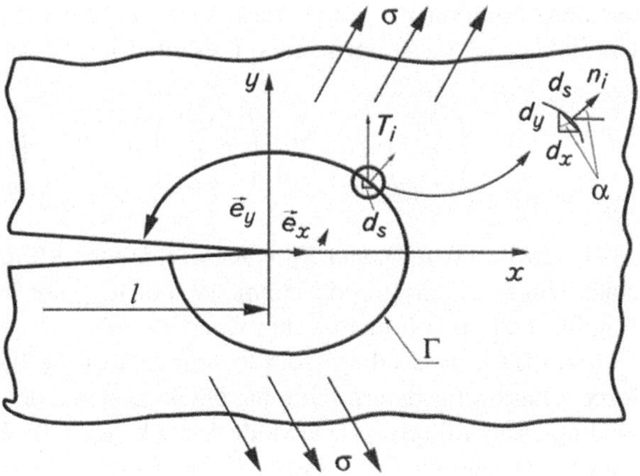

Fig. 1.4 A crack and an integration contour (schematically)

is the displacement in the direction of Ox axis; and dS is the element of contour Γ (Fig. 1.4). A figure in [12, 13] illustrates that for the elastic case, the following equality is valid:

$$G_1 = J. \tag{1.9}$$

This means that the intensity of elastic energy release can also be evaluated by the J-integral. Thus, it is reasonable to assume that at critical values, the equality is not violated. With the development of investigations of the energy criteria based on the balance of the supplied and released energy, the obtained results were systematized in [14, 15].

Local energy criteria that are based on the calculation of the strain energy density for an imaginary cylindrical specimen in the prefracture area with the account of the stress intensity factor (SIF) were proposed in paper [16]. It is assumed that fracture begins when the critical density of the strain energy is attained at a certain critical distance from the crack tip. At present, the application of the energy local and global fracture criteria to the engineering practice appears to be complicated due to the mathematical problems.

1.1.2 Force Criteria

Under the action of the field of tensile stresses applied normally to the crack plane, depending on the specimen thickness near the crack contour, the stress state with various degrees of multi-axiality is formed. For thick specimens, the PDS arises, at which there is no lateral deformation over the specimen thickness, i.e., $\varepsilon_z = 0$, while in thin specimens, the PSS appears when stresses disappear over the specimen thickness, and condition $\varepsilon_z = 0$ is valid. In the coordinate system $Oxyz$ (Fig. 1.5) for the stress components at the crack tip $0 < r \leq l$, we can get approximated equations with taking into account results, presented in [17–19]:

$$\begin{pmatrix} \sigma_x \\ \sigma_y \\ \tau_{xy} \end{pmatrix} = \frac{K_{\mathrm{I}}}{\sqrt{2\pi r}} \begin{pmatrix} f_1(\theta) \\ f_3(\theta) \\ f_5(\theta) \end{pmatrix} + \frac{K_{\mathrm{II}}}{\sqrt{2\pi r}} \begin{pmatrix} f_2(\theta) \\ f_4(\theta) \\ f_6(\theta) \end{pmatrix} + O(1). \tag{1.10}$$

Here, $K_{\mathrm{I}} = K_{\mathrm{I}}(p, l)$, and $K_{\mathrm{II}} = K_{\mathrm{II}}(p, l)$ are the SIF that are the functions of load p, crack length $2l$ and a body geometry, but they are independent of the coordinates r and θ; $O(1)$ is a limited value when $r \to 0$.

Irwin [20], proceeding from the concept of the limiting SIF values, proposed a force criterion for determining the crack start in a deformed solid under quasi-static loading. Accordingly [4], a Mode I crack ($K_{\mathrm{I}} \neq 0$; $K_{\mathrm{II}} = K_{\mathrm{III}} = 0$) begins to grow under a condition: $K_{\mathrm{I}}{}^* = K_{\mathrm{I}}(p^*, l)$, i.e. K_{I} (p, l) for $p = p^*$ at the given point of a contour becomes equal to a certain (constant for this material) value of K_{IC} (or K_C for thin plates)—the experimentally evaluated characteristic of static crack growth resistance of the material [21]. This criterion is written as follows:

$$K_{\mathrm{I}}^* = K_{\mathrm{I}}(p^*, l_0) = K_{\mathrm{I}c}. \tag{1.11}$$

For an infinite plate under stress σ, and which contains a crack of the length $2l$:

$$K_{\mathrm{I}} = \sigma\sqrt{\pi l} \tag{1.12}$$

or for a specimen of the thickness t and width b, Eq. (1.12) is written as follows:

$$K_{\mathrm{I}} = \sigma\sqrt{\pi l} \cdot Y(l/b). \tag{1.13}$$

For brittle fracture, as Irwin showed, equalities (1.1) and (1.11) are equivalent, as well as expressions (1.5) and (1.11).

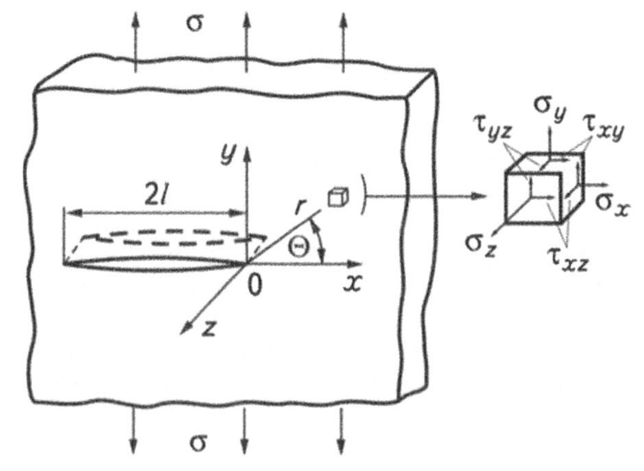

Fig. 1.5 A plate with an internal crack and a system of coordinates

Thus, the Irwin force criterion (1.11) is valid for quasi-brittle bodies with cracks when characteristic size of the prefracture zone is considerably less than that of a macro-crack (MC) (self-similarity condition) and enables one to simplify the complicated calculations of the elastic energy release, true surface energy, and work of local plastic deformations that are used in other approaches and criteria. Based on the obtained dependences, Irwin considered a relation between the force fracture criteria and the stress state at the crack tip, and showed the equivalence of the energy and force approaches:

$$K_{\mathrm{I}c} = \sqrt{\frac{2E\gamma}{1-\nu^2}}. \tag{1.14}$$

Mathematical relationships of self-similarity conditions for this criterion, together with the fracture criterion, form a closed calculation model for evaluating the limiting equilibrium state of the cracked body.

Similar to the Irwin criterion, the condition of Mode II crack start ($K_{\mathrm{I}} = K_{\mathrm{III}} = 0$) can be represented as:

$$K_{\mathrm{II}_*} = K_{\mathrm{II}c}; \quad (K_{\mathrm{II}_*} = K_{\mathrm{II}}(\sigma_*, l)), \tag{1.15}$$

and for Mode III crack ($K_{\mathrm{I}} = K_{\mathrm{II}} = 0$), it is written as follows:

$$K_{\mathrm{III}_*} = K_{\mathrm{III}c}; \quad (K_{\mathrm{III}_*} = K_{\mathrm{III}}(\sigma_*, l)). \tag{1.16}$$

A principal difference of the values in the left-hand and right-hand parts of criterion relationships (1.11), (1.15) and (1.16) is that a principal difference of the values in the left-hand and the right-hand parts of criterion relationships (1.11), (1.15) and (1.16) is that K_I^*, K_{II}^*, K_{III}^* represent the geometrical shape of the cracked body and its loading conditions, while $K_{\mathrm{I}c}$, $K_{\mathrm{II}c}$, $K_{\mathrm{III}c}$ are characteristics of the material which, similarly to the yield strength and ultimate strength, residual elongation, etc., represents a certain property of the material-resistance to crack propagation, namely crack growth resistance.

For a micro-heterogeneous body, in which various fracture mechanisms are simultaneously realized for their particular elements, functional dependence that describes some bounding surface in coordinates (K_{I}, K_{II}, K_{III}) for this material, at which a crack passes from the stable to the unstable state, begins to grow as follows:

$$F(K_{\mathrm{I}}, K_{\mathrm{II}}, K_{\mathrm{III}}, C_i) = 0 \quad (i = 1, 2, 3, \ldots). \tag{1.17}$$

After this, it will take another form:

$$\left(\frac{K_{\mathrm{I}}}{K_{\mathrm{Ic}}}\right)^{m_1} + \left(\frac{K_{\mathrm{II}}}{K_{\mathrm{IIc}}}\right)^{m_2} + \left(\frac{K_{\mathrm{III}}}{K_{\mathrm{IIIc}}}\right)^{m_3} = 1. \tag{1.18}$$

Here, m_1, m_2, m_3 are material constants. Criterion relationship (1.18) in some cases well describes the fracture of a wide range of structural materials.

When the criteria (1.11), (1.15), (1.16) and (1.18) are satisfied, it means that the limiting-equilibrium state of the body with a crack has been attained. However, the crack propagation in this state can be both stable and unstable. In the first case, the crack is unmoving under the constant load and, for the increment of its length or area by an insignificant value, a small increase of the external load is required. It means that if, from the fracture criterion, the relationship between external loading σ and crack length l is found, then for a stable crack, the following inequalities

$$\frac{\mathrm{d}\sigma}{\mathrm{d}l} > 0 \text{ or } \frac{\mathrm{d}K_i}{\mathrm{d}l} < 0 \quad (i = \mathrm{I}, \mathrm{II}, \mathrm{III}) \tag{1.19}$$

are satisfied.

If the K_i factor increases with the growth of loading and crack length, then from the condition of the K_i constancy in the limiting state, the second inequality in (1.19) is satisfied.

In the unstable state, an equilibrium crack starts to propagate when the critical load value, found from the criterion of equilibrium, is attained. In the overcritical region, it can grow under a constant load. The region of unstable equilibrium is characterized by the following inequalities:

$$\frac{\mathrm{d}\sigma}{\mathrm{d}l} < 0 \text{ or } \frac{\mathrm{d}K_i}{\mathrm{d}l} > 0. \tag{1.20}$$

To prevent a complete failure, it is important to know which type of the limiting equilibrium it is attributed to. If it is a stable one, there is no danger of an instant fracture of a component or a structural element, but if the limiting equilibrium is unstable, there should be no crack.

Since the limiting equilibrium state of brittle or quasi-brittle bodies (the bodies that are in brittle or quasi-brittle states, respectively) with cracks is likewise correctly determined both by force and energy fracture criteria, then there should be an unambiguous relationship between their basic parameters (crack growth resistance characteristics). To find this relationship, the elastic energy in a cracked body, e.g., for Mode I crack, should be determined by the K_{I} factor and substituted in the corresponding inequalities under condition $K_{\mathrm{I}} \rightarrow K_{\mathrm{Ic}}$. As a result, we obtain for PDS

$$K_{\mathrm{Ic}}^2 = \frac{E g_{\mathrm{Ic}}}{1 - \nu^2} \tag{1.21}$$

and for PSS

$$K_c^2 = Eg_{\mathrm{Ic}}. \tag{1.22}$$

Here, $g_{IC} = 2\gamma$ for brittle bodies and $g_{IC} = J_{IC}$ for elasto-plastic bodies.

Dependences (1.21) and (1.22) should be used when energy and force criteria are jointly applied to the assessment of the strength of elasto-plastic bodies with cracks by one criterion, while the material crack growth resistance characteristics are assessed by another criterion that is easier to realize from the technical point of view.

In paper [4], the force criteria of strength, accounting for the critical stresses in deformed solids, are synthesized.

1.1.3 Deformation Criteria

When evaluating the crack growth resistance of materials with a large PD at the crack tip (for example, medium- and low-strength steels, etc.) the self-similarity conditions are not satisfied and the Irwin force criterion in this case is incorrect, since it does not take into account the physical and geometrical non-linearity of strain. In this case, the deformation criteria for cracked bodies, accounting the local and global deformations, are used [22, 23].

Considering that the crack propagation begins in an elasto-plastic body at critical stress σ^* (Fig. 1.5), then according to classical deformation strength criterion, we can write

$$\varepsilon_{\max}(l, \sigma^*) = \varepsilon_c, \tag{1.23}$$

where $\varepsilon_{\max}$ is the maximum tensile strain in the process zone, and ε_c is the threshold tensile strain of the material.

Evaluation of the value of $\varepsilon_{\max}$ is very complicated for various reasons. To simplify, we can select an elementary volume of the height of h at the crack contour. Its elongation during material deformation is equal to the crack tip opening displacement (Fig. 1.6). According to [4]

$$\varepsilon_{\max}(l, \sigma) = \delta_{\mathrm{I}}(l, \sigma)/h. \tag{1.24}$$

Assuming that $\delta_c = h\varepsilon_c$, we get

$$\delta_{\mathrm{I}}(l, \sigma^*) = \delta_{\mathrm{Ic}}. \tag{1.25}$$

This critical crack opening displacement criterion (COD criterion) describes the limiting-equilibrium state of the elasto-plastic body with a crack as the moment when the crack opening displacement $\delta_{\mathrm{I}}(l, \sigma)$ attains the critical value δ_{IC} [22]. Consequently, the crack begins to grow, regardless of the specimen geometry and

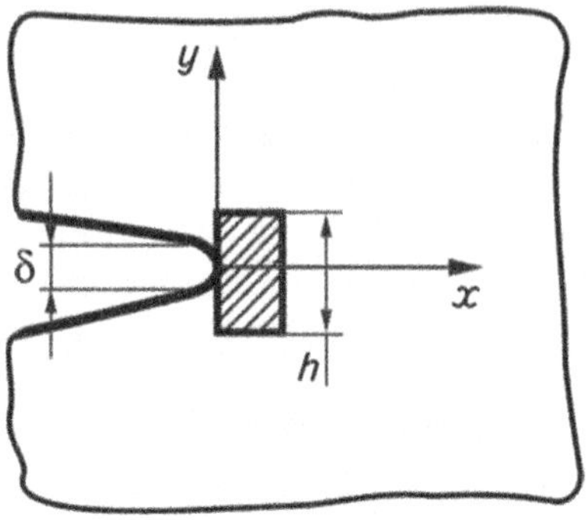

Fig. 1.6 Crack tip calculation (schematically)

the plastic zone size. This criterion forms the basis of models, where cohesive forces in the crack tip are considered [24–32].

Today, the calculation δ_c-model of Leonov-Panasyuk [24–28] is the most widely used. In this model, a crack is considered to be a cut in a linear elastoplastic body, the opposite sides of which are attracted by stresses $\sigma_0 = \text{const}$ for $\delta < \delta_c$, and $\sigma_0 = 0$ for $\delta > \delta_c$, in the region of length d (Fig. 1.7). We can see that the concrete value of stresses causing failure are introduced, and the ultimate plasticity of a material is taken into account, the latter being expressed in terms of the flaw faces displacement δ_c. Using the law of distribution of the intensity of cohesive forces between the neighboring atoms, the values of σ_0 and δ_c are connected by the dependence

$$\sigma_0 \delta_c = 2\gamma, \tag{1.26}$$

where $\sigma_0 = \sigma_{0.2}$ for a perfectly elasto-plastic body.

According to the δ_c–model, a crack in the deformed solid starts to grow under the condition

$$2v_n(l_0, l, p^*, \sigma_0) = \delta_c, \tag{1.27}$$

where $v_n(l_0, l, p^*, \sigma_0)$ is a normal component of the vector of crack faces displacements; l, l_0 are typical final and initial crack sizes, respectively; p^* is the critical external load. For macroscopic cracks, the δ_c–model yields the same results

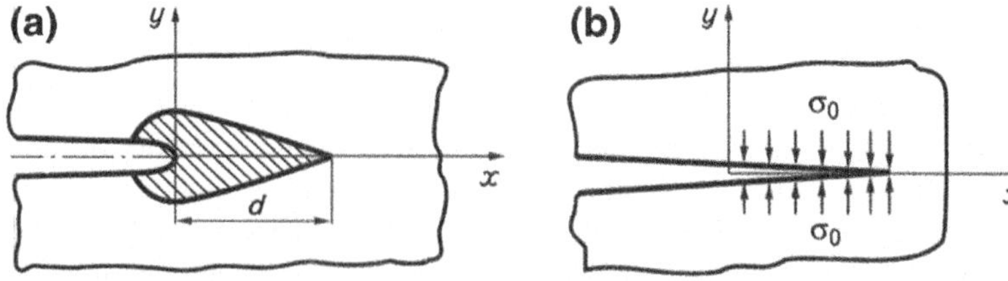

Fig. 1.7 Plastic zone on the crack continuation (**a**) and its calculation model (**b**)

as the Griffith-Irwin model. The main disadvantage of the δ_c–model is that it is obtained for a perfectly plastic material and does not consider the hardening of the material.

Some time later, a similar model was proposed by Dugdale [29]. The length of the plastic zone r_p in a cracked infinite plate is proposed to be evaluated from the equation

$$r_p = l - l_0 = l_0 \left[\sec\left(\frac{\pi\sigma}{2\sigma_y}\right) - 1 \right], \tag{1.28}$$

where σ, σ_y are the applied stress and the yield stress, respectively.

Except for those mentioned above, the local criteria proposed for the first time by Orowan and later developed by McClintock [31] and other researchers [32] are successfully used today. Contrary to the global criteria, they take into account the complicated effect of the geometry of a structural element and its loading mode. The distribution of stresses and strains in these models is determined by calculations using the finite elements method. These models, despite the complicated and expensive calculations, are more and more widely used in engineering.

Thus, in the fracture mechanics of solids, the most important problem is the interpretation of the crack initiation and propagation processes. In this approach, the mechanism of fracture is included, and a crack becomes the tool by which it is carried out. It becomes clear that it is very important to study the problem of a limiting equilibrium state of deformed solids containing crack-like defects. In this connection, it is significant to be able to perform not only a quantitative evaluation of the pre-fractured state of structural materials using fracture criteria, but also to effectively determine the processes taking place in the plastic zone in front of the crack tip. The solution to this problem certainly presupposes the development and use of new, modern, non-destructive methods of physical research, among which the AE-based methods occupy the leading place.

1.2 Micro-Cracking of Solids

All known mechanisms of micro-cracking due to PD are mainly reduced to two groups: cracking as the result of the accumulation of dislocations, and the crossing of twins. Dislocation mechanisms at the initial stage presuppose a blocking of the dislocation movement by barriers: grain boundaries or inclusions. Consequently, if dislocations in a certain slip plane stop in front of a strong enough barrier, they form a group that causes a high stress concentration at the barrier. Mutual repulsion of dislocations of the same sign tends to increase the distance between them, and the tangential stress τ acting in the slip plane presses one dislocation to the other and all of them together to the barrier. If a group contains n dislocations, the force $F \sim n\tau$ acts on the barrier. According to the Zener-Stroh model [33–35], when the critical number of dislocations is attained, head dislocations can approach each other so closely that they join together and form a crack. Stroh has also shown [34] that a group of n dislocations produces a normal stress concentration at the head of the group in the planes located above and below the slip planes, this stress concentration being proportional to the number of dislocations. The critical number of dislocations in the head of the group, which is capable of initiating a crack, is about 300 for metals. Such a model of micro-crack initiation excludes, of course, the decrease of stress concentration due to slipping or twinning in the vicinity of the dislocation group head, and crack formation is the only possible way for energy redistribution.

Cottrell has proposed another mechanism of crack initiation due to the formation of a dislocation group in the *bcc* lattice. He stated [36] that the two dislocations moving in the mutually intersected slip planes should merge to form a sessile dislocation, which becomes a barrier for the motion of other dislocations and causes the nucleus micro-crack initiation. This model is doubted by some researchers [1, 37].

In a number of highly symmetric structures subjected to a significant loading, the deformation twins can be formed. If the boundaries of such twins intersect, high local stress concentration and formation of micro-cracks is possible. This phenomenon was observed in monocrystals of silicon iron [38]. Although there was a great number of twin intersections, cracks were formed only in some of them.

According to the above described mechanisms of micro-cracking, it has been considered to be experimentally proved and theoretically grounded that the stage of a single micro-crack initiation ends when its length is $l_0 \sim 10^{-7}$ m, and then starts the stage of its sub-critical growth [37].

Using the methods of microfractography analysis, it was found that depending on the loading conditions and structural state of the material, crack propagation can occur by the following fracture micromechanisms: transcrystalline cleavage, quasi-cleavage, intergranular cleavage, pitting tearing, laminations along the slip planes, and the formation of grooves during fatigue crack propagation. Each of them leaves its own traces on the fracture surfaces [1, 2, 4, 39, 40].

1.3 Physical Grounds of AE Generation

When investigating the strength of the deformed cracked bodies, it is very important to evaluate the stresses and strains that arise near the defect. The region close to the crack contour, the PD zone, is of utmost interest since in this region alone there is a maximum concentration of stresses and the material failure is almost surely right there [3–9]. The internal mechanisms causing PD of crystals and the change in their internal microstructure have been studied deeply enough [2]; however, nowadays such methods and means do not correspond to the demands because they cannot be used directly for the IO. Therefore, the necessity arose to develop new methodological approaches that make it possible to quantitatively assess the fracture processes exactly in the PD zones using non-destructive testing methods.

The AE method has lately been more and more accepted [41]. This is first of all related to enormous developments in modern electronics, which permit creating highly sensitive and compact facilities with a wide range of functional possibilities, as well as information-measuring systems (IMS) and complexes. There are two well-expressed tendencies of applying the AE method. The first is for NDT and for diagnosing products and elements of structures, and the second, as a high sensitive physical method, for investigating the dynamics of deformation and fracture development in a material.

Over the last few decades, fracture mechanics and physics of strength and plasticity, as fundamental sciences in the field of solid state investigation, have met with great success. Today it is not necessary to prove that the process of PD in a crystal is a macroscopic manifestation of dislocation generation and motion under the applied loading. Still, the problem of the PD stages and micro-crack initiation in a structural material that is characterized by very short duration of their elementary acts during fracture has not been fully studied. Cracks during their initiation and propagation are the sources of the AE signals (AES) generation; this is why the application of the methods based on the AE phenomenon is considered to be very promising for theoretical and experimental research on the processes mentioned.

The data on the application of the AE phenomenon have been published [42–47]. It is seen that the modern concepts of PD processes in metals are based on the idea of mass formation and motion of defects in a crystal lattice under loading. Thus, dislocations make the main contribution to these processes. However, the path from scientific research to the verification of the final version of a modern hypothesis was not easy, and its application greatly contributed to the investigation of these processes, which has been proven by the great number of publications in the literature throughout the world.

Slipping and twinning are considered to be the two determining PD mechanisms. Both are caused by dislocation initiation and propagation and are accompanied by the elastic waves generation, namely by AE [48]. In this study, it is also stressed in general features that the deformation conditions, type, and state of the material, type of crystal lattice, and a number of other factors, significantly affect the AE parameters. Developing studies in this direction, the authors [49] show that the accelerated motion of dislocations and their abrupt arrest is a source of the generation of elastic AE waves. In the last case, additional frictional stresses arise, of which, for example, the Cu-10% Zn crystal are approximately 5 MPa. Actually, this is almost the lower yield stress of the alloy, and the mechanism of AE irradiation arrest is caused by a higher stress than the frictional shear stress. The authors believe that these mechanisms are typical of the PD processes and that in a crystal, which contains many local dislocation fields, the dislocation movement changes from quasi-static to a dynamic one with an increasing average velocity.

Thus, the dislocation mechanisms of the AES generation in monocrystals of fluorine lithium, deformed by the loading applied to the slip plane, were studied in [50]. Displacement of 25×10^{-8} m was accompanied by the AE pulses of rather large amplitudes, while smaller ones of 0.5×10^{-8} m resulted in a synchronous initiation of pulses of insignificant amplitudes that did not always appear during the test, while some durable AE pulses did not correspond to any displacements. Based on these results, the mechanism of the AES occurrence was considered and the dislocation group velocities were evaluated, which agrees with the dislocation velocity measurements in fluorine lithium. The mechanism provides for the interaction between the dislocation groups and the barriers that cause them. The local stress energy around these dislocation groups increases. When the minimum energy required for the breakage and for the accelerated movement of the dislocation group is attained at the head of the dislocation group, the local energy is enough to cause lattice vibrations leading to AES generation. The process of the secondary AE occurrences due to collisions between the coherent groups of the moving dislocations and barriers in their slip planes is also taken into account in the study. Finally, the emission with the alternative type of vibrations arises at high stresses after a huge PD, when stress concentration at the head of a dislocation group causes micro-crack initiation and propagation.

In turn, the authors of [51], investigating the deformation of copper, magnesium, iron monocrystals and polycrystals of brass and Cu-7% Al alloy, concluded that the AE bursts corresponded to the slip bands formation by an avalanche dislocation motion. Using their evaluations, the average value of deformation ε_{av}, per one AE pulse was found—such as for a magnesium monocrystal $\varepsilon_{av} = 8 \times 10^{-7}$. This value corresponds to the specimen elongation by 4×10^{-8} m, while for a monocrystal of manganese, these values are $\varepsilon_{av} = 5 \times 10^{-7}$ and 2×10^{-8} m, respectively. By means of electron microscopy, it was possible to determine the free length of dislocations at the stage of easy slipping, which was 10^{-4} m, and each dislocation source consisted of about 40 dislocations. The authors showed, based on the results, that the strain $\varepsilon_{av} \cong 2.5 \times 10^{-11}$ corresponded to each dislocation source, and during slip band formation, a single AE pulse was generated with the participation of about 2×10^4 dislocation sources. Usually, the results of AE research obtained at that time were very approximate and the determination of the average strain per each AE pulse by dividing a total strain by the total number of pulses was not correct. At present, it is well known that the AE behavior changes at the stage of strain hardening; moreover, the above results show that a significant AE occurs during non-stationary movement of dislocations on a certain part of a free path, which is determined as a distance between dislocation groups. The obtained empiric estimation of the strain value per one AE pulse due to an avalanche movement of dislocations is higher than the similar one for twinning by a few orders [52]. At the same time, the amplitude of the AES pulses, and, consequently, displacement during twinning, is considerably higher in comparison with the dislocation slipping [53, 54]. The results of the authors' experiments [53] argued that the beginning of twinning during deformation of the annealed polycrystalline pure zinc causes an increase in the AE intensity by about one order in comparison with slipping, while an abrupt increase of this value proved the beginning of crack initiation. In [55], the author, considering the nature of a continuous and discrete AE, suggested that the first one arose due to accumulation and breakage of dislocations, and the second one had a different nature in its origin.

It should be noted that in [51], an attempt was made for the first time to evaluate the time of AE pulse rise by the velocity of dislocations at the free path distance;

according to their data, it is about 3 μs. It should be mentioned that at the early stages of AE research, the presented quantitative characteristics of the AE were doubtful due to the non-perfect IMS and facilities. Moreover, some authors did not even provide the technical characteristics of such important parameters of extraction and processing of the obtained results as an amplification factor of the AE channel, sensitivity and frequency properties of primary AE piezo-transducers, threshold level of AES, channel bandwidth, etc. This, in its turn, did not permit comparing the obtained results and diminished their scientific value.

Important results for comprehending the nature of AE sources were obtained in [56–58] in the alkaline-haloid KCl and LiF crystals testing. The authors, using the dislocation etching method, showed [58] on LiF monocrystals that the source of AE at the initial PD stages is dislocation break-off from the fixing points. In annealed crystals, this process stipulates the beginning of a micro-scale yielding, while during further deformation of crystals it transforms into macro-scale yielding when break-off from the strong fixed points begins. According to the investigation's results, it was established that the generation of every AE pulse occurs due to a simultaneous motion of a great number of dislocations. If we sum up their length, the resulting length will be several thousand centimeters.

The above-mentioned studies put forward a hypothesis that the AES intensity depends on the degree of strain, which is to be determined by the function of distribution of dislocation segments' length in a crystal. Thus, there is a simultaneous break-off of many dislocation segments of a similar length at certain moments of time. It is assumed that a few groups of segments with an amount of dislocations from 10^5 to 10^6 in each group can be torn away from the fixing points due to the effect of elastic vibrations during acoustic wave propagation. Taking into account that the energy of AE pulses undergo summarizing, it is possible to observe an initiation of break-off of other shorter segments of dislocations during the motion of stress waves in a crystal. This explains, for example, such PD features as discontinuous yielding in the crystal, formation of slip bands, and low-temperature brittleness of crystals. Their kinetics can be determined by AES amplitude values of discrete (burst) type and the frequency of their arrivals. Somewhat later, a high efficiency of the AE method was also confirmed by experimental results obtained in investigations of deformation processes in molybdenum monocrystals [59].

1.4 Basic Parameters of the AE Signals

Since new state standards were adopted in Ukraine, the interpretation of some terms regarding the AE information parameters was corrected [60] according to the terms widely used in the literature. Therefore, in this section and throughout the monograph, we will use a set of terms used in scientific journals in order to avoid misunderstandings while comparing the investigation's results.

1.4.1 Cumulative Count [60]

In [61], the local PD formation is considered to be accompanied by the energy irradiation in several different directions. The energy is released as thermal, exo-electron, and acoustic emissions. The first two emissions are thought to be quite promising research directions, while a good correlation between AES was found for AE from the aluminum alloy. Thus, a theoretical model was proposed for a movable dislocation group as a function of the PD stresses. The preliminarily developed theoretical model for a plastic volume around a sharp crack in the elastic stress field presupposes that the total count of AES, i.e., the sum of AES pulses exceeding the threshold level set for the measuring path of the AE IMS or any AE facility, should change as the fourth degree of the SIF at the crack tip. The experimental data presented for 7075-T6 aluminum testing concur with this model, as well as with experimental results of [62–66], where AES are analyzed. The latter were recorded under a loading of structural material specimens containing the induced sharp fatigue cracks with the account of SIF from the viewpoint of linear fracture mechanics. It was shown that the AES origin was related to the microscale dynamic processes of dislocation motion. Besides, it was noted that the results of the AE research could be considered comparable only if the experimental conditions, material preloading history, its heat treatment, and stress-strain state (SSS) [65] are identical. Recommendations on the production of specimens aimed at obtaining a good correlation of experimental data with the available theories of micro-plasticity have been proposed [66]. It was suggested to use these approaches while developing NDT methods for materials and products.

1.4.2 AE Count Rate [60]

This parameter in literature is often called the AE intensity. In particular, it was used as an information parameter of AES (a number of AE pulses per time unit) in [67–69]. In [67], using this parameter, a violation of the Kaiser effect (KE) for a polycrystalline zinc was shown for the first time, and in [68] it was confirmed for steels. It was revealed as follows: During repeated loadings following a specimen unloading from a certain level to zero, AE began to appear a bit before attaining the previous force value. In [69], AES were found against the background of sound and ultrasonic noises produced by the working reactor. Measurements were made at frequencies up to 1 MHz to determine the band of low-frequency noises, and their results showed a good correlation between the plastic zone size at the crack tip and the AE count rate values. Developing a concept of direct dependence of the AES count rate on the strain (load) value, the authors [70] tested a number of structural materials and found the relationship between "strain ε—stress σ" and "strain ε—AE pulses rate $\dot{N}$" diagrams. Steels 3, 45, U8A, 40X, 15XGSA, X13, and 1X18N9T, as well as alloys of the nonferrous metals AMc, D16 M, L62, and Vt6S were used in the tests. However, according to paper [71], the count rate bears basic information on AE.

1.4.3 Amplitude Distribution of AES

Using the results of theoretical and experimental investigations performed with good approximation, it is assumed that the AES from one act of deformation should be described in the form of a decaying radio pulse [55]:

$$u(t) = \begin{cases} A \exp\left(\frac{-t-T}{\tau}\right) \sin 2\pi\nu_0(t - T), & t > T, \\ 0, & t < T, \end{cases} \tag{1.29}$$

where A is the amplitude, T is the time of the AES appearance, τ is the beginning of the time of counting determined by the quality factor of an equivalent contour that describes the response of a specimen, piezo AE transducer (AET) and equipment, and ν_0 is the resonance frequency of AET (A and T are random values). According to data in the literature [72], the range of A variation can reach three orders of magnitude, while the analysis of amplitude distribution can provide necessary information on the deformation process. Here it should be noted that macro-crack initiation and propagation are accompanied by AES of a considerably higher amplitude than during the deformation processes.

The amplitude distribution of AES is assumed to be represented by a histogram of the AES envelope maxima. In the case of discrete AE, when the signals do not overlap, it really yields distribution of pulse amplitudes A, and in the case of continuous AE, when separate pulses are densely superposed, the amplitude distribution is of a different meaning.

1.4.4 Spectral and Energy Distribution of AES

The AES spectrum contains the information about the duration of the AE source action and allows to analyze a detailed time structure of the source, which is important for theoretical research as well as making it possible to divide the contributions in the AE generation of deformation and fracture and to distinguish useful signals from noise. When analyzing the spectral data, a leading role is attributed to the correlation between the effective duration of spectral density of the mechanical loading pulse, caused by a deformation jump, namely by its cut-off frequency ν^* and duration τ_0 of this pulse [55]:

$$\nu^* \approx \tau_0^{-1}. \tag{1.30}$$

As concerns AE, Stephens and Pollock [73] were the first to establish the above-mentioned, having assumed the Gaussian shape of a pulse. The value of τ_0 also determines the effective width of the correlation shape of a pulse. In papers [74–77], experimental results of the AES spectral characteristics were obtained for commercially pure aluminum [74], aluminum alloys 2024 (4.6 Cu, 1.3 Mg) and 5046 (4.9 Mg) [75], and also for polycrystalline Al 99.99% with a different grain size d = 0.2…3.6 mm [76] as well as for monocrystalline Al 99.99% [77].

1.4.5 *Identification of AES by the Waveform Type*

The first attempt to identify AES by this factor and by the amplitude-frequency characteristics (AFC) was made in [78]. The authors confirm that during plastic deformation, continuous AES are emitted with their own vibration frequency of about 20 kHz, which are characterized by a sporadic emission of decaying amplitudes. Fracture in the compliance area is accompanied by discrete AES with a vibration frequency of about 50 kHz. Brittle fracture also generates discrete AES in the range up to 250 kHz. Therefore, AE well represents the processes that correlate with the mechanical parameters of crack growth, namely with the crack propagation rate and dynamic changes in the crack tip region.

Recognition of the mechanisms of AE sources by means of the energy spectrum analysis is presented in study [79]. Theoretical developments of such AE research methods are proposed, and the results are compared with the experimental data obtained on specially prepared zirconium specimens that belong to reactor alloys.

Thus, as we see from the analysis of the above-mentioned literature sources, until the middle of the 1980s, the main concepts of the research directions were formed, and certain methodological approaches were built with the use of such information parameters of AES as amplitude and its distribution, cumulative count, rate of count, spectral distribution, waveforms, energy distribution and others. With the development of engineering equipment, the AE research was carried out at a new, higher level, with preservation of their conceptual orientation that is proved by subsequent research. The elements of stochastic approach to the AES [80, 81] analysis, as well as the methods of their probabilistic evaluation, were worked out [82].

1.5 Basic Analytical Dependences Between the Fracture Parameters and the AE Signals

Most of the papers published so far deal with the problem of material fracture study at a macrolevel using the AE phenomenon. Firstly, this is conditioned by the fact that during macro-crack propagation, the elastic energy that is released is considerably higher than the energy caused by a fracture at a microscale level. Besides, the irradiation spectrum of the first one has powerful low-frequency components [59, 83, 84]; this facilitates its recording. Secondly, the AE phenomenon is successfully used for NDT and technical diagnostics (TD) of various materials, products, structures, building constructions, etc. At present, a number of IMS and other facilities of a different complexity, which are used in combination with the specially created methods, have already been developed.

Many researchers have made an attempt to find a correlation between the AES parameters and cracks. A huge amount of data has been obtained in this field of investigation, and its analysis shows that in order to establish such correlations, the AE parameters presented in the previous subsection were used. Thus, in [85] it was found that the cumulative count of AES is proportional to the fourth degree of SIF: $N \sim K^4$ or $N \sim l^2$, where $2l$ is the crack length. The $N \sim K^2/E$ expression (E is the Young's modulus) was obtained in [86]. It was also noted that the results of other authors showed that the power value can be within the limits of 1…8. Authors explained this by the various types of facilities used in experiments, as well as by the various conditions of recording and processing of the AE information. A dependence was also obtained that relates the value of the total crack tip opening displacement with the AE pulses number: $\delta_\Sigma \sim (1\text{-}\nu)N(0.2\ \sigma_y)^{-1}$, where ν is the Poisson's ratio and σ_y is the yield strength.

A number of studies have been published, for example [87, 88], in which authors set a correlation between the cumulative count of AES and newly-formed area of the macro-crack, and the information on every discrete increment is suggested to be obtained using the amplitudes of AES. It is shown in [89] that every AE pulse corresponds to the macro-crack jump, and the values of the AES amplitudes are linearly correlated with the energy that accompanies its growth. Authors [87] argued that the AES amplitude is proportional to the micro-crack area, while the square of the sum of the amplitudes of AE pulses, which accompany a jump-like macro-crack growth, is, in the authors' opinion [88], proportional to its area.

An interesting result was obtained in [90], where, by solving the non-stationary problem of crack propagation with the account of crack tip interaction with the stress field, a relationship between spectral characteristics of AES and the macro-crack length was found. These results were confirmed experimentally in [91].

The results of a review of the literature sources on the fracture studies by AES are presented in Table 1.1. It should be noted that the relationships in this table were in most cases obtained empirically, basing them on the very schematic reasoning that caused their obvious incorrectness in some cases. Moreover, when interpreting the experimental results that were obtained, most authors did not differentiate the obtained AE data by the mechanisms of their origin, a priori giving an advantage to one of them. This also substantially reduces the scientific value of such studies, and in some cases even makes it unusable.

Therefore, taking into account modern-day knowledge of the sources of AES generation at various stages of micro- and macro-crack initiation and propagation, it is possible to state the absence of scientific grounds and unity of the above results. It

Table 1.1 Relationships of the AES parameters with the fracture parameters of materials

Dependence	Author	Literature
SIF(K)		
$N \sim K^n$, $n = 4$	Dunegan, Harris, Tatro	[85]
$N \sim K^n$ $n = 1\ldots11.9$	Liptai, Harris, Engle, Tatro, Hartbower, Reuter, Morais, Chimmins, Greshnikov, Drobot	[86, 91, 92]
$N \sim K^2/E$	Hartbower, Reuter, Morais, Chimmins	[86]
$N = -\beta \frac{K_C^2}{K_I^2 + K_{IS}^2}\left(\frac{K_I^2 - K_{IS}^2}{K_C^2} + \ln \frac{K_C^2 - K_{IS}^2}{K_C^2 - K_I^2}\right)$	Andreykiv, Lysak, Skalskyi, Serhiyenko	[93, 94]
Crack opening displacement (δ)		
$N \sim a\delta^n$, $n = 0.34\ldots1.2$	Palmer, Brindley, Harrison Indham, Bentley	[95]
$\delta_\Sigma \sim (1\text{-}\nu)N(0.2\sigma_y)^{-1}$	Liptai, Harris, Engle, Tatro	[96]
Macro-crack length and increment (l, Δl)		
$N = Dl\left\{\sec\left(\frac{\pi\sigma}{2\sigma_y} - 1\right)\right\}$	Palmer, Heald	[97]
$l = c_1 T$	Maslov L, Bolotin, Greshnikov, Drobot	[90, 98]
$\Delta l \sim (\sum_i A_i)^2 E/K_I^2$	Hartbower, Gerberich, Liebowitz	[88, 99]
$\Delta l \sim A$	Radon, Pollock	[89]
$\Delta l_\Sigma \sim \sum_i A_i^{2/3}$	Bolotin, Maslov L, Polunin	[100]
Area increment (S, S_m) for macro- and micro-cracks		
$S \sim N$, $\Delta S_m \sim A$	Gerberich, Alteridge, Lessar	[87]
$S \sim \alpha l$, $\alpha = 2\ldots44$ counts/mm^2	Sinclair, Connors, Formby	[101]
Plastically deformed volume (Vp)		
$N \sim V_p$	Dunegan, Harris, Tatro	[85]
$A^2 \sim V_p^2 \sim r_p^4 \sim (B\, r_p^2)^2 \sim E_T$	Kishi, Kuribayashi	[102]
J integral (J)		
$\Delta E_T/\Delta l \sim J$	Ishikawa, Kim	[103]
Length of the PD zone (r_p)		
$N \sim r_p^2$	Palmer	[104]
Strain Rate ($\dot{\varepsilon}$)		
$A \sim \sqrt{\dot{\varepsilon}}$, $\dot{\varepsilon} = \rho/v$	Imanaka, Sano, Shimizu, Fudjimoto	[105]
AE Energy of PD of one grain (E_g)		
$\Delta E_g = d^2 nb/2(\sigma + \sigma_f)$	Tetelman	[106]
Tensile stress (σ)		
$N \sim \sigma^m \ln(\sigma/\sigma_{AE})$, $m = 2\ldots5$	Tetelman, Chow, Lottermoser, Waschkies, Zenner, Voss, Gotz	[107]
Density of distribution of the AES amplitudes during macro-crack jump (m)		
$m(a,t) = ck(t)/A^{c+1}$ $c = 0.4\ldots2.2$	Pollock	[108, 109]

means that modern research in fracture by the AE method should be carried out at a higher scientific level, and with the involvement of novel theoretical developments.

It should be noted that in real structural materials such as metals, polymers, composites, etc., fracture occurs by crack propagation. As shown above, plastic deformation of a great amount of the material at a crack tip, i.e., formation of the PD zone as an integral part of the process zone, is a very important stage during macro-crack initiation and propagation. The maximum PD volume is often commensurable with the crack size. Therefore, a quantitative evaluation of the material damaging in a plastically deformed volume (PDV) is very important for diagnostics and for the prediction of the safe operation of structures and products.

References

1. McClintok F, Argon I (1970) Deformaziya i razrushenie materialov (Deformation and fracture of materials). Mir, Moskva
2. Honeycombe RWK (1972) Plasticheskaya deformaziya metallov (The plastic deformation of metals). Mir, Moskva
3. Dahl V, Anton V (eds) (1986) Staticheskaya prochnost' i mechanika razrusheniya staley (Static strength and fracture mechanics of steels). Metallurgia, Moskva
4. Panasyuk VV, Andreikiv AY, Parton VZ (1988) Osnovy mechaniki razrusheniya materialov (Bases of fracture mechanics of materials). In: Panasyuk VV (ed) Mechanika razrusheniya i prochnost' materialov (Fracture mechanics and strength of materials), vol 1. Naukova Dumka, Kiev
5. Neimitz A (1998) Mechanika pękania. Wydawnictwo Naukowe PWN SA, Warszawa
6. Griffith AA (1920) The phenomena of rupture and flow in solids. Phil Trans Roy Soc London A 221(1):163–198
7. Irwin GR (1957) Analysis of stress and strain near end of a crack traversing a plate. J Appl Mech 24(3):361–364
8. Felbek DK, Orowan EI (1955) Energy criteria of fracture. Weld J Res Suppl 34:157–160
9. Morozov EM (1971) Energeticheskiy kriteriy razrusheniya dlya uprugo–plasticheskich tel (Energy fracture criterion for elasto-palstic bodies). Konzentraziya napryazheniy (Stress Concentration) 3:85–90
10. Morozov EM (1985) Mechanika uprugoplasticheskogo razrusheniya (Elasto-plastic fracture mechanics). Nauka, Moskva
11. Cherepanov GP (1974) Mechanika khupkogo razrusheniya (Brittle fracture mechanics). Nauka, Moskva
12. Rice J (1968) Nezavisyaschiy ot puti integral i priblizhennyy analiz konzentrazii deformaziy u vyrezov i treschin (Path-independent integral and approximate analysis of deformation concentration at notches and cracks). Prikl. mechanika. Ser. E (Appl Mech Ser E) 35 (4):340–349
13. Rice J (1976) Matematicheskie metody v mechanike razrusheniya (Mathematical methods in fracture mechanics). Razrushenie (Fracture), vol 2. Mir, Moskva, pp 204–209
14. Libowitz H, Eflis L (1971) On non–linear effects in fracture mechanics. Eng Frac Mech 3 (2):267–281
15. Eshelby JD (1956) A continuum theory of lattice defects. On progress solid state physics, vol 3. Academic Press, New York, pp 79–85
16. Sih G (1974) Strain energy density factor applied to mixed mode crack problems. Int J Fract 10(3):124–129

17. Weighardt E (1907) Über das Spalten und Zerressen elestischer Körpes. Z Math Und Phys 50:60–103
18. Sneddon I (1966) Preobrazovanie Fur'e (Fourier transformation). Inostrannaya Literatura, Moskva
19. Williams ML (1957) On the stress distribution at the base of a stationery crack. J Appl Mech 24(1):109–114
20. Irwin GR (1958) Fracture. Handbuch der Physik, vol 6. Springer, Berlin, pp 551–590
21. (1985) GOST 25.506–85. Raschety i ispytaniya na prochnost'. Metody mechanicheskich ispytaniy metallov. Opredelenie charakteristik treschinostoykosti (vyazkosti razrusheniya) pri staticheskom nagruzhenii. Vved. v deystvie 27.03.1985 g. (State Standard 25.506–85. Calculations and tests for strength. Methods of mechanical testing of materials. Determination of crack growth (fracture toughness) characteristics under static loading. Approved 27.03.1985). Izdatel'stvo standartov, Moskva
22. Wells AA (1961) Unstable crack propagation in metal: cleavage fast fracture. In: Symposium on Crack Propagation, College of Aeronautics, Cranfield, 1961
23. Cottrell AH (19610 Theoretical aspects of radiation damage and brittle fracture in pressure vessels steel. Iron Steel Inst Spec Report 69:261–296
24. Leonov MYa, Panasyuk VV (1959) Rozvytok naydribnishych trischyn v tverdomu tili (Development of the smallest cracks in a solid). Prykladna Mekhanika 5(4):391–401
25. Leonov MYa, Panasyuk VV (1961) Rozvytok trischiny, yaka v plani maye formu kruga (Developemnt of a crack, which in plane has a shape of a circle). Dopovidi AN URSR (Reports of the Academy of Sciences of the Ukrainian SSR) 2:165–168
26. Panasyuk VV (1960) Do teoriji poshirennya trischyn pry deformazii krychkogo tila (To the theory of crack distribution during brittle body deformation). Ibid 9:1185–1189
27. Panasyuk VV (1968) Hranychna rivnovaha til z trischynamy (Limiting equilibrium of bodies with cracks). Naukova Dumka, Kiev
28. Vitvitski PM, Panasyuk VV, Yarema SYa (1975) Plastic deformation around crack and fracture criteria. Eng Fract Mech 2:305–319
29. Dugdale DS (1960) Yielding of steel sheets containing slits. J Mech Phys Solids 8(2):100–108
30. Orowan E (1948) Cleavage fracture of metals. Rep Prog Phys 12:185–189
31. McClintock F (1976) Plasticheskie aspekty razrusheniya (Plastic aspects of fracture). Razrushenie (Fracture), vol 3. Mir, Moskva, pp 67–262
32. Ritchie RO, Knott JF, Rice JR (1973) On the relationship between critical tensile stress and fracture toughness in mild steel. J Mech Phys Solids 21(3):359–410
33. Zener C (1949) Micromechanism of fracture. Fracturing of metals. ASM, Novelty, Ohio, pp 3–31
34. Stroh AN (1954) The formation of cracks as a result of plastic flow I. Proc Roy Soc A223:404–414
35. Stroh AN (1955) The formation of cracks in plastic flow II. Ibid A232:548–560
36. Cottrel AH (1958) Theory of brittle fracture in steel and similar metals. Trans AIME 212:192–203
37. Panasyuk VV (1991) Mechanika kvazichrupkogo razrusheniya materialov (Quasi-brittle fracture mechanics of materials). Naukova Dumka, Kiev
38. Hull D (1960) Twinning and fracture of single crystals of 3% silicon iron. Acta Met 8:11–18
39. Romaniv OM, Zyma YuV, Karpenko GV (1974) Elektronna fraktografiya zmiznenich staley (Electron fractography of strengthened steels). Naukova Dumka, Kiyv
40. Romaniv ON et al (1990) Fatigue and fatigue crack growth resistance of structural steels. In: Panasyuk VV (ed) Fracture Mechanics and strength of materials: reference book, vol 4. Naukova Dumka, Kiev
41. (2003) DSTU 4227-2003 Rekomendaziyi schodo akustyko–emisiynogo kontrolyu ob'yektiv pidvischenoyi nebezpeky (State Standard of Ukraine 4227-2003 Recommendations on the acoustic-emission control of objects of a special danger). Derzhspozhyvstandart Ukrainy, Kyiv

42. Skalskiy VR, Andreykiv OY, Serhiyenko OM (2003) Doslidzhennya plastychnogo deformuvannya materialiv metodom akustichnoyi emisiyi. Oglyad (Investigation of plastic deformation of materials by the method of acoustic emission. A review). Fizyko–chimichna mechanika materialiv (Physico-Chem Mech Mater) 1:77–94
43. Nazarchuk ZT, Koshoviy VV, Skalskiy VR (2001) Neruynivni metody kontrolyu materialiv i technichna diagnostyka (Nondestructive test methods of materials and technical diagnostics). Fiziko–mechanichniy institut: postup i zdobutki (Physico-Mechanical Institute: progress and achievements). Prostir, Lviv, pp 171–214
44. Skalskiy VR, Demchyna BH, Karpukhin II (2000) Ruynuvannya betoniv i akustichna emisiya (Oglyad). Povidomlennya 1. Statychne navantazhennya i vplyv temperaturnogo polya (Failure of concretes and acoustic emission (A review). Report 1. Static loading and temperature field effect). Tekhnicheskaya diagnostika i nerazrushayushchii control (Tech Diagn nondestructive testing) 1:12–23
45. Skalskiy VR, Demchyna VR, Karpukhin II (2000) Ruynuvannya betoniv i akustichna emisiya (Oglyad). Povidomlennya 2. Koroziya zalizobetonu. Aparaturni zasobi. AE – kontrol' ta diagnostyka budivel'nich sporud (Failure of concretes and acoustic emission (A review). Report 2. Static loading and temperature field influence). Ibid 2:9–27
46. Skalskiy VR, Demchyna BH (2000) Stan i perspektyvy rozvytku akustiko–emisiynogo diagnostuvannya budivel'nich konstrukziy (The state and prospects of development of acoustic – emission diagnostics of building constructions) In: Mechanika i fizyka ruynuvannya budivel'nich materialiv i konstrukziy (Mechanic and physics of failure of building materials and constructions), is 4. Lviv, pp 520–531
47. Andreykiv OY et al (2000) Rozvitok doslidzhen' prozesiv ruynuvannya iz zastosuvannyam yavyscha akustichnoyi emisiyi u FMI im. HV. Karpenka NAN Ukrayiny (Development of researches of fracture processes with application of the phenomenon of acoustic emission in Karpenko Physico-Mechanical Institute of NASU). Preprint, NAN Ukrayiny, Fizyko-mechanichnyi instytut, 1(2000), L'viv
48. Fitzgerald ER (1960) Mechanical resonance dispersion and plastic flow in crystalline solids. J Acoust Soc Am 32(10):1270–1289
49. Ookawa A, Yazu I (1963) The energy radiated from a dislocation by an accelerated motion through impurity fields. J Phys Soc Jpn 18(A):36–43
50. Engle RB (1966) Acoustic emission and related displacements in lithium fluoride single crystals. Dissertation, Michigan State University
51. Fisher RM, Lally LS (1967) Microplasticity detected by an acoustic technique. Canad J Phys 45(2):1147–1159
52. Boiko VS et al (1970) Zvukovoe izluchenie dvoynikuyuschich dislokaziy (Sonic irradiaton of twinning dislocations). Fizika tviordogo tela (Phys Solid) 12(6):1753–1755
53. Melekhin VP, Mints RI, Kugler LM (1971) Vliyanie mechanizmov plasticheskoy deformazii zinka na akusticheskuyu emissiyu (The effect of the mechanisms of plastic deformations of zinc on acoustic emission). Izvestiya VUZov. Zvetnaya metallurgiya (Reports of High Schools. Non-ferrous metallurgy) 3:128–131
54. Schofield BH (1972) Research on the sources and characteristics of the acoustic emission. In: Acoustic emission, ASTM STP 505. Baltimor, pp 11–19
55. Yudin AA, Ivanov VI (1985) Akusticheskaya emissiya pri plasticheskoy deformazii metallov (obzor) (Acoustic emission during plastic deformation of metals (A review)). Probl Prochn (Strength Mater) 6:92–107
56. James DR (1971) The source of acoustic emission in deforming single crystals. In: International conference on mechanical behaviour of materials, vol 3. Kyoto, pp 960–961
57. Segwick RT (1968) Acoustic emission from single crystals of LiF and KCl. J Appl Phys 39 (3):1728–1740
58. James DR, Carpenter SH (1971) Relationship between acoustic emission and dislocation kinetics in crystalline solids. J Appl Phys 42(12):4685–4697

59. Gusev OV (1982) Akusticheskaya emissiya pri deformirovanii monokristallov tugoplavkich metallov (Acoustic emission during deformation of monocrystals of refractory metals). Nauka, Moskva
60. (1994) DSTU 2374-94 Rozrachunky na miznist' ta viprobuvannya technichnich vyrobiv. Akustichna emisiya. Termini ta viznachennya (State Standard of Ukraine 2374-94 Calculations of strength and testing of technical products. Acoustic emission. Terms and definitions)
61. Dunegan HL, Harris AT (1968) Acoustic emission: and new nondestructive testing tool. In: Proceedings of the third annual symposium on nondestructive testing of welds and materials joining, Los Angeles, California, 12 Mar 1968
62. Dunegan HL (1969) Ultrasonic acoustic emission from materials. IEEE Trans Sonics Ultrason 1:16–32
63. Engle RB, Dunegan HL (1969) Acoustic emission: stress wave detection as a tool for nondestructive testing and material evaluation. Int J Nondestr Test 1(7):109–125
64. Dunegan HL, Harris DI, Tetelman AS (1970) Detection of fatigue crack growth by acoustic emission techniques. Mater Eval 28(10):221–227
65. Dunegan HL, Green AT (1971) Factors affecting acoustic emission response from materials. Mater Res Stand 11(3):21–24
66. Dunegan HL (1969) Ultrasonic acoustic emission from materials. IEEE Trans Sonics Ultrason 16(1):32–35
67. Borchers H, Tensi H-M (1960) Untersuchung von Vorgangen in Metallen bei Mechanischer Beanspruchung und bei Phasenanderung. Z für Metallkd 51(4):212–218
68. Kerawalla JN (1965) An investigation of the acoustic emission from commercial ferrous materials subjected to cyclic tensile loading. Dissertation, University of Michigan
69. Eisenblatter J (1971) Schallemissionsanalyse als zerstorungsfreies Prufverfahren zur Feststellung von Fehlern in Kernreaktordruckgefassen. In: Proceedings of the first international conference on structural mechanism in reactor technology, vol 4. Battelle–Institut eV, Frankfurt am Main, Germany, pp 529–530
70. Greshnikov VA et al (1971) Primenenie emissii voln napryazheniy dlya nerazrushayuschego kontrolya i technicheskoy diagnostiki kachestva materialov i izdeliy (Application of emission of stress waves for nondestructive testing and technical diagnostics of the quality of materials and products). In: Materialy seminara "Novye metody nerazrushayuschego kontrolya kachestva materialov i izdeliy", Chabarovsk, 2 aprelya 1971 g. (Trans. Seminar "New methods of nondestructive testing of the quality of materials and products", Khabarovsk, April 2 1971), Khabarovsk, 1971
71. Averbukh II, Vainberg VYe (1973) Zavisimost' akusticheskoy emissii ot deformazii v raznych materialach (Dependence of acoustic emission on deformation in different materials). Defectoskopia 4:25–32
72. Spanner JC (1974) Acoustic emission: technique and applications, vol 12. Intex publication, Co, Evanston, Illinois
73. Stephens RWB, Pollock AA (1971) Waveforms and spectra of acoustic emission. J Acoust Soc Am 50(3):904–910
74. Hatano H (1971) Quantitative measurements of acoustic emission related to its microscopic mechanisms. Ibid 57(3):639–645
75. Hatano H (1976) Strain–rate dependence of acoustic–emission power and spectra in aluminum alloys. J Appl Phys 47(9):3873–3876
76. Fleischmann P, Lakestani F, Baboux JC (1977) Analise spectrale et energétique d'une source ultrasonore en mouvement. Application a l'émission acoustique de l'aluminium soumis d'éformation plastique. Mater Sci Eng 29(1):205–212
77. Rouby D, Fleischmann P (1978) Spectral analysis of acoustic emission from aluminium crystals undergoing plastic deformation. Phys Stat Solidi (A) 48(2):439–445
78. Hutton PH et al (1968) Crack detection in pressure piping by acoustic emission. In: Nuclear Safety Quarterly Report–July, August, September, October, 1967 for Nuclear Safety Branch

of USAEC Division of Reactor Development and Technology. BNWL–754. – Battelle-Northwest, Richland, Washington. June 1968
79. Nosov VV, Potapov AI (1996) Strukturno–imitazionnaya model' parametrov akusticheskoy emissii (Structural simulation model of acoustic emission). Defektoskopia 6:31–38
80. (1986) Akusticheskaya emissiya geterogennych materialov (Acoustic emission of heterogeneous materials). A.F. Ioffe Physico-Technical Institute, Leningrad
81. Borodin YuP, Gulievskyi IV (1980) Statisticheskaya model' akusticheskoy emissii defektov v materialach i konstrukziyach pri nagruzhenii (Statistical model of acoustic emission of defects in materials and structures under loading). Uchenye zapiski Zentral'nogo aerogidrodinamicheskogo instituta (Scientific notes. Central Areodynamic institute) 11 (2):86–95
82. Robsman VA (1996) Nelineynaya transformaziya veroyatnostnych raspredeleniy signalov akusticheskoy emissii pri evolyuzii ansamblya defektov v tverdom tele (Nonlinear transformation of probabilistic distributions of acoustic emission signals during evolution of an ensemble of defects in a solid). Akusticheskiy zhurnal (J Acoust) 42(6):846–852
83. Andreykiv OYe et al (1990) Metodicheskie aspekty primeneniya metoda akusticheskoy emissii pri opredelenii staticheskoy treschinostoykosti materialov (Methodical aspects of application of the acoustic emission method in evaluation of static crack growth resistance of materials). Preprint, NAN Ukrayiny, Fizyko-mechanichnyi instytut, 165(1990), L'viv
84. Takahashi H et al (1981) Acoustic emission crack monitoring in fracture–toughness tests for AISI 4340 and SA 533B steels. Exp Mech 21(3):89–99
85. Dunegan HL, Harris DO, Tatro IA (1967) Fracture analysis by use of acoustic emission. Eng Fract Mech 1(1):105–122
86. Hartbower CE et al (1972) Acoustic emission for the detection of weld and stress–corrosion cracking. In: Acoustic emission, ASTM STP 505. Baltimore, pp 187–221
87. Gerberich WW, Alteridge DG, Lessar JE (1975) Acoustic emission investigation of microscopic ductile fracture. Met Trans A 6(2):797–801
88. Gerberich WW, Hartbower CE (1967) Some observations on stress wave emission as a measure of crack growth. Int J Fract Mech 3(3):185–192
89. Radon IC, Pollock AA (1972) Acoustic emission and energy transfer during crack propagation. Eng Fract Mech 4(2):295–310
90. Maslov LA (1975) Investigation of acoustic pulses in cracking. Dissertation, University of Novosibirsk
91. Greshnikov VA, Drobot YuB (1976) Akusticheskaya emissiya. Primenenie dlya ispytaniy materialov i izdeliy (Acoustic emission. Application in material and products testing). Izdatel'stvo standartov, Moskva
92. Collaquot R (1989) Diagnostika povrezhdeniy (Damage diagnostics) (trans: Babayevsky PG). Mir, Moskva
93. Baranov VM (1990) Akusticheskie izmereniya v yadernoy energetike (Acoustic measuring in nuclear power engineering). Energia, Moskva
94. Andreykiv OYe et al (1987) Teoreticheskie konzepzii metoda akusticheskoy emissii v issledovanii prozessov razrusheniya (Theoretical concepts of acoustic emission method in the fracture processes investigations). Preprint, NAN Ukrayiny, Fizyko-mechanichnyi instytut, 137(1987), L'viv
95. Palmer IG, Brindley BJ, Harrison RP (1974) The relationship and crack opening displacement measurements. Mat Sci Eng 14(1):3–6
96. Liptai RG et al (1971) Acoustic emissions technique in materials research. Int J Nondestr Test 3:215–275
97. Palmer IG, Heald PT (1973) The application of acoustic emission measurement to fracture mechanics. Mat Sci Eng 13(11):181–184
98. Bolotin YuI et al (1975) Analiz akusticheskoy emissii, vyzvannoy rostom treschiny v pryamougol'noy plastine (Analysis of acoustic emission caused by crack growth in a rectangular plate). Izmeritelnaja Tekhnika (Measuring Equip) 1:54–57

99. Hartbower CE, Gerberich WW, Liebowitz H (1968) Investigation of crack growth stress wave relationships. Eng Fract Mech 1(2):291–308
100. Bolotin YuI, Maslov LA, Polunin BI (1975) Ustanovlenie korelyaziy mezhdu razmerom treschiny i amplitudoy impul'sov akusticheskoy emissii (Establishment of correlation between crack sizes and acoustic emission pulse amplitude). Defektoskopia 4:119–122
101. Sinclair ACE, Formby CL, Connors DC (1975) Acoustic emission from defective C/Mn steel pressure vessel. Int J Press Vessels Pip 3(3):153–174
102. Kishi T, Kuribayashi K (1978) Akustichna emisiya u viprobuvannyach na v'yazkist' ruynuvannya (Acoustic emission in the tests on fracture toughness). Kindzoku 48(2):56–60
103. Ishikawa K, Kim HC (1974) Stress wave emission and plastic work of notched specimens. J Mat Sci 9(5):737–743
104. Palmer AG (1973) Acoustic emission measurements on reactor pressure vessel steel. Mater Sci Eng 11(4):227–236
105. Tetelman AS, Chow R (1972) Acoustic emission testing and microcracking processes. In: Acoustic emission, ASTM STP 505. Baltimore, pp 30–40
106. Tetelman AS (1972) Acoustic emission and fracture mechanics testing of metals and composites. In: Proceedings of the U.S.–Japan joint symposium on acoustic emission—Tokyo, Japan, 4–6 July 1972
107. Lottermoser J et al (1982) Report No 780236–TW. Institut für Zerstorungs freie Prufvertahren
108. Pollock AA (1973) Acoustic emission—2. Acoust emission amplitudes. Non-Destr Test 6 (5):264–269
109. Pollock AA (1976) Acoustic emission. A review of recent progress and technical aspects. In: Stephens RW, Leventhall HG (eds) Acoustic and vibration progress, vol 1(1). Charman and Hall, London, pp 51–84

Chapter 2
Propagation of Elastic Waves in Solids

Elasticity is a solid's most important property for restoring its shape and volume after the termination of the action of the external forces applied to it, while for liquids and gases, only volume is restored. Therefore the medium, whose typical feature is elasticity, is referred to as "elastic medium." Accordingly, *elastic vibrations* are vibrations of mechanical systems, elastic medium, or its part that arises under mechanical disturbances. *Elastic* or *acoustic waves* are mechanical disturbances that reproduce in an elastic medium. A partial case of acoustic waves is a sound, which is audible to man; thus the term "acoustics" (from the Greek "*akustikos*," which means "auditory") was given to this phenomenon. In the widest sense, acoustics involves the study of elastic waves, and in the narrowest, it is often used to define their sound range only.

Elastic vibrations and acoustic waves are widely used in nondestructive testing and technical diagnostics of materials and products, in various engineering devices and equipment. For example, powerful ultrasonic vibrations are used for the local fracture of brittle high-strength materials (ultrasonic crushing); dispersion (fine crushing of solid or liquid bodies in any medium—for example, fats in water); coagulation (enlargement of particles of a substance, such as smoke); and for other purposes. Elastic vibrations and waves are very important for the investigation of the processes of initiation and propagation of the volume damaging and fracture of solids; this it what has made it possible to use them widely in fundamental and applied scientific studies of these processes from the viewpoint of fracture mechanics.

2.1 Types of Elastic Waves

2.1.1 Some General Ideas on Elastic Strain

Elastic vibrations in liquids and gases are characterized by one of the following parameters: change in pressure p or density $\boldsymbol{\rho}$; particle shift from an equilibrium

Z. Nazarchuk et al., *Acoustic Emission*, Foundations of Engineering Mechanics,
DOI 10.1007/978-3-319-49350-3_2

state $\boldsymbol{u}$; vibration motion velocity $\boldsymbol{v}$; or shear potential χ, i.e., vibration velocity φ. It is essential to distinguish the change in pressure or density caused by acoustic wave propagation from their statistical (average) value. All the above-mentioned parameters are interconnected, for example: $\boldsymbol{u} = \text{grad}\chi$; $\mathbf{v} = \text{grad}\varphi$; $\mathbf{v} = \partial u/\partial t$; $p = \rho\partial\varphi/\partial t$, where ρ is the medium density, and t is the time [1].

Unlike liquids and gases, the acoustic field in a solid is of a more complicated nature, because a solid possesses not only the volume elasticity as liquids and gases do, but also the elasticity of their shape, i.e., shear elasticity. The concept of stress is introduced for solids instead of pressure, i.e., the force related to a surface unit.

In the mechanics of a deformed solid there are normal (tensile or compressive) σ_{xx}, σ_{yy}, σ_{zz} and tangent or tangential (shear) σ_{xy}, σ_{yz} stresses, etc. In general, the stress state of a solid is characterized by a third-rank tensor σ_{ij}, where indices i and j take the values of coordinate axes x, y, z. The first index indicates the coordinate toward which the force acts, and the second index defines the plane perpendicular to the direction of the coordinate indicated by the second index, to which this force is applied. This tensor is symmetric, i.e., $\sigma_{ij} = \sigma_{ji}$.

Since in liquids and gases there is no elasticity in a shape, there are no tangential components of a stress tensor, and normal components are equal to each other and to a pressure with a reverse sign. The pressure is negtive when it creates tensile stresses that are considered to be positive; on the contrary, pressure is assumed to be with a plus when it creates compressing stresses.

A typical feature of vibrations in a solid is the change in stresses σ_{ij}, displacement of its particles u_i and shear potential. The concept of vibration velocity is rarely used. More often, vibrations are characterized by deformation ("*deformation*" in Latin means "distortion"), i.e., the change in mutual location ∂u of the body points. This change is compared to the primary distance between points; as a result, the deformation is a dimensionless quantity. If points are displaced along a segment connecting them, the tension-compression deformation occurs in a solid, and if they are displaced perpendicular to this segment direction, we have a shear deformation, which is why the deformation is written as a tensor ε_{ij}, similar to a stress tensor. Herein, $\varepsilon_{xx} = \partial u_x/\partial x$ is the tension-compression deformation along the x axis. For the symmetry of strain tensor, its components are written as $\varepsilon_{xy} = (\partial u_x/\partial u_y + \partial u_y/\partial x)/2$. Other shear components of deformation are written similarly. The value of $\varepsilon = \varepsilon_{xx} + \varepsilon_{yy} + \varepsilon_{zz}$ implies the change of the volume $dxdydz$ of an elementary cube. For liquids and gases, there are no shear deformations while the tension-compression deformations are the same in all directions.

In this chapter we consider an isotropic medium only. Isotropy (from the Greek "*isos*", which means "equal" or "identical," and "*tropos,*" which means "direction") means that physical properties of a medium do not depend on the direction chosen in it. Media, whose properties depend on the direction, are called "anisotropic" (from the Greek "*anisos,*" which means "unequal").

Hook's law relates a proportional dependence between stresses and strains. In a generalized form, it looks like

$$\sigma_{ij} = \delta_{ij}\Lambda\varepsilon_{ii} + 2\mu\varepsilon_{ij}, \tag{2.1}$$

where $\delta_{ij} = 1$, if $i = j$, and $\delta = 0$, for $i \neq j$; Λ and μ are Lame constants. In engineering practice, instead of the latter, the modules of the normal elasticity E and shear G are used. They are related by the following dependence [2]:

$$E = \mu(3\Lambda + 2\mu)/(\Lambda + \mu), \quad G = \mu. \tag{2.2}$$

Alongside these modules, another important elastic constant, i.e., the Poisson's ratio v, is also used in calculations. It is defined as the ratio of compression to elongation of the tensioned bar:

$$v = \frac{\Lambda}{2\,(\Lambda + \mu)} = \frac{E}{2G} - 1. \tag{2.3}$$

In all cases, a couple of independent elastic constants characterize the elastic properties of an isotropic solid.

2.1.2 A Wave Equation for a Solid

It is derived by using the second Newton's law for an elementary volume d*x*d*y*d*z*. The difference of forces applied to its opposite faces is equated to the product of mass and acceleration. As a result, we get for axis x

$$\rho\frac{\partial^2 u_x}{\partial t^2} = \frac{\partial\sigma_{xx}}{\partial x} + \frac{\partial\sigma_{xy}}{\partial y} + \frac{\partial\sigma_{xz}}{\partial z}. \tag{2.4}$$

By analogy, it is possible to write the equation for axes y and z.

By substituting strains from (2.1) instead of stresses, the equation of wave propagation in an elastic medium is obtained:

$$\rho\frac{\partial^2 u_x}{\partial t^2} - (\Lambda + \mu)\frac{\partial_\varepsilon}{\partial_x} - \mu\nabla^2 u_x = 0, \tag{2.5}$$

where $\nabla^2 = \partial^2 / \partial x^2 + \partial^2 / \partial y^2 + \partial^2 / \partial z^2$ is the Laplace operator. The wave equations (2.5) include second-order time, and coordinate derivatives with different signs with respect to some variable. Using a vector analysis, the equation of (2.5) type for all coordinates can be written as one expression:

$$\rho\frac{\partial^2 \mathbf{u}}{\partial t^2} = (\Lambda + \mu)\ \text{grad div}\,\mathbf{u} + \mu\nabla^2 u. \tag{2.6}$$

When $\mu = 0$, and assuming the displacement $u_x = u_y = u_z = u$ to be the same in all directions (scalar), Eq. (2.6) is transformed to a wave equation for a liquid or a gas:

$$\frac{\partial^2 \mathbf{u}}{\partial t^2} = c^2 \nabla^2 u, \tag{2.7}$$

where $c = \sqrt{\Lambda/\rho}$ is the velocity of propagation of elastic waves. The same equations are valid for other elastic values, i.e., pressure, potential, etc.

2.1.3 *Main Ideas of the Wave Process*

The solution of Eq. (2.7) for the potential of velocity:

$$\frac{\partial^2 \varphi}{\partial t^2} = c^2 \nabla^2 \varphi. \tag{2.8}$$

Displacements and vibration velocities for liquids and gases are the same in all directions, so they can be considered to be scalars u and v. For simplicity, assume that function φ depends only on the coordinate x: $\nabla^2 \varphi = \partial^2 \varphi / \partial x^2$. It is known from the theory of differential equations in partial derivatives that the solution of such an equation looks like $\varphi = \varphi_1(x - ct) + \varphi_2(x + ct)$, where φ_1 and φ_2 are arbitrary functions that are differentiated twice. The first element is a wave that propagates along the x axis in the positive direction, and the second is a wave that propagates in the opposite direction. Hereinafter, as a rule, we will take into consideration a direct wave and will omit the second term; therefore, before t there should be a minus if there is a plus before x.

In the case of harmonic wave propagation, we can write (2.9):

$$\varphi = \varphi_1 = \Phi \cos\left[\frac{\omega}{c}(x - ct)\right] = \Phi \cos(kx - \omega t). \tag{2.9}$$

Here, Φ is the amplitude, $kx - \omega t$ is the phase, $\omega = 2\pi f$ is the circular frequency, f is the vibration frequency, $T = 1/f$ is the period of vibrations, $k = \omega/c = 2\pi/\lambda$ is the wave number, and λ is the wavelength, i.e., the distance that a wave passes for a period of vibrations. If t changes by a period or x changes by a wavelength, a phase will change by 2π, and consequently the cosine value will be the same.

The other harmonic wave presentation is

$$\varphi = \mathrm{Re}\left[\Phi e^{j(kx - \omega t)}\right], \tag{2.10}$$

where $j = \sqrt{-1}$. Symbol Re indicates that the real part of the complex function, which is in square brackets, is taken. Since $e^{ja} = \cos a + j \sin a$, expressions (2.9) and (2.10) coincide. The Re sign is not usually written, but it is actually implied.

In a plane, perpendicular to the x axis, the wave phase is the same (this plane is a wave front). A wave with a plane front is called a "plane wave," and a direction perpendicular to the front is called a "ray."

The harmonic plane wave of an arbitrary direction can be written as [3]:

$$\varphi = \mathbf{\Phi} e^{j(\mathbf{kx} - \omega t)}. \tag{2.11}$$

Here, $\mathbf{kr} = k_x x + k_y y + k_z z$ is a scalar product of the radius-vector of a point in space $\boldsymbol{r}$ and the vector $\mathbf{k} = \mathbf{n}\omega / c$, where $\boldsymbol{n}$ is a unit vector that characterizes a wave direction, and k_x, k_y, k_z are vector components. For a plane wave that propagates along an x axis, we have $k_x = k$; $k_y = k_z = 0$; as a result, we get a formula (2.10).

For the harmonic (or monochromatic) plane waves mentioned above, i.e., the waves that have one frequency of vibrations and an infinitely extended plane front, we write formulas for the correlation of the main values that characterize vibrations in this wave:

$$v = jk\varphi; \quad p = j\omega\rho\varphi; \quad u = -\varphi/c; \quad p = \rho c v. \tag{2.12}$$

In the practice of acoustic testing that includes the AE testing methods, the wave processes limited in time and space are used. Instead of monochromatic vibrations, pulses are most often applied. A pulse (from "*impulses*" in Latin = "impact" or "push") is the time-limited vibration process. The amplitude of vibrations in a pulse changes from zero to the final value according to the law that describes the pulse shape [4]. The time during which the amplitude exceeds 0.1 of its maximum value is considered to be the pulse duration τ. The product $c\tau$ is called a spatial duration of a pulse. It determines a space region occupied by a pulse. Using the formulas of spectral analysis, the pulse is represented as a frequency integral of monochromatic vibrations of a different frequency, i.e., it is expanded in the harmonic vibrations spectrum.

No success was achieved in obtaining a limited wave in the form of a bundle of parallel rays. For example, by cutting out a part of the plane wave front by a diaphragm, a complicated wave field is obtained. In practice, however, the weakly scattered ray bundles are used. A wave with an arbitrary front can be represented as an assembly of plane waves by decomposing in the Fourier integral over a wave vector k. For a durable enough acoustic pulse that propagates in a direction of the weakly scattered ray bundle, formulas (2.12) are used, although in this case, as approximate ones [1].

The Laplace operator ∇^2 in Eq. (2.8) can be represented not only in Cartesian but also in cylindrical or spherical coordinates. Accordingly, the simplest solutions of Eq. (2.8) will have the form not of a plane but of cylindrical or spherical waves. A harmonic spherical wave that starts from the origin of coordinates looks like

$$\varphi = \frac{\Phi}{\mathbf{r}} \mathrm{e}^{\mathrm{j}(\mathbf{kr}-\omega t)}. \tag{2.13}$$

Here, $\boldsymbol{r}$ is the radius vector that radiates from the origin of coordinates, and the direction of $\boldsymbol{k}$ coincides with $\boldsymbol{r}$, $k_x = k_y = k_z$. A surface with a constant phase, i.e., the wave front, has the form of a sphere for this wave, and the rays radiate in the direction of the radii. The wave amplitude decreases inversely proportional to the distance along a ray. At large distances r, an insignificant part of a spherical wave front can be considered to be a quasi-plane wave.

In the case of a sphere-like emitter of radius a that pulsates by volume changing with constant frequency and amplitude of vibration velocity, the pressure in an outgoing spherical wave is written as [5]:

$$p = \frac{\mathrm{j}a^2|v_0|\rho\omega}{\mathbf{r}} \mathrm{e}^{\mathrm{j}(\mathbf{kr}-\omega t)} \approx \mathrm{j}\frac{|p_0|}{\lambda \mathbf{r}} 2\pi a^2 \mathrm{e}^{\mathrm{j}(\mathbf{kr}-\omega t)} \tag{2.14}$$

where p_0 is the pressure near the sphere, calculated approximately by formula (2.12).

Elastic wave energy consists of the kinetic energy of medium particles motion and internal (potential) energy of deformation. The density of the kinetic energy is equal to $\rho|\mathbf{v}|^2 / 2$ [1]. In a running wave, the internal energy density is equal to the density of the kinetic energy, so the total energy density is $E = \rho|\mathbf{v}|^2$. The density of the energy flux is

$$W = cE = \rho c|v|^2 = |pv| = |p|^2 / \rho c. \tag{2.15}$$

The time average value of the energy flux density is called the elastic wave intensity. For a plane harmonic running wave, the intensity is determined as

$$J = |p|^2 / 2\rho c = \rho c|v|^2 / 2. \tag{2.16}$$

In a spherical wave, the intensity decreases inversely proportional to the distance squared:

$$J = |p_0|^2 2\pi^2 a^4 / (\lambda^2 \rho c r^2). \tag{2.17}$$

Acoustic waves attenuate when passing in the real media, and this is not accounted for by Eqs. (2.6) and (2.7). As a result, a wave number becomes a complex one: $k = k' + \mathrm{j}\delta$, where δ is the attenuation factor. A plane wave that propagates along the x axis with the account of attenuation is written as

$$\varphi = \Phi \mathrm{e}^{-\delta x} \mathrm{e}^{\mathrm{j}(k'x-\omega t)}. \tag{2.18}$$

2.1.4 Spatial Elastic Waves

By using Eq. (2.5) or (2.6), it is possible to show that in the infinite solid medium there are two types of waves that propagate with different velocities [3]. It is known from the vector analysis that any vector field can be represented as a sum of two vectors, one of which has a scalar potential and the other a vector one:

$$\mathbf{u} = \mathbf{u}_l + \mathbf{u}_t = \text{grad}\varphi + \text{rot}\boldsymbol{\psi}. \tag{2.19}$$

Taking into consideration that $\text{rot}\, u_l = \text{div}\, u_t = 0$ and substituting (2.19) into (2.6) with the application of operations rot and div, we get

$$\partial^2 \mathbf{u}_l / \partial t^2 - c_l^2 \nabla^2 \mathbf{u}_l = 0; \quad c_l = \sqrt{(\Lambda + 2\mu)/\rho}, \tag{2.20}$$

and

$$\partial^2 \mathbf{u}_t / \partial t^2 - c_t^2 \nabla^2 \mathbf{u}_t = 0; \quad c_t = \sqrt{\mu/\rho}. \tag{2.21}$$

Equations (2.20) and (2.21) are of the wave type and are similar to Eq. (2.7). Hence, the vector $\boldsymbol{u}$ in a solid decomposes into two waves propagating with different velocities.

The wave $\boldsymbol{u_l}$ is called a longitudinal wave or a wave of expansion-compression (Fig. 2.1), because the vibration direction in it coincides with its propagation direction. For the volume strain ε, the same Eq. (2.20) is satisfied.

The wave $\boldsymbol{u_t}$ in which the direction of vibrations is perpendicular to the direction of wave propagation and in which deformations are shear is called a transversal or shear wave (Fig. 2.2). There are no transversal waves in liquids and gases due to the absence of shape elasticity in these media. Strictly speaking, in liquids, there are waves similar to transversal ones, in which vibrations are transferred due to ductile forces; however, they decay quickly [1]. The ratio of velocities of longitudinal and transversal waves depends on the Poisson's ratio ν of the medium. For example, in metals, where $\nu \approx 0.3$, it is possible to get $c_t/c_l \approx 0.55$ (see Table 2.1).

When a transversal wave propagates in an infinite isotropic medium, all directions of transversal vibrations are equal. If there is a limiting surface, to which a transversal wave propagates parallel or at an angle, the question arises regarding the direction of vibrations in the transversal wave with respect to this surface. The wave, in which the direction of vibrations is parallel to the limiting surface, is called "horizontally polarized" (TH-wave). If vibrations occur in the plane perpendicular to the separating surface, such a wave is called "vertically polarized" (TV-wave). This type of wave is more frequently used in non-destructive testing. Therefore, if no special notations are given, by transversal wave we mean a vertical polarized wave [1].

In ultrasonic testing, a longitudinal wave is usually excited by special piezoelectric transducers that cause tension: compression deformation at a certain part of

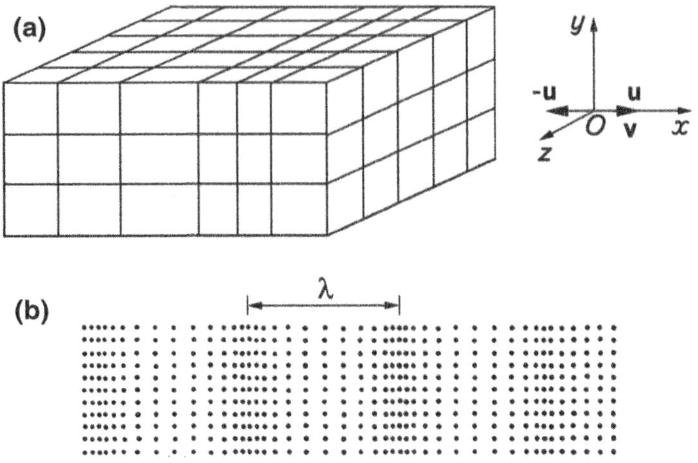

Fig. 2.1 Propagation of longitudinal wave vibrations (schematically)

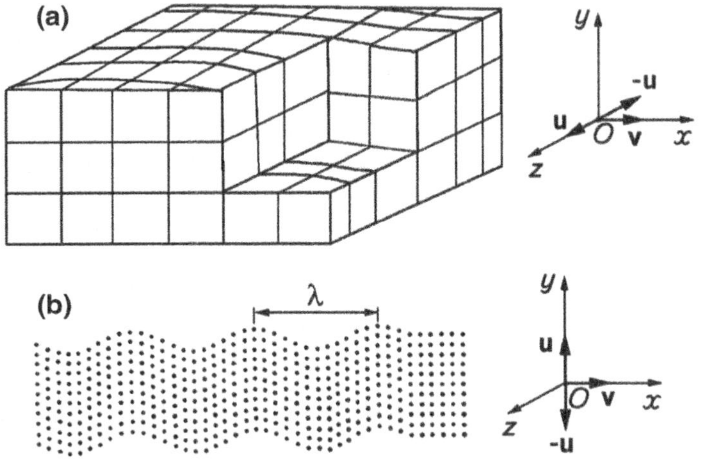

Fig. 2.2 Distribution of transversal wave vibrations (schematically): **a** is TH-wave; **b** is TV-wave

the IO surface, and a transversal wave—by shear deformation. Most often, a vertical polarized wave inclined to the surface is excited by a longitudinal incident wave at a certain angle to the IO surface from an external medium. Thus, the incident longitudinal wave is transformed into a transversal one. The external medium, from which a longitudinal wave falls at a certain angle, is called a "prism of a transducer" [6].

Table 2.1 Elastic properties of some media [1]

Material	Density ρ, 10^3 kg/m^3	Velocity of wave propagation c, 10^3 m/s			Wave resistance for longitudinal waves, z, 10^6 Pa × s/m
		Longitudinal	Transversal	Surface	
Metals					
Aluminum	2.7	6.35	3.08	2.80	17.1
Beryllium	1.82	12.8	8.71	7.87	23.3
Bronze	8.5…8.9	3.5…3.8	2.3…2.5	2.1…2.3	31…33
Bismuth	9.80	2.18	1.10	1.03	21.4
Tungsten	19.3	5.18	2.87	2.65	100
Duralumin	2.7…2.8	6.25…6.35	3.1	2.9	17.2…17.5
Iron	7.8	5.91	3.23	3.0	46.1
Gold	19.3	3.24	1.20	1.12	62.5
Cadmium	8.6	2.78	1.5	1.4	27.0
Brass composition-metal	8.5	4.43	2.12	1.95	37.7
Lithium	0.53	3.00	–	–	1.6
Magnesium	1.74	5.77	3.05	2.875	10.1
Copper	8.9	4.72	3.72	3.52	42.0
Molybdenum	10.1	6.29	3.35	3.11	63.5
Nickel	8.8	5.63	2.96	2.64	49.5
Niobium	3.9	4.10	1.70	1.58	35.3
Tin	7.3	3.32	1.67	1.56	24.2
Platinum	21.4	3.96	1.67	1.57	84.6
Mercury	13.6	1.45	–	–	19.8
Lead	11.4	2.16	.0.70	0.63	24.6
Silver	10.5	3.60	1.59	1.48	38.0
Steel corrosion-resistant	8.03	5.73	3.12	2.90	46.6
Carbon steel	7.80	5.92	3.28	3.01	46.1
Titan	4.50	6.00	3.50	3.20	27.0
Uranium	18.7	3.37	1.94	1.8	63.0
Zinc	7.1	4,17	2.41	2.22	29.6
Zirconium	6.5	4.65	2.25	2.15	30.2
Cast-iron	7.2	3.5…5.6	2.2…3.2	–	25…40
Non-metals					
Araldite	1.18	2.5	1.1	–	3.0
Capron	1.1	2.64	–	–	2.9
Quartz melted	2.2	5.93	3.75	3.39	13.0
Nylon, perlon	1.1…1.2	1.8…2.2	–	–	1.8…2.7

(continued)

Table 2.1 (continued)

Plexiglass	1.18	2.65…2.73	1.12…1.13	1.05	3.0…3.22
Aluminum oxide	3.7…3.9	10	–	–	37…39
Polystyrene	1.1	2.37	1.12	1.04	2.61
Rubber					
Raw	1.3…2.1	1.5	–	–	1.9…3.1
Vulcanized	0.9…1.6	1.5…2.3	–	–	1.3…3.7
Resin acrylic	1.18	2.6	1.12	–	3.2
Window glass	2.6	5.7	3.4	3.1	14.5
Textolite	1.2…1.3	2.6	–	–	3.1…3.9
Fluoroplastic	2.2	1.35	–	–	3.0
Porcelain	2.4	5.3…5.35	3.5…3.7	–	12.8
Ebonite	1.2	2.40	–	–	2.9
Epoxy resin hard	1.15…1.3	2.5…2.8	1.1	–	2.8…3.7
Liquids (20 °C)					
Acetone	0.792	1.192	–	–	0.4
Water	0.998	1.90	–	–	1.49
Glycerin	1.265	1.923	–	–	2.42
Kerosene	0.825	1.295	–	–	1.07
Acetic acid	1.05	1.84	–	–	1.45
Oil					
Diesel	0.88…1.02	1.25	–	–	1.1…1.3
Machine	0.89…0.96	1.74	–	–	1.5…1.7
Transformer	0.9…0.92	1.38…1.40	–	–	1.25…1.27
Alcohol					
Methyl	0.792	1.123	–	–	0.89
Ethyl	0.789	1.180	–	–	0.93
Gases (0 °C)					
Hydrogen	0.9×10^{-4}	1.248	–	–	1.1×10^{-4}
Air	1.3×10^{-3}	0.331	–	–	4.3×10^{-4}

2.1.5 Rayleigh Surface Wave

A specific type of waves propagate along the solid surface. For an unloaded (free) surface, the existence of some of them is proved in the following way [6]: We can assume a priori that there is a wave running along the solid boundary (along the x axis) and it consists of a linear combination of longitudinal and transversal vibrations, the amplitudes of which depend on the depth y under the surface (Fig. 2.3). For this purpose, the velocities of propagation of transversal vibrations should be the same and equal to some value c_s, while the wave number should be

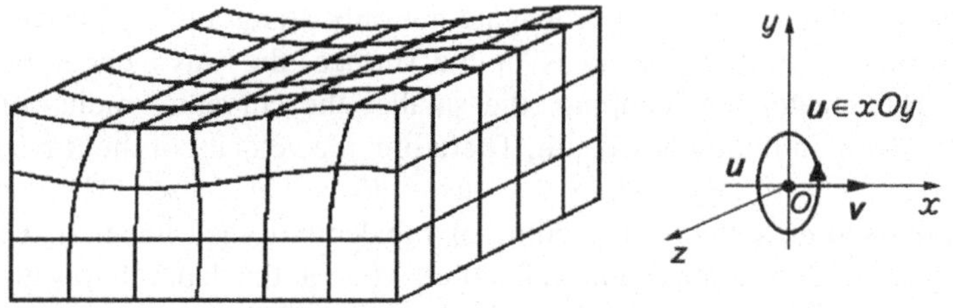

Fig. 2.3 Schematic presentation of the Rayleigh wave propagation on the free surface of a solid

equal to $k_s = \omega / c_s$. These predicted solutions are substituted into wave equations (2.20) and (2.21), which satisfy the conditions:

$$u_l = A\exp\left(-\sqrt{k_s^2 - k_l^2 y}\right)\exp(-\mathrm{j}k_s x),$$

and

$$u_t = B\exp\left(-\sqrt{k_s^2 - k_t^2 y}\right)\exp(-jk_s x). \qquad (2.22)$$

Here, and further, a multiplier $e^{-\mathrm{j}\omega t}$ is omitted. Thus, the obtained expressions correspond to the class of heterogeneous waves. In this wave, the amplitude changes in the front direction (along the y axis).

Functions (2.22) are to be substituted in boundary conditions of the equation: normal and tangential stress components vanish on a free surface, i.e., $\sigma_{yy} = 0$, $\sigma_{xy} = 0$. Two unknowns k_s / k_t and a ratio A/B of amplitudes are found from these two equations. For $(k_s / k_t)^2$, the 3rd degree equation is obtained. One real and positive root proves that an a priori assumption is correct and that the wave that was sought really exists.

An approximate formula for evaluating the velocity of the Rayleigh surface wave is known [1]:

$$c_s \approx c_t(0.87 + 1.12\nu)(1 + \nu)^{-1} \approx 0.93c_t \qquad (2.23)$$

for condition ν = 0,3. The absence of the imaginary part in a root for k_s indicates a weak decay of the surface wave caused just by an ordinary spatial wave decay, which is why the Rayleigh wave can propagate over a large distance along the solid surface. Its penetration underneath the surface of the body is small: at the wavelength λ_s, the intensity is about 5% of the intensity on the surface of the body.

During the surface wave propagation, the body particles move, revolving around ellipses (Fig. 2.3) with a large axis perpendicular to the surface forming a TV-type wave, and the extension of the ellipse increases with the depth. These conclusions yield from formulas (2.22), in which the greater the decrease of multipliers with depth, the greater the difference $k_s^2 - k_{l,t}^2$. A wave similar to the Rayleigh wave, i.e., a quasi-Rayleigh wave, can propagate not only along a plane but also along a distorted surface. On concave areas of the surface, the wave undergoes further damping. The greater the damping, the smaller the curvature radius due to the energy irradiation into the body depth. Therefore, the velocity of the Rayleigh wave decreases on concave areas, while on convex areas it increases. The wave selectively responds to the defects, depending on the depth of their location. The defects on the surface produce a maximum reflection, while at the depth that is greater than the wavelength, they are practically undetected.

2.1.6 Head (Creeping) Wave

When analyzing the solution to the problem on excitation of elastic waves upon a certain area of a solid surface, we can see that a wave propagates along the surface with the velocity practically equal to the velocity of a longitudinal wave. In [3, 6] it is referred to as a quasi-homogeneous wave because the amplitude along its front changes slowly. However, in the literature, such a wave is usually called a "head" or "creeping wave."

At every point on the surface along which a head wave propagates, a transversal wave is excited at an angle θ_t to the surface which is determined as $\theta_t = \arcsin(c_t / c_l)$. As a result, the head wave decays rapidly. Such waves, with a velocity smaller than that of a generating wave, are called "side waves." The combination of a head wave and side wave provides the situation with the stresses being equal to zero on the free surface of the body. Waves that consist of surface and spatial components, in which a surface component is continuously transformed into a spatial one, belong to the type of leaky waves [1, 6].

The head wave is usually excited by a longitudinal wave that falls inclined from the external medium (a prism) onto a partial area of the IO surface at an angle $\beta = \arcsin(c_0 / c_l)$ (Fig. 2.4). The bundle of longitudinal waves radiates from this area of the surface, with one of the rays passing along the surface—this ray actually being the head wave. The ray that propagates at an angle of 10…15° to the surface has the maximum energy of irradiation. The fronts of transversal waves *T*, generated by the head wave, are shown by lines whose width expands with an increase of the depth that corresponds to the wave amplitude increase. This takes place due to an increase of the number of surface points, which contribute to the formation of a side transversal wave.

Concurrent with the excitation of waves in IO, the waves in a prism are also excited. In seismic acoustics, a side wave is called a "*B* wave," which corresponds to a side wave in a prism. In seismic acoustics, it is referred to as a "head wave."

A quasi-homogeneous (head) wave is almost insensitive to surface defects and to a surface roughness. Nevertheless, it can be used for locating surface defects in a layer, beginning with the depth of 1…2 mm. by these waves. It is difficult to test thin structures by the side transversal waves that reflect from the opposite surface of IO and give off false signals [1].

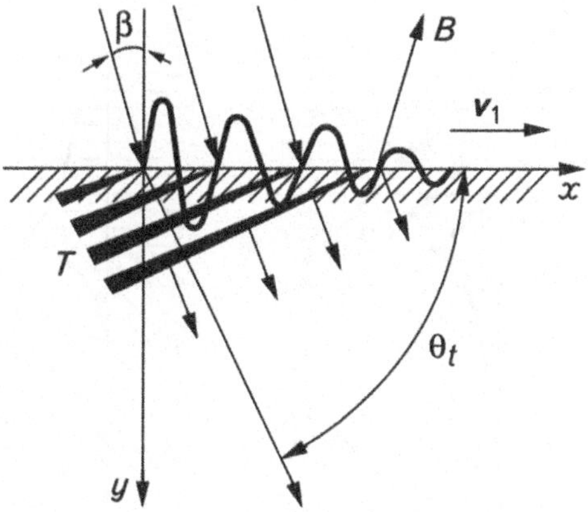

Fig. 2.4 Propagation of a head (creeping) wave on the free surface of a solid (schematically) [6]

2.1.7 Waves at an Interface of Two Media

The waves on the free surface of a solid are discussed above. An external medium on a limited area of the surface was introduced just to explain the wave excitation mechanism. When the solid surface is loaded with a liquid or hard medium, specific types of waves appear [1, 3, 6–8].

For a solid-liquid interface, with the sound velocity in a liquid $c_p < c_s$, a surface wave along the interface generates a side wave in a liquid and thus decays (Fig. 2.5a). For a steel-water interface, a side wave amplitude decreases e times at a distance of 10 λ_s. This feature classifies the wave as a leaking one.

Moreover, a wave can predominantly propagate in a liquid with a velocity less than c_p (Fig. 2.5b). In a solid, it is located in a layer of a thickness $\lambda_p/2\pi$, while in a liquid it is located in the layer of a thickness much larger than λ. The wave is used for testing the solid material surface by an immersion method. Similarly to the Rayleigh wave, it decays very slowly with the distance along the surface.

For an interface of two solid media (Fig. 2.5c) whose elastic modules and density differ insignificantly, the Stoneley wave propagates along the interface. It consists of something like two Rayleigh waves, each existing in its own medium, but both with identical propagation velocities that are lower than the velocities of spatial waves in both media. In each medium, the wave is localized in a layer of the thickness of a wavelength order and is vertically polarized. Such waves are used for the inspection of bi-metal joints.

Transversal waves propagating along the interface of two media and being horizontally polarized are referred to as the "Love waves" (Fig. 2.6). They arise when there is a layer of solid material on the solid half-space surface, with the velocity of transversal waves propagation less than in a half-space. The depth of the wave penetration in a half-space increases with a decrease of the layer thickness. When there is no layer, the Love wave in a half-space transforms into a spatial one, i.e., into a horizontally polarized plane transversal wave. The Love waves are used for testing the quality of coatings on the IO surface.

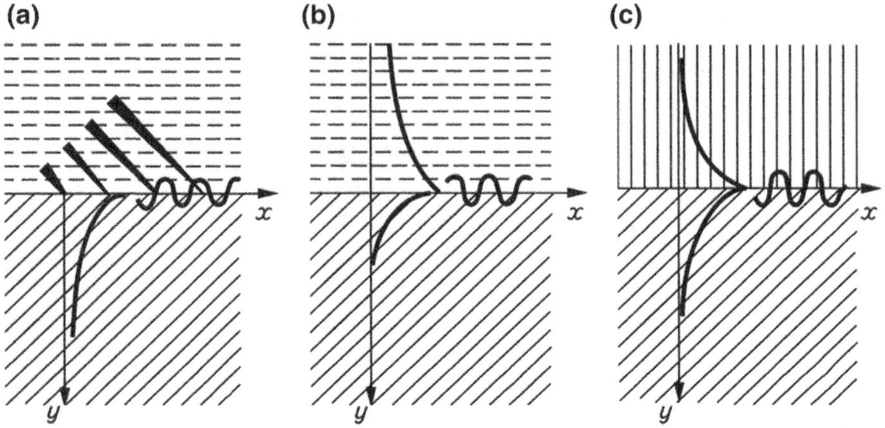

Fig. 2.5 Types of waves at the interface of two media: **a** is the decaying Rayleigh wave at the solid-liquid interface; **b** is the weak decaying wave at this interface; **c** is the Stoneley wave at the two solids interface

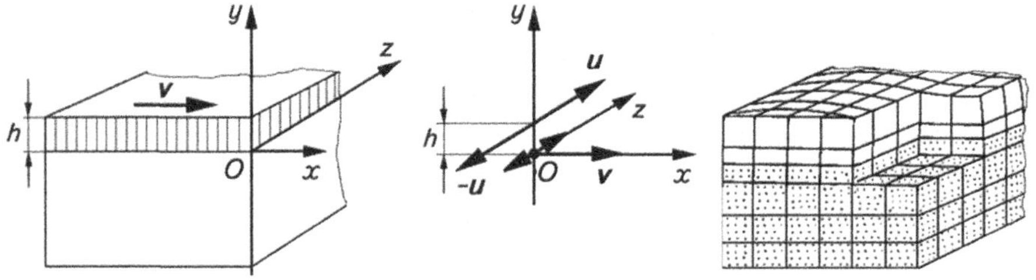

Fig. 2.6 Schematic representation of the Love wave propagation in a layer—half-space system

2.1.8 Waves in Layers and Plates

In bodies containing two free surfaces (a plate), specific types of elastic waves can propagate [1, 3, 6, 7]. They are called the "waves in plates" or the "Lamb waves" and are attributed to the normal waves, namely the waves propagating (transporting the energy) along a plate, a layer or a bar, and stationary waves (where energy is not transported) in a perpendicular direction. The solution to the wave equation for a plate with boundary conditions, when stresses on both surfaces are equal to zero, gives a system of two characteristic equations for the wave number k_p. It has two or more positive real roots, depending on the product of a plate thickness and frequency. A certain type of the wave (a mode) in the plate corresponds to each root.

To illustrate the physical essence of the waves in plates, let us consider the formation of normal waves in a liquid layer. Let a plane longitudinal wave fall outside onto a layer of thickness h (Fig. 2.7) at an angle β. Line AD shows the incident wave front. Because of the refraction at the interface, the wave with a CB

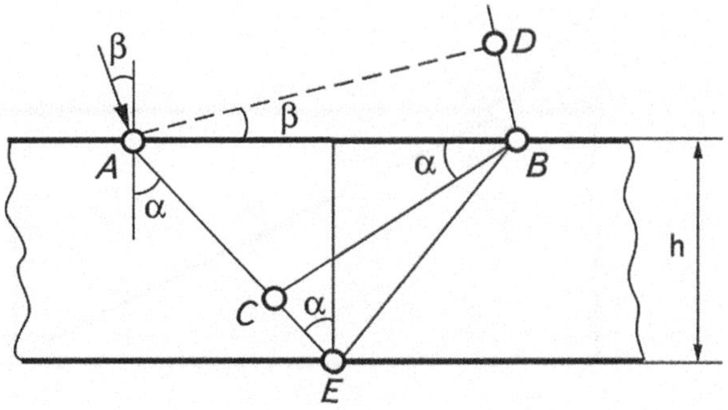

Fig. 2.7 Formation (schematically) of normal waves in a layer of liquid

front arises in the layer. The wave propagates at an angle α and reflects many times in the layer. At a certain angle of incidence, the wave reflected from the lower surface coincides in a phase with a direct wave that propagates from the upper surface. We can define the angles β or α, $\sin\beta / c_1 = \sin\alpha / c_2$, where c_1 and c_2 are velocities of the sound in a media, at which this phenomenon is observed.

As is known, the difference phases of the direct and the reflected waves are determined as

$$2\pi\left(\frac{AEB}{\lambda_2} - \frac{DB}{\lambda_1}\right) = 2\pi\left(\frac{2h}{\lambda_2\cos\alpha} - \frac{2h\,\mathrm{tg}\alpha\,\sin\alpha}{\lambda_1}\right) = \varphi. \tag{2.24}$$

Here, λ_1 and λ_2 are the wavelengths in an upper medium and in a layer. The condition of phase coincidence is attained if $\varphi = 2n\pi$, where n is an integer. Hence,

$$h\cos\alpha = n\lambda_2/2. \tag{2.25}$$

Thus, a wave in a layer is caused by the interference of the waves propagating in different directions. Interference (from the Latin "*inter*," meaning "mutually," and "*ferio*" meaning to "strike" or "hit") is the process of combining two or several waves in space. A monochromatic wave that spreads zigzag-like along a layer under condition (2.25) can be considered to be a wave that envelopes all the sections of a layer and moves along it. It differs from the above-mentioned waves by the velocity that varies, depending on frequency, i.e., velocity dispersion takes place.

Using a wave in a layer as an example, it is convenient to consider the concept of phase and group velocities. A group velocity characterizes the velocity of energy propagation in the direction of a wave motion. A wave pulse is a typical energy carrier. Since a pulse in a layer propagates zigzag-like, the energy propagation velocity by this wave along a layer is (Fig. 2.8) [1]:

$$c_g = c_2\sin\alpha. \tag{2.26}$$

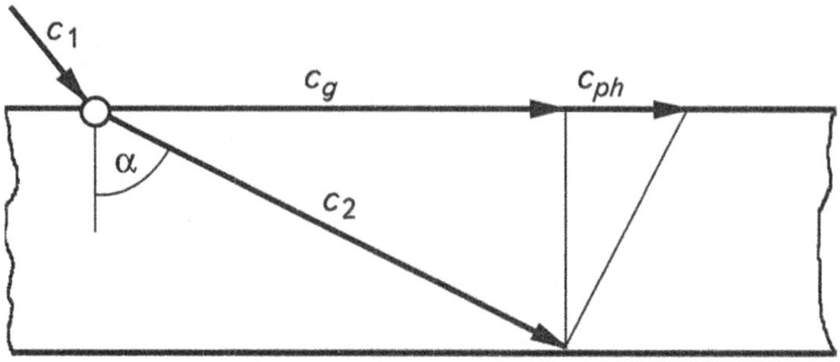

Fig. 2.8 Relationship between spatial c_2, group c_g and phase c_{ph} velocities of waves in a medium

A phase velocity determines the velocity of phase propagation in the direction of wave propagation. It is equal to the velocity of the phase variation of an incident wave along the layer, i.e., it is determined from the sine law:

$$\sin\beta/c_1 = \sin\alpha/c_2 = 1/c_{\text{ph}}; \quad c_{\text{ph}} = c_2/\sin\alpha. \tag{2.27}$$

Since a newly formed wave moves along the layer, its refraction angle is 90°. The angle α is evaluated from (2.25). As a result, we get

$$c_g = c_2\sqrt{1 - (n\lambda_2/2h)^2}; \quad c_{\text{ph}} = c_2/\sqrt{1 - (n\lambda_2/2h)^2}. \tag{2.28}$$

Thus, the phase and group velocities of normal waves depend on the frequency of elastic vibrations and the layer thickness. The dispersion curves, i.e., the graphs of the c_{ph}/c_2 versus h/λ_2 dependence are shown in Fig. 2.9. At the points where $h/\lambda_2 = 1/2$; 1; 3/2, etc., phase velocities tend to infinity. It means that the whole surface vibrates simultaneously, and c_g in these cases is equal to zero.

When $h/\lambda_2 \to \infty$, for all values of n, velocities c_g and c_{ph} of normal waves tend to c_2, i.e., the spatial wave velocity. The waves with the odd values of n are called symmetric because the motion of particles within these waves is symmetric about the layer axis, and the waves with even values of n are antisymmetric.

The formula for the relation of phase and group velocities is known

$$\frac{1}{c_g} - \frac{1}{c_{\text{ph}}} = -\frac{f}{c_{\text{ph}}^2}\frac{dc_{\text{ph}}}{df}. \tag{2.29}$$

Dependences (2.27) and (2.28) satisfy this equation.

Regarding solid layers (plates), it should be noted that the essence of the phenomenon, i.e., the formation of standing waves along a plate thickness due to interference of normal waves, remains here, although the conditions of formation of normal waves are more complicated due to the presence of longitudinal and transversal waves. Due to reflection, these waves are partly transformed into one another, and the wave phase can vary by an aliquant π number.

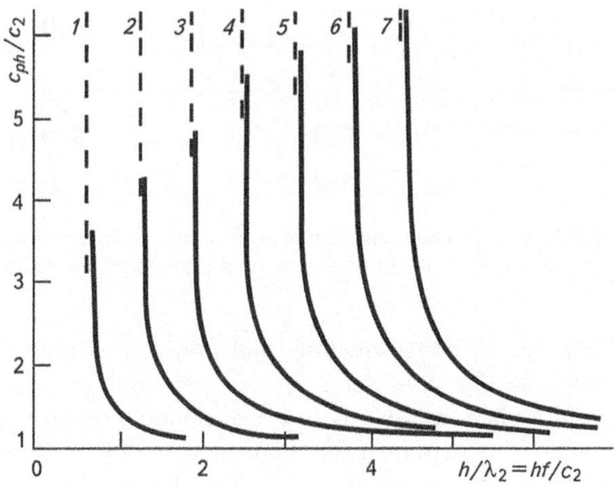

Fig. 2.9 Dispersion curves of normal waves propagating in a liquid layer

In Fig. 2.10, a system of dispersion curves for the phase velocity of waves in a steel plate is plotted. Zero indices indicate the modes that transform into a surface wave with an increase of the plate thickness. These waves exist at any frequency and for any thickness of plates. A zero symmetric mode s_0 (Fig. 2.11a) corresponds to the wave of extension-compression, while a zero antisymmetric mode a_0 under condition $h \ll \lambda$ (Fig. 2.11b) corresponds to the bending wave. For a wave s_0 under condition $fh \to 0$, phase and group velocities are the same since there is no dispersion.

In the considered modes of normal waves, the medium particles vibrate in their propagation plane. They are caused by the interference of longitudinal and transversal vertical polarized waves. In a plate, the waves can also be formed due to the interference of transversal horizontally polarized waves. When the waves are reflected from the plate boundaries, the waves with horizontal polarization are not

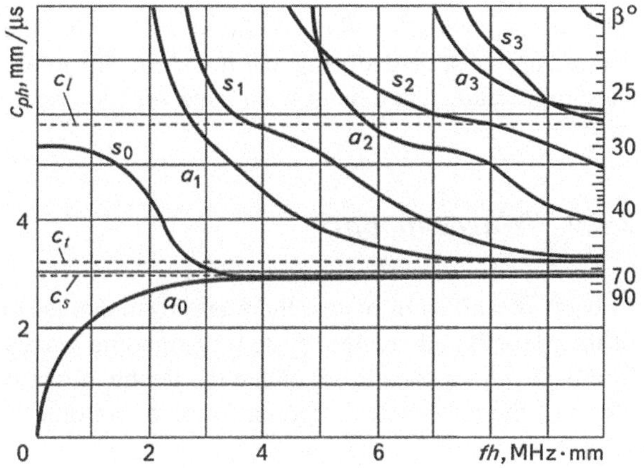

Fig. 2.10 Dispersion curves of the Lamb waves in a steel plate: c_l, c_t and c_s are the velocities of longitudinal, transversal and surface waves; $a_0, a_1, a_2\ldots$ are antisymmetric modes; $s_0, s_1, s_2\ldots$ are symmetric modes of vibrations

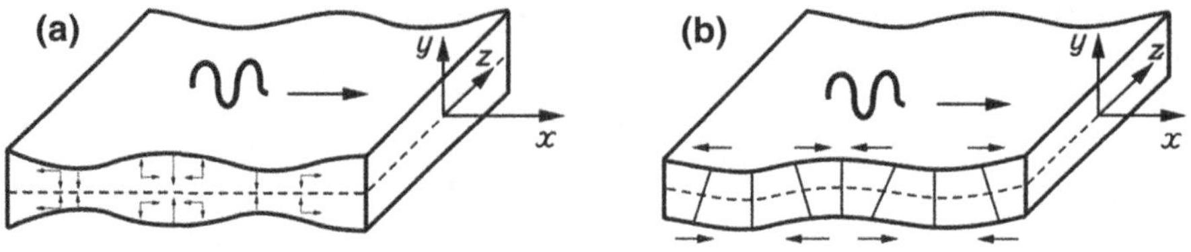

Fig. 2.11 Schematic representation of normal symmetric (**a**) and antisymmetric (**b**) waves: x is wave propagation direction (*arrows* indicate directions of displacements along the x and y axes)

subjected to transformation, and a system of dispersion curves is similar to that shown in Fig. 2.9.

In a plate, as well as in a waveguide, normal waves propagate over large distances. This permits using them successfully in testing the sheets, shells, and pipes that are 3…5 mm or less thick. The change of the waveguide cross-section and the presence of heterogeneity (defects) in it, cause the reflection of normal waves. It should be noted that the changes in the wave propagation conditions in a waveguide are caused not only by transversal but also by longitudinal defects—for example by the laminations located along the wave propagation direction. The defects located along the spatial wave propagation direction are poorly detected; this feature is useful in flaw detection.

As has already been shown above, the Love waves are very important surface waves that propagate in a layered media. The phase velocity of a Love wave is higher than the velocity of a transversal wave in a layer, although it is lower than their velocity in a half-space. From the solution of the known equations [7], it follows that the dispersion of Love waves is due to the presence of a dimensional parameter, i.e., a layer thickness.

Strictly speaking, Love waves are not true surface waves because they are caused by layered media. Various roots of a dispersion equation characterize the corresponding normal modes, and their number is larger than the product of the wave number k_2 (transversal wave) and the layer thickness h. Sometimes, Love waves are interpreted simply as the lower mode of vibrations that exists on all layer thicknesses, including the condition $k_2h \rightarrow 0$. These waves, like the Rayleigh waves, are observed during earthquakes, since the earth's crust has a lamellar structure. Lately, they have been used for the creation of dispersion delay lines.

2.1.9 Waves in Bars

In bars, as well as in plates, there are normal waves that propagate in the direction of their length and create a system of standing waves in a transversal cross-section. These waves are sometimes called the Pochhammer waves. For bars with a different shape of the cross-section (round, square and other), special systems of dispersion curves take place that indicate symmetric and asymmetric modes. The velocity of a mode s_0 in a bar is less than in a plate.

Except for the already mentioned symmetric and asymmetrical waves, a twist wave can propagate in a bar or a pipe. Vibrations in a twist wave arise due to rotation about an axis of a certain cross-section of a bar or a pipe. Different modes of normal waves in a bar are excited by an off-normal incidence of a longitudinal wave from an external medium or by electromagnetic-acoustic method (a twist wave).

2.1.10 Other Types of Waves

Besides the waves described above, there are other types of surface waves in solids that can be used in technical diagnostics and nondestructive testing of objects. First of all, the surface waves in crystals must be mentioned [3, 5, 7]. Today, the existence of surface waves in most directions of any cross-section of crystals is proved. Owing to the anisotropy of their elastic properties, a plane surface wave has three components of displacement, and its wave vector does not coincide with the direction of the group velocity vector. Only for crystal symmetric directions, the vectors of the group and phase velocities are collinear, and the paths of particles lie in the plane that passes through the wave vector and a normal to the surface. These surface waves are very similar to the Rayleigh waves in an isotropic solid, and so they are sometimes called the "Rayleigh-type waves" [5].

A typical example of such a wave is the wave that propagates in the *ZY* direction —the cross-section of the lithium niobates piezoelectric crystal. It is worth noting that in piezo-crystals, the surface wave is usually accompanied by a quasi-static electric field, and this feature is used in different acoustic-electronic devices of signal processing.

A piezoelectric effect in a number of cases leads to the formation of purely shear surface waves known in the literature as the "Guliaev-Bleustein" waves. They, unlike the Rayleigh waves, are weakly heterogeneous and decrease with depth at a certain distance depending on the coefficient of electromechanical coupling. Hence, they depend less on the surface irregularities and are used in the acoustic-electric devices that operate at super-high frequencies. The waves of this type can exist in non-piezoelectric materials at the presence of external fields: electric (due to a resulting piezoelectric effect), and magnetic (the action of the Lorentz force on electrons in metals). In such cases, it is possible to inspect the depth of the wave localization by applying a field of specific value [7].

There is an analogy, however, only with the electromagnetic waves in a metallic comb, for shear surface waves that propagate along a periodically unequal solid boundary (Fig. 2.12) [7]. Wave equations and critical conditions in both cases are identical. Therefore, it is possible to reach the conclusion, based on the known solution for electromagnetic waves, that in the system under discussion a surface wave can propagate with the phase velocity of

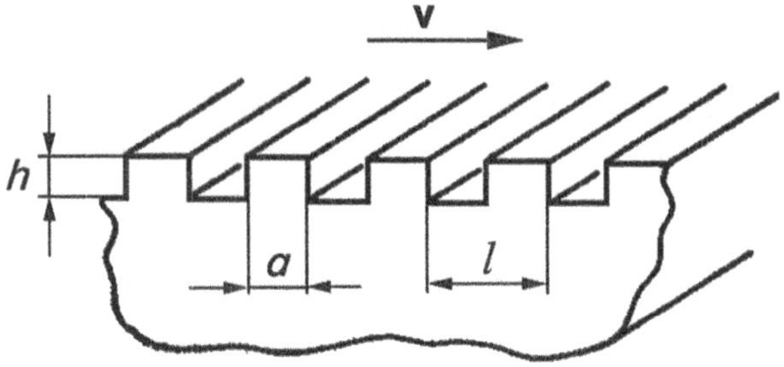

Fig. 2.12 A comb-like structure on a solid surface

$$c = c_t\left[1 + (a/l)^2 \mathrm{tg}^2(kh)\right]^{-1/2}, \tag{2.30}$$

where a, l and h are the geometrical parameters of a medium of propagation of waves, k is the wave number.

Equation (2.30) is valid under condition $kl \ll 1$. Thus, it yields that at certain values of a, l, and h, the surface wave velocity c can be significantly less than c_t.

A number of surface waves are defined purely by geometrical factors. On a convex cylinder surface of a solid, except for the Rayleigh type waves, there should also be non-Rayleigh surface waves with polarization in the plane that passes through a wave vector and a normal to the surface. A component of longitudinal motion in those waves behaves like the motion in the Rayleigh wave, decreasing with depth by an exponential dependence. A shear component, decaying with depth, oscillates; such waves are called "mixed-type" waves. Their velocity is somewhat higher than the velocity of a shear wave, and it is attained asymptotically with an increase of the cylinder radius. In convex cylinders, there are purely shear surface waves that are polarized parallel to the surface. Since the reflection of horizontally polarized shear waves is similar to the reflection of waves in a liquid, such surface waves do not differ from mixed-type ones.

Other linear waves, which propagate along the elastic wedge edges, are also known to exist (Fig. 2.13) [7]. They represent another class of wave motions of a continuum that concentrates near the surface borders. The waves are similar to the surface ones, but they have not been studied comprehensively. They are analyzed by numerical methods, and calculations illustrate that an arbitrary field at the wedge edge, similar to any other waveguide, can be represented as the sum of vibration modes that propagate along the edge with their inherent phase velocities.

From the practical viewpoint, the most important modes of a wedge are the lower anti-symmetric ones. Their amplitudes also quickly decrease with depth, so that practically all the energy of a wave remains concentrated near the edge line. Owing to this fact, such waves are called linear or wedge waves. They are not dispersion waves, because the wedge is characterized only by an opening angle α_k rather than by linear sizes. The velocity of wedge waves diminishes with a decrease

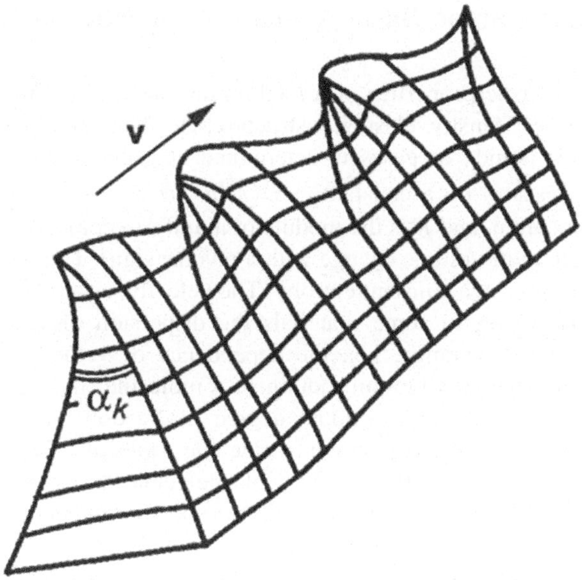

Fig. 2.13 Anti-symmetric vibrations of the wedge edge (schematically)

of α_k and can be less than the velocity of a spatial transversal wave if $\alpha_k \approx 5\ldots10^{\mathrm{o}}$ by one order of magnitude. Physically this is clear, because anti-symmetric waves in a sharp wedge are similar to those in a thin plate. The latter ones are also called the "lower anti-symmetric Lamb" modes or "bend waves."

Finally, we should mention Sesave waves and the waves caused by the mechanisms of the field non-locality that are related to the medium microstructure. The first ones arise in the system: layer half-space, similar to the Love waves under condition of a small thickness of a layer (Fig. 2.6). In this case, there is only one mode of vibration, which transforms into a Rayleigh wave when $kh \rightarrow 0$. The following mode of such vibrations is a "Sesave" wave. It occurs when the velocity of transversal waves in a layer is larger than the transversal velocity of a sound in a half-space. These waves, as well as the Lamb waves, are widely used in acoustic electronics for making the surface wave waveguides operate according to a general principle of open waveguides [6].

The second ones cannot be explained from the position of continuum mechanics. The use of non-local continuum models permits predicting the occurrence of strongly heterogeneous surface waves whose amplitude also decreases at the depth of the order of atomic spacing in a lattice.

The surface wave types in solids discussed in this chapter do not include all the types of waves occurring in nature. This concerns, first of all, the waves in layered media [3, 4] and the others caused by various mechanisms.

2.2 Some Basic Acoustic Properties of Media

Acoustic properties of a medium are determined by its physico-mechanical properties: density, elasticity, structure, etc. The velocity of acoustic waves propagation for liquids or gases is determined at a given state of the medium (temperature, pressure) by a constant $c = \sqrt{(\partial p / \partial \rho)_s} = \sqrt{K\rho}$, where p is the pressure in the medium, and K is the module of a uniform compression (the ratio of pressure to the strain of volume change with a reverse sign). The index s shows that the derivation is done at a constant entropy. The velocity, as a rule, does not depend on frequency. However, in some materials the dispersion of velocity is observed in a certain frequency range, i.e., the dependence of velocity on the number of degrees of freedom of vibration motion of a molecule. In the frequency range that has been mentioned, an additional degree of freedom should be involved in vibrations: mutual motion of atoms inside the molecule. Investigation of the properties of materials and the kinetics of molecular processes by measuring velocity and damping of acoustic waves is the subject of molecular acoustics. Isotropic solids are characterized by the velocities of elastic wave propagation determined by formulas (2.20) and (2.21) (see Table 2.1). These two values of velocities can be used as a pair of elastic constants instead of the Lame coefficients or modules of elasticity.

A crystal solid is an anisotropic medium. Its properties vary depending on the directions, and the maximum possible number of independent elastic constants is 21. However, the presence of a crystal symmetry diminishes the number of independent elastic constants for the majority of them. In a crystal, three elastic waves can propagate with different velocities in every direction. In an isotropic solid, longitudinal wave and two transversal waves with mutually perpendicular directions of vibrations correspond to them. The velocities of these transversal waves are identical. In a crystal, a displacement vector in each wave has the components, which are both parallel and perpendicular to the propagation direction, i.e., each wave will be neither purely longitudinal nor purely transversal [1].

The group velocity that determines the direction of the energy flow $\boldsymbol{J}$, due to the anisotropy of crystal characteristics, does not coincide with the direction of the wave vector $\boldsymbol{k}$, although the wave fronts remain perpendicular to $\boldsymbol{k}$. An angle between $\boldsymbol{J}$ and $\boldsymbol{k}$ is spatial and can be of tens of degrees.

The acoustic wave propagation velocity is a function of temperature. This dependence is characterized by a change of velocity by one degree of temperature. For gases, this value is positive, while for liquids and solids, it is negative—an order of 0.01…0.05%.

2.2.1 *Impedance and Wave Resistance of a Medium*

The term "impedance" (from the Latin "*impedio*," meaning "oppose") means "resistance." Acoustic impedance is determined as a ratio of a complex sound pressure to a spatial vibration velocity. This concept is used, in particular, for the description of acoustic wave propagation. In the extended medium, the concept of specific acoustic impedance is introduced by using the correlation of sound pressure with the vibration velocity (not spatial). Since only this case is considered here, the term "specific" is omitted (a mechanical impedance differs from an acoustic) in what follows.

If a plane harmonic wave propagates in a liquid medium, then according to formula (2.12), acoustic impedance is equal to $z = p/v = \rho c$. This value characterizes a medium in which the wave propagates. It is called the "wave resistance" of a medium or its characteristic impedance. This concept of impedance is also used for a solid (for longitudinal and transversal waves), defining it as a ratio of the corresponding mechanical stress to the vibration velocity of a medium particles taken with a reverse sign (see Table 2.1).

The origin of the term "impedance" is related to the system of electromechanical analogies, in which voltage is compared with pressure, and current with velocity. From the physical viewpoint, acoustic and mechanical impedance shows how difficult it is to "loosen" the system or the degree of the system's incompliance with vibrations.

2.2.2 *Decay of Elastic Waves*

A decrease in the plane harmonic wave amplitude as a result of its interaction with a medium takes place according to the law $e^{-\delta x}$, where x is the distance that the wave passes in the medium and δ is the damping coefficient. In what follows, the term "damping" will refer only to the wave amplitude decrease that is taken into account by an exponential multiplier, contrary to a decrease in the amplitude related to the wave front extension, e.g., in a spherical wave.

The value that is the reverse of the damping coefficient illustrates the distance at which the wave amplitude decreases e times, where e is the Napier number; therefore, the dimension of the damping coefficient is m^{-1}. In the literature [1], this unit is sometimes written as Np/m. However, this unit is not foreseen by the State Standard of Ukraine. The damping coefficient is often expressed by the number N of negative decibels, by which the wave amplitude decreases at a unit distance of its propagation $x = 1\text{m} \times N = 20\lg e^{-\delta l} = -8.68\ \text{dB/m}\ (1\ \text{m}^{-1} = 1\ \text{Np/m} = -8.68\ \text{dB/m})$.

The damping factor consists of the coefficients of absorption δ_1 and dissipation δ_2: $\delta = \delta_1 + \delta_2$. During absorption, a sound energy transforms to thermal, and during dissipation the energy remains sonic, but is emitted by a wave whose propagation is directional. Absorption is conditioned by ductility, elastic hysteresis (i.e., by a different elastic dependence during extension and compression), and heat conductivity. The last absorption mechanism is conditioned by the fact that the process of acoustic wave propagation is considered to be adiabatic. The extension or compression of an elementary volume accompanied by the temperature alteration is so short that the process of temperature equalization cannot be taken into account. In fact, there is a heat conductivity that assists in the loss of vibration energy. There are also other mechanisms of absorption revealing themselves at frequencies higher than those used in technical diagnostics and nondestructive testing of components and structural elements.

The dissipation of waves occurs due to the medium heterogeneities whose wave resistance differs from a medium one; their sizes are commensurable with the wavelength. A difference in the wave resistances causes the reflection of waves. Small sizes and a large number of heterogeneities specify a statistical character of a dissipation process. Such heterogeneities can be represented, for example, by the inclusion of various types in alloys, solid particles, or bubbles of air in water. In gases and liquids, which do not contain foreign particles, there is no dissipation, and the damping is determined by absorption. The absorption coefficient is proportional to the square of frequency. Therefore, a value $\delta' = \delta/f^2$ is introduced as a characteristic of sound absorption in liquids and gases.

The absorption coefficient in solids is proportional to the frequency f (glass, biological tissues, metals, some plastics), or f^2 (rubber, majority of plastics). For the same medium, the absorption of transversal waves at $f = \text{const}$ is less than for longitudinal ones; this is because transversal vibrations are not conditioned by the change in a volume and there are no heat conductivity losses.

There is no dispersion in homogeneous amorphous solids such as glass and plastic. A weak dispersion in them can arise due to internal stresses that cause the change in elastic wave velocity and their refraction. In heterogeneous materials such as cast iron, granite, concrete, etc., dispersion is very large. Considerable dispersion is also observed in most metals, even at a high degree of homogeneity [1].

As is known, metals have a polycrystalline structure and consist of a great number of crystals (grains)—single crystals that are not clearly facetted. All crystals are most often randomly oriented; at the transition of elastic waves from one crystallite into another, their velocity can vary to a greater or lesser degree due to their anisotropy. As a result, a partial reflection, refraction, and transformation of wave types appears that determine the mechanism of dispersion. The greater the elastic anisotropy of crystals, the larger the dispersion. Anisotropy is also characterized by the parameter of elastic anisotropy. In a cubic crystal, it is a measure of relative resistance of crystals to two types of shear deformation. A large anisotropy is typical of copper, zinc, and austenitic stainless steel; a small elastic anisotropy is typical of tungsten and aluminum. Alpha iron and carbon steels belong to intermediate materials according to the value of elastic anisotropy and dispersion.

The value of the dispersion coefficient in a medium is greatly affected by the correlation of the average size of heterogeneities and the average distance between heterogeneities with the elastic wavelength. In metals, the medium parameter that significantly affects the dispersion of crystals is the average size of a particle $\bar{d}$. In

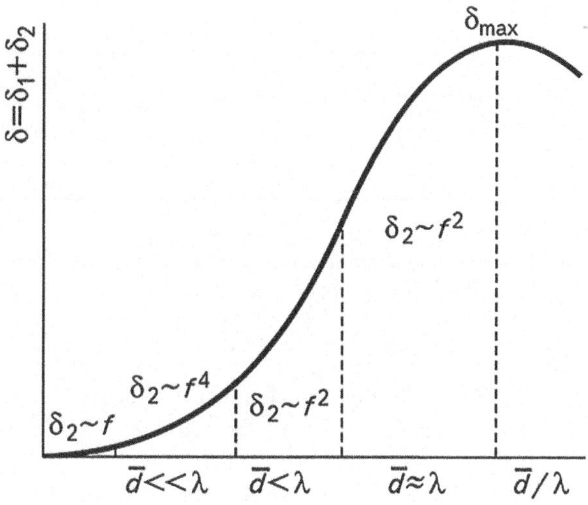

Fig. 2.14 Schematic dependence of the damping coefficient on the correlation of average grain diameter and the elastic wavelength

the case of $\bar{d} \gg \lambda$, coefficient δ_2 is proportional to f^4 (Rayleigh dispersion) (Fig. 2.14). Then, total damping is calculated by the equation

$$\delta = Af + Bf^4\overline{d^3}, \tag{2.31}$$

where A and B are the constants, and f is the frequency of vibrations. The term Af is caused by absorption, and it dominates at low values of f. In the region $4 \le \lambda / \bar{d} \le 10$, coefficient δ_1 is proportional to the product $\bar{d}f^2$.

In metals with different grain size, the power index at f varies from 2 to 4, and maximum damping is observed when $\lambda \approx \bar{d}$. In carbon steel, grains consist of a great number of fine plates of iron and cementite (Fe_3C). Their sizes are considerably smaller than the average size of grain $\bar{d}$. This is probably the reason that in a wide range of frequencies up to the values of $f = 4\ldots5$ MHz in fine-grained carbon steels, the damping is determined by absorption, i.e., it is proportional to frequency. In austenite steel welded joints, the crystal orientation is ordered.

In general, a wave field generated by the AE source consists of elastic waves of different types: spatial longitudinal and transversal waves that, in their turn, generate the surface Rayleigh, Lamb, and other waves (Fig. 2.15). These waves propagate with various velocities and damp according to different laws. Thus, for instance, the damping of spatial longitudinal waves is proportional to R^{-2}, where R is the distance from the source to the observation point. Surface waves, such as Rayleigh and Lamb waves, damp proportionally to R^{-1} and are capable of carrying information on the AE source character at the distance greater than spatial waves.

In paper [10], a problem concerning the evaluation of elastic displacements in Rayleigh waves excited by the vertical harmonic point force source, which acts at a depth h under the surface of an elastic half-space, is solved. It shows that with the growth of h, the amplitude of waves decreases, and under condition

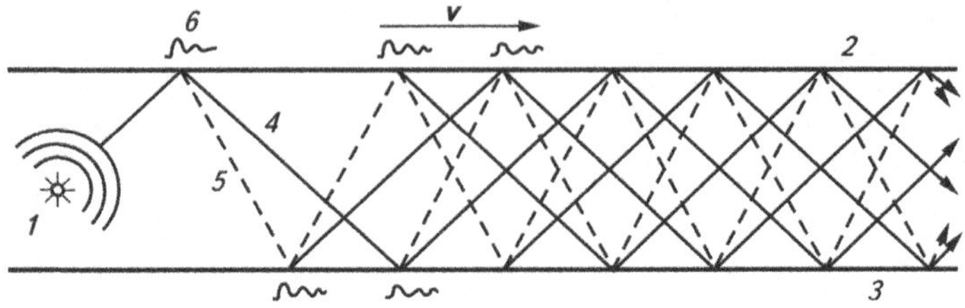

Fig. 2.15 Illustration of the elastic wave propagation in a plate (passing of one ray): *1* is a source, *2* is the upper surface of a plate, *3* is the bottom surface of a plate, *4* is a longitudinal wave, *5* is a transversal wave, *6* is a surface wave [9]

$h \geq (2...3)\lambda$ (λ is the wavelength), it tends to zero. The authors of [11] showed that due to the difference in the damping of spatial and surface waves, at large distances from the AE source, surface waves prevail even when the source mainly generates spatial waves. They stated the prevalence of that or other types of waves depending mainly on mutual locations of an AE source and an AET mounted on the IO. Therefore, in the case of their locating on one side of a plate, the Rayleigh wave prevails, while in other cases, when a source is inside a plate or when a source and a receiver are on different sides of the plate, the sequence of spatial longitudinal and transversal waves prevails in a waveform.

In paper [12], the propagation of elastic waves in pipe specimens of 17G1SU steel with various wall thicknesses was investigated using the method of "sounding." Wide-band AETs were used for emitting and sensing elastic waves. To study the effect of water on the character of elastic waves propagation, experiments were performed on water-filled specimens. It was found that after filling the specimen with water, the velocities of elastic waves did not change, and only the dynamic parameters of the wave field underwent changes: the amplitude of spatial waves diminished, and vibrations arose between the appearance of longitudinal and transversal waves. The effect of the pipe specimens filling with water on the waveform was studied for various wall thicknesses h of specimens, taking into account the dimensionless parameter kh, where $k = 2\pi / \lambda$ is the wave number and λ is the wavelength. The physical essence of this parameter is that it determines the conditions of elastic wave propagation in pipe specimens: $kh \approx 1$ corresponds to the conditions of wave propagation in a plate, $kh \gg 1$ is for wave propagation in an infinite medium. The variation range of kh is from 2 to 10. At the values of $kh > 10$, the effect of water on the damping of elastic waves was not observed (Fig. 2.16).

Authors [12] suppose that the main contribution to the energy of wave package is given by Lamb waves, whose amplitude is by one order of magnitude higher than the amplitude of longitudinal spatial waves. The filling of a pipeline with water causes distortion of the wave forward front structure, producing an error during the AE signal location in the hydrotesting of pipelines.

Figure 2.17 shows the effect of pipe steel degradation in the feeding pipelines of supercritical pressure power units of thermal power stations on the damping of AE elastic waves. A knee pipe about 1130 mm long, cut out of the pipeline and not

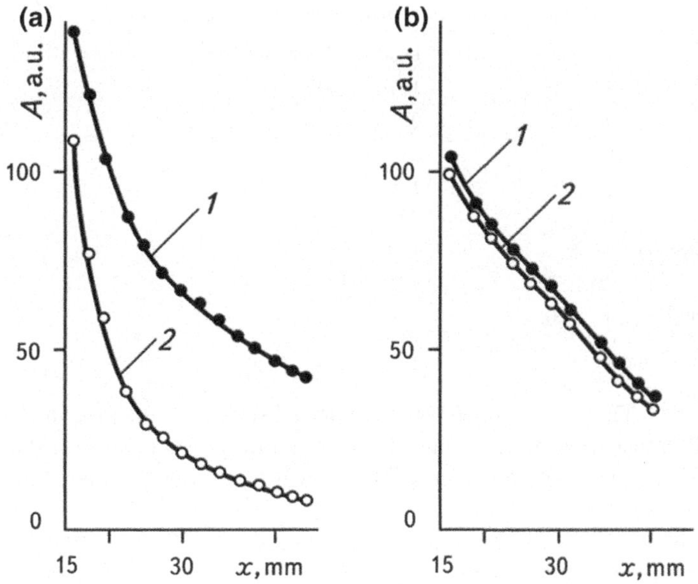

Fig. 2.16 Dependence of elastic wave amplitudes damping A (arbitrary units) on the value of *kh* and a passing distance *x*: **a** $kh = 2$; **b** $kh = 10$; *1* is for a specimen without water, *2* is for a specimen with water

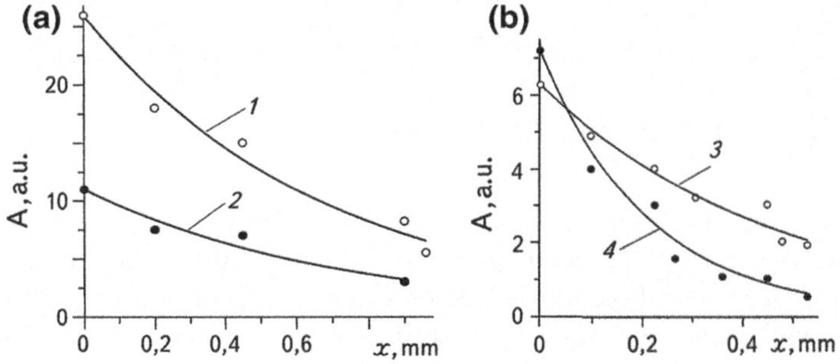

Fig. 2.17 Elastic AE waves damping while passing through the degraded material of a pipe bend made of 16GS steel, recorded by the resonance (**a**) and a wide-band (**b**) TAE: resonance frequency is 260 kHz (*1*); 320 kHz (*2*); 560 kHz (*3*); 810 kHz (*4*)

filled with the service environment, was investigated. For AE sounding, a probe-simulator of the AE pulses was used. The operating time of large-scale equipment was over 120,000 h.

Mechanical pulses were excited by electric rectangular pulses of voltage 12.5 V and frequency of 0.4 Hz. In order to select and record the AE signals, the AETs (see Chap. 1) were used as a set of serial AE devices that have various gain-frequency

Table 2.2 Damping of elastic waves in some liquids and in air at 20 °C [1]

Medium	Frequency *f*, MHz	Damping coefficient α/f^2, 10^{14} $s^2/m = 10^5$ m^{-1} MHz^{-2}
Air	1.1...1.4	1670...2000
Water	7...250	2.5
Glycerin	0.5...4	250
Kerosene	6...21	170
Vinegar acid	0.5	9000
Oil (transformer)	1...5	130
Mercury	20...50	1.2...1.3
Alcohol ethyl	1...220	5.4

characteristics. They are mounted on the external surface of the pipe through a layer of acoustic transparent oil at certain distances from the stationarily installed sound simulator. The external diameter of the knee pipe was 410 mm, and its wall thickness was 55 mm.

As follows from the results of investigations (see also Table 2.2), the AE signals damp in a different manner depending on their frequency range. The design features of AET significantly affect this parameter as presented below:

Thus, the character of the wave field depends on the mutual arrangement of an AE source and a receiver, the source type and IO, frequency characteristics of the primary transducer. This means that in each case it is necessary to carry out experimental studies regarding the effect of the above-mentioned factors on the formation of the wave field that irradiates an expected or known AE source.

2.2.3 *Diffraction of Elastic Waves*

Diffraction (from the Latin "*diffractus*" meaning "broken down") of waves is a deviation of waves from the geometrical laws of propagation when interacting with obstacles. Hence, the diffraction of a sound (and ultrasound) is a deviation of the sound behavior from the laws of geometrical (ray) acoustics, conditioned by the wave nature of a sound. The elastic fields created by an initial wave diffraction on the obstacles are called "scattered" or "diffracted" waves [1].

We should consider the diffraction of elastic waves on the objects of a regular geometrical shape simulating real defects. Usually, it is very difficult to get the exact solution to most problems on elastic waves diffraction. Therefore, approximate methods are used for this purpose.

According to Young's theory, the field that arises due to a wave diffraction is caused by the interference of waves that propagate according to geometrical laws as well as by diffracted waves that arise at the critical points with discontinuous boundary conditions. The boundaries of the obstacles—edges on their surfaces—are the locus of such points. According to the Fresnel theory, a diffraction field appears due to the action of fictitious secondary sources excited by an incident wave on the obstacle during the reflection or outside it (during passing). For the same conditions, the calculation results obtained using both methods coincide.

Diffraction on a slot edge and on a strip. An infinitely thin slot simulates real defects, such as lamination, extended crack, or faulty fusion, and its edge is the edge of the corresponding defect. The diffraction field calculation in this case is usually done using the Sommerfeld method, which is actually Young's theory development for plane obstacles. From each point of the edge (Fig. 2.18), a diffraction wave propagates as a cone, one of its generatrices being a continuation of the incident longitudinal wave ray; the transformed transversal wave forms another cone. Moreover, there are waves that spread along the slot surface. Amplitudes of all these waves are proportional to the amplitude of the incident wave, though by 1...2 orders less.

If a slot is not semi-infinite and is of a finite width, there are two edges, on each of which the diffraction waves similar to those considered in Fig. 2.18 appear. When a limited bunch of rays of an elastic wave covers both edges of the strip, and a mirror reflection from the strip does not reach the receiving transducer, its signal is determined by interference of diffracted waves from the edges. The points on the reflecting edges (similar to real defects), on which diffraction waves appear giving maximal contribution to the scattered wave field, are called "bright points."

Diffraction on a hollow disk. Small-sized, crack-like defects are simulated by a disk. A reflector is considered to be a hollow one if stresses on its boundaries are equal to zero. A problem on diffraction upon a disk is among the problems on the diffraction of elastic waves upon the small-sized objects in comparison with a wavelength. The exact solution of similar problems consists of expanding the incident and scattered waves into a series by functions close to the object shape. Oblate spheroidal functions are used for a disk, i.e., eigenfunctions of a wave equation in the system of oblate spheroidal coordinates that coincide with the disk surface.

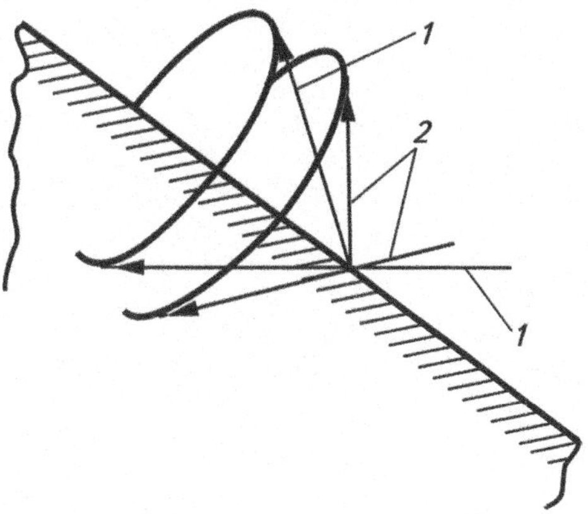

Fig. 2.18 Formation of diffraction cones of longitudinal (*1*) and transversal (*2*) waves at an inclined incidence of a longitudinal wave on the slot edge

Expansion into a series is carried out, assuming that the value of $k_1 b < 2\pi$, where k_1 is the wave number for a transversal wave, and b is the disk radius. In the incident wave, coefficients of the series terms are known, while in the scattered longitudinal and transversal waves they are unknown. They are to be found from the boundary conditions: the normal and tangential stresses on a hollow disk surface are equal to zero. These conditions should be satisfied for the terms of identical powers in a series for the incident and scattered waves. Figure 2.19 shows the solution for perpendicular incidence of a longitudinal wave on a hollow disk in an aluminum body.

As we can see, the amplitude A_d of a scattered longitudinal wave at small angles of observation grows in a non-monotonous manner with the growth of $k_1 b$. At the same time, the directional diagram of the scattered wave narrows.

Diffraction on a cylinder, sphere, or ellipsoid. These objects simulate the real defects, such as pores, slag inclusions of various shapes, etc. They have a smooth convex surface. From the viewpoint of diffraction theory, they differ from the slot edge, strip, and disk by the absence of bright points, and the diffraction waves appear in every point on their surface.

A problem concerning diffraction on a cylinder is solved by expanding the scattered and incident wave potentials into a series by cylindrical functions. If kd is assumed to be of small value, where k is the wave number for longitudinal or transversal waves and d is the cylinder diameter, the solution is called a "long-wave approximation," and if $1/kd$, it is considered to be a "short-wave approximation."

An energy (ray) approximation (d ≫ λ) yields the following expression for the amplitude of reverse reflection from a cylinder $B_c = 0.5\sqrt{d/\lambda}$. The long-wave approximation for $d \ll \lambda$ gives the value $B_c = 7.4(d/\lambda)^2$. The change in B_c for intermediate values of d/λ shows significant differences in the diffraction of various wave types (Fig. 2.20).

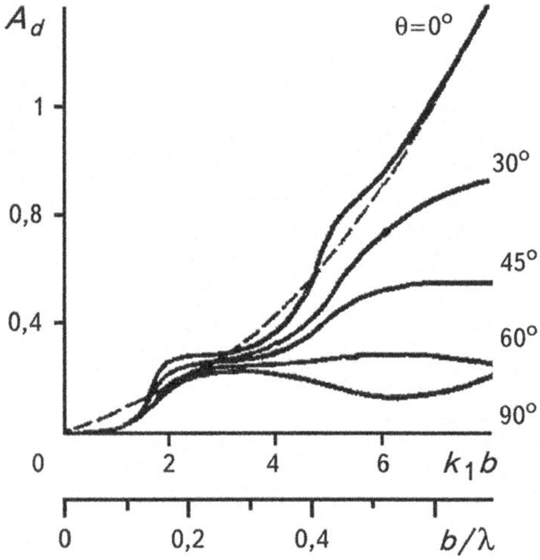

Fig. 2.19 Amplitudes of the longitudinal wave scattered on a hollow disk in aluminum at different observation angles θ [13]: *solid lines* correspond to a model for a solid; the *dashed line* is the Kirchhoff approximation for $\theta = 0$

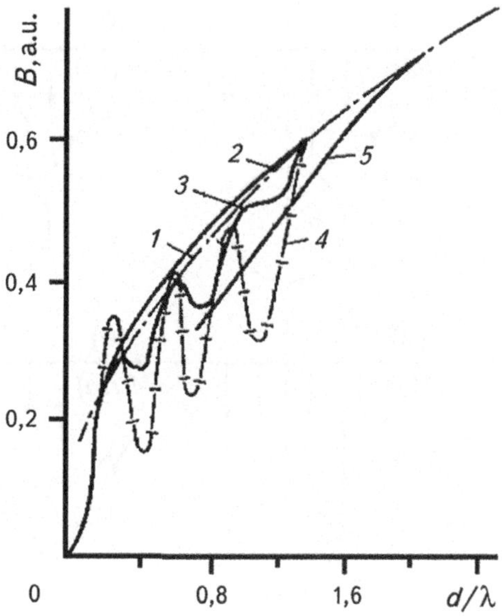

Fig. 2.20 Amplitude of a wave reflected from a hollow cylinder [13]

Curve *2* for a longitudinal wave rather precisely coincides with the energy theory data (curve *1*) up to the values of $d/\lambda \geq 0.2$, and the reflection amplitude of a transversal wave coincides only up to $d/\lambda \geq 2$. At smaller value of d/λ, the theory predicts the occurrence of oscillations for a transversal wave reflection (curves *3*, *4*). These oscillations are especially large for waves with vertical polarization (perpendicular to the cylinder axis, curve *4*). However, the results show (curve *5*) that for a pulse character of irradiation theoretically calculated for continuous irradiation, there are either no oscillations of function *5* or they are very small, even though curve *4* is lower than for the energy theory.

The origin of oscillations and their smoothing can be explained by a short-wave approximation [13]. It is shown that in the case of a transversal vertical polarized wave incidence (Fig. 2.21), various types of waves are formed: transversal (Fig. 2.21a) and longitudinal (Fig. 2.21b), transformed from a transversal. A ray of the incident transversal wave that touches the cylinder surface excites the heterogeneous wave of a transversal type (Fig. 2.21c), which envelops the surface of the cylinder (enveloping wave).

Figure 2.21d shows the same wave that envelops the cylinder in a reverse direction (for other waves, which are examined below, the variants of reverse enveloping are not shown). These two waves generate transversal waves of sliding that starts from every point of a cylinder tangential to its surface.

A ray that falls on the cylinder surface at the third critical angle (Fig. 2.21e) generates a longitudinal enveloping wave (a head wave) which, in turn, generates the wave of sliding also of a transversal type that begins at the third critical angle. The ray of the incident transversal wave that passes close to the cylinder (Fig. 2.21f) creates an enveloping wave of the Rayleigh type, which also

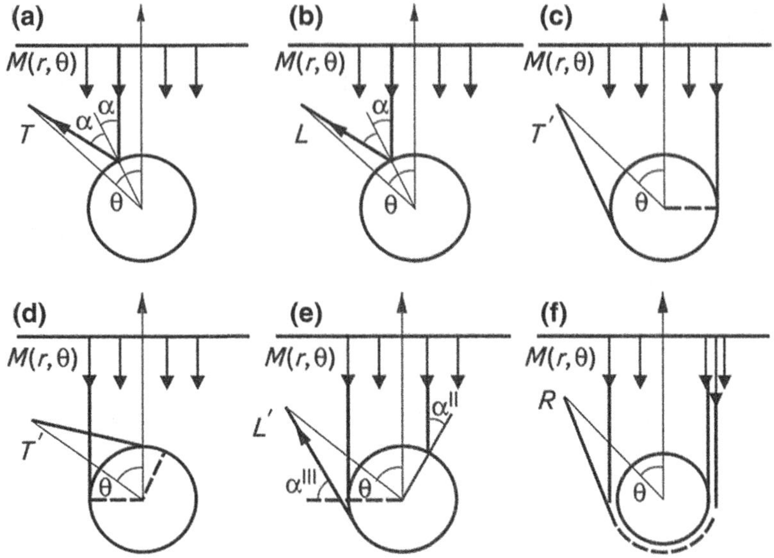

Fig. 2.21 Initiation and propagation of waves during diffraction of a transversal perpendicular polarized wave on a cylinder (a short-wave approximation)

re-irradiates the energy into space in the form of a transversal sliding wave, because it propagates along a concave surface.

Thus, except for the directly reflected elastic waves, three more transversal waves emerge in the observation point. These waves are generated by the enveloping waves of transversal, longitudinal and Rayleigh types. The amplitudes of the enveloping longitudinal waves are much lower than those of transversal and Rayleigh waves.

For a large diameter cylinder, the mirror reflected signal and a series of pulses of enveloping waves are clearly observed. With a decrease in the cylinder diameter, the pulses come closer, and they merge for the condition $d/\lambda \approx 1.5$, and their combined interference brings about oscillations. The condition of the lack of oscillations $d/\lambda > 1.5$ is close to the condition of coincidence of the experimental curve with the energy approximation curve ($d/\lambda \approx 2$).

Diffraction waves on a sphere are formed according to the same laws as on a cylinder. The difference is that during the longitudinal wave incidence on a sphere, the difference of amplitudes of the diffracted and mirror-reflected signals are fewer than for a cylinder. Due to an axial symmetry of the problem for a sphere, the rays that envelop a cavity at different directions arrive at the observation point simultaneously. This condition is satisfied only for a combined type of sounding ($\theta = 0$).

When a transversal wave falls on the sphere, the oscillations at various areas of its front are oriented differently with respect to the sphere surface; they can be expanded into vertical and horizontal polarized vibrations, the diffraction of which occurs by various laws. The intensive enveloping waves are typical only of the vertical polarized vibrations, whose portion for a spherical object is less than for a cylinder.

For an elastic wave incidence on an ellipsoid or an elliptical cylinder, a diffraction field is formed; it has the features typical of diffraction on spatial objects, such as sphere or cylinder, and plane, such as disk or band. The prevailing type of diffraction depends on the degree of the ellipse oblateness determined by the ratio of the axes.

2.2.4 *Refraction of Elastic Waves*

Refraction—in the wide sense—is the same as the deflection of rays. Regarding acoustic waves, by refraction we mean a continuous variation of the acoustic ray direction in a heterogeneous medium. The velocity of the waves therein depends on the coordinates. This phenomenon is observed in a layered-heterogeneous and anisotropic medium, whose velocity varies according to a particular law. This medium is assumed to consist of an infinite number of infinite thin layers; in each layer, the elastic wave velocity is constant, but it varies step-wise at the boundaries of the layers. The sine law should be applied at the boundary of these two layers in order to determine the ray behavior: $\sin\alpha/c = \cos\gamma/c = \text{const}$, where $\gamma = 90^\circ - \alpha$ is the sliding angle. The variation of velocity c causes the deviation of the rays from their linear propagation direction, thus forming both acoustic shadows and zones of energy concentration in which acoustic surfaces arise.

We should mention some examples of the media having variable velocity of wave propagation, such as welded joints of austenitic steel, transversal isotropic non-metal materials, surface-hardened components, rolls of cold rolling, axes and bushes of some mechanisms, etc. A special heat treatment mode improves hardness of the surface layers while the internal layers of metal remain unhardened, ductile. They are often called "raw" metals. Thus, in the non-destructive testing of materials, components, and structures, specific features of a material structure should be considered.

2.3 AE Sources

Since most objects inspected by means of nondestructive testing and technical diagnostics are made of metallic alloys, the classification of the AE sources, which generate AE signals during initiation and propagation of a fracture therein, is of special interest to researchers. According to [14] they can be classified as follows: dislocation processes of metal plastic deformation, phase changes, second-phase particles fracture, magnetic effects, surface phenomena and external effects, and material failure. Let us briefly consider some of them, since they were described in the previous chapter.

Dislocation processes. It is shown in [15] that even in aluminum monocrystals of a high degree of cleanness, the AE was generated at stresses lower than the threshold of macroscopic yielding of metal. Later on, a group of researchers discovered a similar effect for other metals. Thus, in paper [16], the AE generation is recorded in dependence on the defect energy of a crystalline structure of copper alloy monocrystals. Using the terms of redistribution and motion of dislocations, the effect of Bauschinger that also causes the AE generation is explained [17]. It is thought that mobile screw dislocation generates AE during motion from one low-energy state to the other, at this point eliminating the oscillation in a lattice [18] that is confirmed by certain theoretical calculations [19–21]. It should be noted that the energy of an AE signal, which is conditioned by a single dislocation, is very low and is therefore difficult to detect. However, a cooperative motion of many dislocations was recorded without much difficulty by modern devices [22].

Annihilation of dislocations. In [23–25], models of the AE origin during annihilation of dislocations on the free surface or during their crossing the interface between two sections of the metal with a different elasticity module were described. The validity of these models is proved by the fact that AE irradiates during the electrochemical removal of the Al_2O_3 film from aluminum specimens [26].

Uniform motion of dislocation groups. There have been studies in which it has been indicated that the number of dislocations involved during co-operative motion varies from several hundred to a few thousand pairs [27, 28]. For the continuous AE, these values can be considerably smaller—from 10 to 100. There is a hypothesis that even irradiation by dislocations of discrete AE is conditioned by uniform motion of a great number of dislocations [29–31]. However, these statements have been corrected, and have shown that the AE is generated only during accelerated or decelerated motion of dislocation groups [22, 28, 32].

Action of a dislocation source. During the action of any dislocation source for a short interval of time on a slip plane in the same direction, a great number of similar dislocations is released and a dislocation avalanche is formed. In [22, 33] it is established, from the analysis of literary sources, that the action of the Frank-Read source causes the AE from LiF monocrystals and metals. It is emphasized that on the silicon steel, the sources of dislocations appear at stresses lower than the macroscopic elasticity limit. The same is set for monocrystals of copper and zinc.

Dislocation breaking from the pinning points [22]. There are models in which the AE bursts are explained by the concepts of sudden pressure (avalanche) of newly formed free dislocations [33–35]. This was experimentally confirmed in papers [36–38] on various metals. For the development of the mentioned concepts, a hypothesis [39] was proposed, according to which dislocations move in the slip bands packages, each of them having a certain number (n) of dislocations. With the beginning of an unsteady plastic flow, the n is large and the AE irradiation is substantial. The behavior of other AE parameters can be explained by the fact that the density of dislocation packages is described by the Weibull distribution as a function of strain.

Formation of slip bands. Studies are known in which the dependences between the slip bands' initiation and the formation of discrete AE in Mg, Cu, and Fe monocrystals were investigated [35]. It was found that the appearance of the first AE pulse corresponds to the origination of the first slip band. The authors believe that the process of formation of the first slip band is simulated by a primary plastic flow caused by a creeping avalanche. It should be noted that such a model, although having the right to exist, is very blurred.

Grain boundary slipping. Since Kaiser's studies [40], a hypothesis has been proposed that says that AE is conditioned by the motion of grain boundaries. However, modern achievements in AE studies refute these assertions, leaving them partly for description of the AE parameters generated in the case of deformation of lead and its alloys. Thus, we can disregard the viewpoint that the mechanisms of grain boundaries sliding are crucial or important sources of the AE [14].

Microvoids coalescence. It is commonly known that development of plasticity in metals includes the process of the formation of cavities or pores and their further development and coalescence. The mechanism of final coalescence can cause a spontaneous failure of ligaments between cavities. In this case, the AE pulses are also generated, and especially considerable AE arises during ligament breakage [22].

Phase transformation. A sudden transition of one metastable phase into another is, possibly, the first case of AE best revealed by man. The famous "cry of tin" [14, 22] is the most widely used example to illustrate the processes of AE generation. Such processes are basic AE sources in some metals under certain conditions of their deformation, and rarely occur when other AE sources are comparatively small. In this case, AE is an exceptionally accurate and useful factor for investigating transformation processes.

Strain twinning. The twinning of tin and zinc is the phenomenon known in metallurgy, which in the case of tin loading generates such AE that it is usually perceived by the human ear as a series of crunches [41]. Similar effects of AE generation are observed in titanium [42] and its alloys, and in zirconium [43] and zinc [44]. This proves that in these metals, the twinning is the basic source of AE, while the remaining mechanisms, including dislocation sliding, are the secondary ones.

Elastic twinning. The AE pulse [45] accompanies the initiation and disappearance of a single twin induced by the application of stresses to a calcite. It was established in [14] that the initial stages of a twin development and sliding are very similar processes. Therefore, the same model can be used to explain them [45]. This assumption today is insufficiently grounded to provide a clear distinguishing of the mentioned above mechanisms as the AE sources.

Twinning expansions. The AE was observed in zirconium with the expansion of twins [39]. The phenomenon has not yet been studied well enough and needs special investigations.

Martensite transformations. The AE appears during transformation of austenite into martensite, beinite, pearlite, and ferrite [46]. The most important characteristics of the AE parameters are as follows:

- The AE beginning coincides with the first martensite formation;

- The cumulative count of the AE attains the maximum at a martensite final temperature, i.e., the AE accompanies a transitional process;
- The cumulative count of AE increases with an increase of carbon content;
- The count rate is maximal at the temperature lower by 50 °C than the temperature of the beginning of martensite transformation. Thus, this temperature can be evaluated by the AE data;
- The cumulative count of AE per volume unit of martensite formation is very sensitive to carbon content, which is determined by the martensite morphology; and
- The lamellar formations of martensite irradiate the AE of a greater energy than the strip martensite. It is not caused simply by the relative value of the volume of the formations, but can be caused by high rates of a lamellar martensite growth.

These ideas are confirmed in papers [47–49], where the alloys of non-ferrous metals and steel were experimentally investigated. Some metastable steels irradiate AE at relatively high and uniform rates of plastic deformation [50–53]. This formed a basis for the study of the processes of martensite formation using the AE parameters.

Beinitic, eutectoid reactions and other phase transformations are also accompanied by the AE [54] generation, which to some extent can be used for their qualitative or quantitative estimation.

Fracture of secondary phase particles. In many cases, metal ductility depends on the fracture easiness of either a particle or the boundary between the particle and the matrix, as well as on the volume fraction of these particles, their distribution, shape, etc. Therefore, it is reasonable to expect that such failures are accompanied by AE, as they are the processes of local stresses relaxation. Every industrial alloy has its inherent concentration of the secondary phase particles. It can change from extremely fine-grained required precipitations in, for instance, areas of recrystallization in duralumin, to relatively large, undesirable inclusions, such as silicates in steels. The nature, size, amount, and the effect of every particle varies within a wide range, hence the AE generation is specific for every metal.

Non-metal inclusions in steels. In steels, the amount of inclusions reaches 10^{12}... 10^{15} per tone of metal for relatively pure grades. Their composition, sizes, shape, properties, etc., are usually different, but they can be grouped as aluminates, silicates, or sulfides. The failure of such inclusions under mechanical stresses is accompanied by the AE [28, 53]. This occurs at almost all stages of deformation. In study [54], the mechanisms related to inclusions are distinguished as [54]:

- Decohesion of MnS streaks along the interface that is located parallel to the rolling plane;
- Fracture of the silicates streaks or MnS across or along the direction of rolling;
- Decohesion of MnS and failure of SiO_2 particles of a spherical shape;
- Failure of the piled-up dislocation groups at an inclusion during the break or decohesion of the last; and
- Friction between the surfaces of the failed brittle particles.

Carbides in steels. Some authors have devoted a number of papers to the problem of carbide fracture in steels [54]. In some of the papers it is proved that the majority of AEs are caused by failure of carbide plates in pearlite or at the ferrite grain boundaries. However, spheroidized carbides in pearlite emit a relatively low amount of AES. There is a scientifically sound assertion that grain boundary films fail under very insignificant deformations. Plates of pearlite fail at greater deformations depending on their size, and spheroidal particles fail within the strain range from 30 to 40%. There is a tendency of these processes to depend on the size, shape, and orientation of the particles in a metal.

Precipitations in aluminum alloys. The action of the AE sources in aluminum alloys is described in detail in [55], and the appearance of two AE peaks is vividly illustrated in paper [37], in which the aluminum alloy went through the yielding point at the early stages of deformation. It is shown that the first AE peak is caused by dislocation mechanisms, and the second by failure of the fine brittle intermetallic inclusions. These peaks split under the action of tensile strain, which was later verified by other researchers [54]. It is seen from the above that AE in most aluminum alloys is mainly the result of the fracture of solid particles. It should be emphasized that a dependence of fracture toughness was found on the brittle particles' location in front of the macrocrack tip in aluminum alloy.

Fibers and fibrous composites. Mechanisms of initiation and development of fracture of different fibers and fibrous composites are described in detail in [56]. The AE caused by fiber failure are highly energetic and are well detected by equipment. Although individual features of the AE generation in materials differ, in the case of composites it is possible to quite easily distinguish the AE sources related to the material fracture mechanisms by the AE parameters [56].

Slag inclusions in welded joints. The fracture of slag inclusions takes place during a weld cooling, which is accompanied by the initiation of mechanical stresses. As a result, the AE emits (a detailed description is given in [57]).

Magnetic effects. Ferromagnetic materials subjected to mechanical stresses tend to order their domains so that the magnetostriction deformation is in the direction of the applied loading. The Barkhausen effect is observed in iron alloys caused by the turning of the domain walls. It is known [58–61] that such motion of domain walls under tensile deformation causes the AE generation during the elastic deformation of the material. This phenomenon forms the basis for investigations of a number of structural materials in order to determine their physical properties [54]. Recently, the methods of AE investigations, comprising the Barkhausen effect, have been more widely used in the non-destructive testing of products and constructions.

Surface effects. The AE is also generated during cracking or delamination of oxide layers from the surface of steel specimens. In other cases, the AE sources are also the traces of sliding lines; the formation of dimples that collapse as well as pores, cracks, or dislocations that come out on the surface are considered to be the AE sources [54]. It is also known [62–64] that the processes of surface friction are accompanied by the AE.

Fracture of materials. Fracture is a complex process that includes a number of possible mechanisms and phenomena that have already been described above as the

AE sources. It mainly concerns the formation and development of the processes of plastic deformation in front of the crack tip. Therefore, when investigating the AE during the fracture of metals, it is essential to clearly define the origin of AE signals, and to find their correspondence to proper mechanisms that occur in a metal at different stages of fracture (see Chap. 1). The most important of them are related to the processes that accompany metal plastic deformation, such as dislocation processes, initiation, and propagation of pores, inclusion failure, and cracking.

The start of a macro-crack and the stages of a macro-crack sub-critical growth are accompanied by discrete high-amplitude AE, which is described in detail in Chap. 5 of this monograph. These fracture stages have been quite well studied, which is attested to by the research results published in a number of monographs [65–71] and review papers [72–78]. Here we will mention only a few of them:

Stress corrosion cracking. This phenomenon is one of the most active AE sources. A number of high-strength steels under stress corrosion cracking in an NaCl environment generate the AE with energy that is proportional to the crack area. The AE during intergranular fracture is energetically higher than during transgranular cracking. A similar situation is observed for stress corrosion cracking of coarse-grained and fine-grained steels [54, 79]. The AE method presented above is successfully used for the estimation of the threshold stress intensity factor under stress corrosion cracking of steels [80, 81].

Hydrogen embrittlement. During the fracture of a hydrogen-induced embrittlement of steel, a high amplitude level of the discrete AE is also observed [54, 80, 82–84]. The energy level of AE signals during such processes is higher than under stress corrosion cracking. Taking this into consideration, the AE method has proved to be very useful in detecting and observing the delayed fracture in forgings and welded joints [57]. The studies have shown a good correlation between the rate of hydrogen-induced cracking and the AE parameters [85–87].

Summing up the above, Table 2.3 presents the most typical AE sources during initiation and development of fractures in metals.

It is worth noting that the AE is also generated during fracture initiation and propagation in other non-metal materials. Numerous publications in the world's scientific journals are devoted to this phenomenon. Since there is a certain analogy in the propagation of elastic waves and in the AE sources action, we will not discuss it.

When conducting non-destructive testing of objects in industrial or in field conditions, the sources of AE can be gusts of wind, rain, snow, hail, etc., that cause the propagation of elastic AE waves as noise or background level of noises. This phenomenon should be taken into account when interpreting the results of non-destructive testing and diagnostics of the state of products and constructions. In Tables 2.4 and 2.5, the quantitative AE parameters and ranges of their variations are presented for the most characteristic sources of AE.

Table 2.3 Some AE sources in metals

Group	Source	Comment
1	2	3
Dislocation mechanisms under plastic deformation	Motion of one dislocation[1]	Below the susceptibility threshold of serial equipment
	Annihilation of dislocations on the free surface[2]	
	Self-annihilation of a couple of dislocations[3]	Below the susceptibility threshold of serial equipment
	Fracture of a dislocation loop[4]	Below the susceptibility threshold of serial equipment
	Uniform motion of a group of dislocations[5]	Below the susceptibility threshold of serial equipment
	Action of the Frank-Read source	Possible source of AE, but not prevailing
	Action of the source of dislocations located at the grain boundary	Possible source of AE, but not prevailing
	The avalanche of the dislocation—dipoles capture phenomena that occur at the grain edge	Possible source of AE applied in fatigue testing at small deformation
	Non-fixing of dislocations or breaking of dislocations[6]	Source is proved soundly
	Formation of a slip band	It is not the basic mechanism
	Slipping in the grain lattice	Possible only in lead alloys at 20 °C
Phase changes	Deformation at which twinning takes place	Possible source in certain metals and alloys
	Formation of elastic twins	Possible, but does not refer to metals
	Extension of twins	
	Martensite transformations	Steel, brass
	Beinitic resistance in steels	Requires investigations
	Phase transformations in tin	Not identified
Particles of the second phase	Non-metal inclusions in steel	Obviously play very important part in industrial steels
	Carbides in steels	Obviously play an important part in industrial steels
	Phase precipitation in aluminum alloys	Dominant source
	Fibers in composites	Dominant source
	Slag inclusions in seals	Play an important part
Magnetic effects	Motion of magnetic domain wall (the Barkhausen effect) Superconductivity	Very important, but not in most cases
Surface effects	Surface oxides	A source is very effective, sometimes is used successfully
	Surface coatings	

(continued)

Table 2.3 (continued)

Group	Source	Comment
1	2	3
Fracture	Formation of cavities	Important part in fatigue testing
	Failure of bonds	Dominant sources of AE
	Breakage of surfaces	High-energy sources of AE
	Micro-crack formation	Plays an important part in most cases, including corrosion and welding
	Crack propagation	
	Stress corrosion	
	Hydrogen embrittlement	

Energy of the AE source [J]: $^{1}10^{-23}$; $^{2}4 \times 10^{-18}$; $^{3}5 \times 10^{-24}$; $^{4}{\cong}10^{-24}$; $^{5}10^{-23}$; $^{6}10^{-19}$

Table 2.4 Ranges of change in some AE parameters during its fail-safe recording on IO

Parameter	Range of the parameter change		Notes
	Electric signal	Mechanical signal	
Cumulative count of AE	$0\ldots10^{7}$ counts	–	During the time of specimen tension to failure
Count rate	$0\ldots10^{5}$ counts/s	–	
Amplitude	$10^{-7}\ldots10^{-2}$ V	$10^{-7}\ldots10^{-14}$ m	Single signal of discrete AE
Signal energy	$10^{-11}\ldots10^{-5}$ J	–	
Pulse duration	$10^{-4}\ldots10^{-8}$ s	–	–

Table 2.5 Parameters of acoustic emission signals for some sources [57]

Type of source	Amplitude or pulse energy of the AE, J	Signal duration, μs	Width of a signal spectrum, MHz
Dislocation source of Frank-Read	$(10^{-7}\ldots10^{-8})G^{*}$	$5\ldots 50 \times 10^{3}$	≤1
Annihilation of dislocations of length of $10^{-6}\ldots10^{-4}$ cm	$4(10^{-18}\ldots10^{-16})$	5×10^{-5}	Hundreds
Micro-crack initiation	$10^{-10}\ldots10^{-12}$	$10^{-3}\ldots 10^{-2}$	≤50
Disappearance of twins of volume ≈1 mm^{3}	$10^{-2}\ldots10^{-3}$	1×10^{4}	–
Plastic deformation of a volume $\approx10^{-3}$ mm^{3}	1×10^{-4}	$\leq 1 \times 10^{3}$	≤0.5
Energy of thermal noises	4.2×10^{-4} J/Hz	–	Uniform spectrum to 10^{7}

Taking into consideration the factors that cause the AE origination, and according to the characteristics of a qualitative effect on the formation of AE sources, they can be grouped as follows.

Factors that cause the initiation of the AE signals with large amplitude	Factors that cause the initiation of the AE signals with small amplitude
• High breaking strength	• Low breaking strength
• High rate of loading	• Low rate of loading
• Anisotropy	• Isotropy
• Heterogeneity	• Homogeneity
• Dense materials	• Porous materials
• Twinning	• Absence of twinning
• Fracture along the weld plane (neck of welding)	• Shear deformations
• Low temperatures	• High temperatures
• Materials with cracks	• Materials without cracks
• Martensitic phase transformations	• Diffusive controlled transformations
• Cracks propagation	• Plastic deformation
• Cast structures	• Forged structures
• Size of a large grain	• Size of a small grain

Thus, in summary, it is possible to state that various elastic waves initiate and propagate in materials under fracture initiation and further propagation. Their sources are also formed through a variety of mechanisms. Therefore, when using non-destructive test methods based on the AE phenomenon, it is necessary to take this variability into consideration and choose the optimum facilities as well as the modes of selection and recording of the AE signals.

References

1. Yermolov IN et al (1991) Akustichestke metody kontrolya (Acoustic methods of testing). In: Nondestructive testing, vol 2. Vysshaya Shkola, Moskva
2. Novatskiy V (1975) Teoriya uprugosti (Elasticity theory). Mir, Moskva
3. Aki A, Richards PD (1983) Kolichestvennaya seysmologiya. Teoriya i metody (Qualitative seismology. Theory and methods), vol 1. Mir, Moskva
4. Chebanov VY (1986) Lazernyy ul'trazvukovoy kontrol' materialov (Laser ultrasonic testing of materials). Izdat. Leningrad. U-ta, Leningrad
5. Mazon U, Tereton R (eds) (1973) Fizicheskaya akustika (Physical acoustics), vol 6. Mir, Moskva, p 132
6. Viktorov IA (1981) Ul'trazvukovye poverchnostnye volny v tverdych telach (Ultrasound surface waves in solids). Nauka, Moskva
7. Krasilnikov VA, Krilov VV (1984) Vvedenie v fizicheskuyu akustiku (Introduction to physical acoustics). Nauka, Moskva

8. Brehovskikh LM, Godin OA (1989) Akustika sloistych sred (Acoustics of layered media). Nauka, Moskva
9. Birchon D (1979) Cries of stress. Spectrum 165:5–8
10. Budenkov GA et al (1981) Vozbuzhdenie voln Releya istochnikom tipa garmonicheskoy sosredotochennoy sily, deystvuyuschey nad poverchnost'yu uprugogo poluprostranstva (Excitation of the Raleigh waves by a source of the harmonic concentrated force type acting over the surface of an elastic half-space). Defektoskopia 2:37–42
11. Vakshys E, Botava E (1982) Rasprostranenie uprugich voln pri akustiko-emisionnom kontrole yadernych reaktorov (Propagation of elastic waves during acoustic emission testing of Nuclear reactors). Materialy 10-y Mezhdunarodnoy konferenzii po nerazrushayuschemu kontrolyu (Proceedings of the 10th international conference on nondestructive testing), vol 4. Moscow, pp 33–37
12. Stryzhkov SA, Vynkler ON (1988) Issledovanie charaktera rasprostraneniya uprugich kolebaniy v trubach pri akustiko-emissionnom kontrole (Investigation of the character of elastic vibrations propagation in pipes during acoustic emission testing). In: Nerazrushayuschiy kontrol' i diagnostika truboprovodov (Nondestructive testing and disgnostics of pipelines). Nauka, Moskva, pp 15–21
13. Alioshyn NP (ed) (1989) Metody akusticheskogo kontrolya metallov (Methods of acoustic testing of metals). Mashynostroyenie, Moskva
14. Jaffrey D (1979) Sources of acoustic emission (AE) in metals. A review. Aust Chem Eng 20 (9–10): 9–17, 21
15. Skalskiy VR, Andreykiv OY, Serhiyenko OM (2003) Doslidzhennya plastychnogo deformuvannya materialiv metodom akustichnoyi emisiyi. Oglyad (Investigation of the materials plastic deformation by the acoustic emission method. A review). Fizyko-chimichna mechanika materialiv (Physico-chemical mechanics of materials) 1:77–94
16. Schofield BH (1961) Acoustic emission under applied stress. Contract AF-33(616)—5640. Progress Report, vol 11. Lessells and Associates, Inc., Boston, Massachusetts
17. Hatano H (1976) Strain-rate dependence of acoustic-emission power and spectra in aluminum alloys. J Appl Phys 47(9):3873–3876
18. Siegel E (1977) Burst acoustic emission during the Bauschinger effect in FCC and HCP metals and alloys. Acta Metall 2S(4):383–394
19. Eshelby JD (1949) Dislocations as a cause of mechanical damping in metals. Proc Roy Soc L A 197(1050):396–416
20. Eshelby JD (1956) A continuum theory of lattice defects. Prog Solid State Phys 3:79–85
21. Kosakevych AM, Margvelashvili IG (1967) Izluchenie elektromagnitnych i zvukovych voln dislokaziey, ravnomerno dvizhuscheysya v ionnom kristalle (Irradiation of electromagnetic and sound waves by dislocation moving uniformly in ion crystal). Izvestiya AN SSSR. Seriya Fizika (Reports of the Academy of Sciences of the USSR. Ser. Physics) 31(5):848–850
22. Cortellazzi G et al (1973) A lattice-dynamics approach to the acoustic signal by a uniformly moving dislocation. J Appl Phys 44(4):1518–1523
23. Natsyk VD (1968) Izluchenie zvuka dislokaziey, vychodyaschey na poverchnost' kristalla (Irradiation of a sound by dislocation, appearing on the crystal surface). Pisma zhurnal eksperimentalnoi i tekhnicheskoy fiziki 8(6):324–328
24. Graff KF (1975) Wave motion in solids. Clarendon Press, Oxford
25. Natsyk VD, Chyshko KA (1972) Zvukovoe izluchenie pri annigilyazii dislokaziy (Sound irradiation caused by annihilation of dislocations). Fizika tverdogo tela (Phys Solids) 14 (11):3126–3132
26. Schofield BH (1972) Research on the source and characteristics of acoustic emission. In: Acoustic emission. ASTM STP 505. American Society for Testing and Materials, Philadelphia, Pennsylvania, pp 11–19
27. Kim HC (1976) Atomic structure and mechanical properties of metals. In: Proceedings of the International School of Physics "Enrico Fermi", 8–10 July 1976
28. Mirabile M (1975) Acoustic emission energy and mechanisms of plastic deformation and fracture. Non-Destr Test 8(2):77–85

29. Dunegan HL, Harris D, Tatro CA (1968) Fracture analysis by use of acoustic emission. Eng Fract Mech 1(1):105–122
30. Gillis PP (1972) Dislocation motions and acoustic emission. In: Acoustic emission. ASTM STP 505. American Society for Testing and Materials. Philadelphia, Pennsylvania, pp 20–29
31. Malen I, Bolin LA (1974) Theoretical estimate of acoustic-emission stress amplitudes. Physica Status Solidi (B) Basic Res 61(2):637–645
32. Kim HC (1975) Characterization of mechanical properties by acoustic emission using an energy criterion. In: Ultrasonic symposium proceedings. Institute of Electrical and Electronics Engineers, Inc., Los Angeles, California. 22–24 Sep 1975
33. Sedgwick RT (1968) Acoustic emission from single crystals of LiF and KCl. J Appl Phys 39 (3):1728–1740
34. Day CI (1969) An investigation of acoustic emission for defect formation in stainless steel weld coupons. Richlai, Washington
35. Fisher RM, Lally LS (1967) Microplasticity detected by an acoustic technique. Canad J Phys 45(2):1147–1159
36. James DR, Carpenter SH (1971) Relationship between acoustic emission and dislocation kinetics in crystalline solids. J Appl Phys 42(12):4685–4697
37. Carpenter SH, Higgins FP (1977) Sources of acoustic emission generated during the plastic deformation of 7075 aluminum alloy. Met Trans A 8A(10):1629–1632
38. Carpenter SH (1976) Report No. AFML-TR-75-212. Rockwell Science Center, Thousand Oaks, Jan 1976
39. Tetelman AS (1972) Acoustic emission and fracture mechanics testing of metals and composites. In: Proceedings of the U.S.–Japan Joint symposium on acoustic emission, 4–6 July 1972
40. Kaiser J (1953) Erkenntnisse und Folgerungen aus der Messung von Gerauschen bei Zugbeanspruchung von Metallischen Werkstoffen. Arch Eisenhuttenwesen 1/2:43–45
41. Woodward B, McDonald NR (1975) Flaw identification using acoustic emission. In: Proceedings of the 3rd international conference on structural mechanics in reactor technology, London, England, 1–5 Sep 1975
42. Tanaka H, Horiuchi R (1975) Acoustic emission accompanied by twin formation in titanium and titanium alloy. Bull Inst Space Aeronaut Sci Univ Tokyo 11(2A):427–435
43. Toronchuk JP (1977) Acoustic emission during twining of zinc single crystals. Mater Eval 35 (10):51–53
44. BoikoVS Gaber RI, Kryvenko LF (1974) Zvukovaya emissiya pri annigilyazii dislokazionnogo skopleniya (Sound emission during annihilation of dislocation group). Fizika tverdogo tela (Phys Solids) 16(4):1233–1235
45. Boiko VS et al (1970) Zvukovoe izluchenie dvoynikuyuschich dislokaziy (Sound irradiation of twinning dislocations). Ibid 12(6):1753–1755
46. Speich GR, Schwaeble AJ (1975) Acoustic emission during phase transformation in steel. In: Spanner JC, McElroy JW (eds) Monitoring structural integrity by acoustic emission, ASTM STP 571. Pennsylvania, Philadelphia, pp 40–58
47. Speich GR, Fisher RM (1972) Acoustic emission during martensite formation. Acoustic emission, ASTM STP 505. American Society for Testing and Materials. Philadelphia, Pennsylvania, pp 140–151
48. Liptai RG, Dunegan HL, Tatro CA (1969) Acoustic emission generated during phase transformations in metals and alloys. Int J Nondestr Test 1:213–221
49. Pascual R et al (1975) Acoustic emission and the martensitic transformation of β-brass. Scripta Metall 9(1):79–84
50. Hartman WF, Kline RA (1977) Variations in frequency content of acoustic emission during extension of HF-I steel. Mater Eval 35(7):47–51
51. Palmer IG, Holt J, Goddard DJ (1974) Acoustic emission measurements on type 316 stainless steel. In: EWGAE—ISPRA: proceedings of third meeting of the european working group on acoustic emission, ISPA, Italy, 25–26 Sept 1974, pp 11–26

52. Ono K (1974) Acoustic emission and microscopic deformation and fracture processes. In: Proceedings of the second acoustic emission symposium, Japan, 2–4 Sep 1974, pp 1–63
53. Tetelman AS, Chow R (1972) Acoustic emission testing and microcracking processes. In: Acoustic Emission, ASTM STP 505. American Society for Testing and Materials. Philadelphia, Pennsylvania, pp 30–40
54. Jaffrey D (1979) Sources of acoustic emission (AE) in metals. A review. Australian Chemical Eng 20(11):9–11, 20(12):13–17
55. Andreykiv OYe, Skalskiy VR, Serhiyenko OM (2001) Viznachennya ob'yemnoyi poshkodzhenosti alyuminiyevych splaviv za sygnalamy akustichnoyi emisiyi (Determination of volume damaging of aluminium alloys by the acoustic emission signals). Fizyko-chimichna mechanika materialiv (Physicochem Mech Mater) 3:77–90
56. Andreykiv OY, Skalskiy VR, Serhiyenko OM (2001) Akustiko-emisiyni kryteriyi dlya ekspres ozinky vnutrishnich poshkodzhen' kompozitnich materialiv (Acoustic-emission criteria for express estimation of internal damages in composite materials). Ibid 1:91–100
57. Ivanov BI, Belov VM (1981) Akustiko–emissionnyy kontrol' svarki i svarnych soedineniy (Acoustic emission testing of welding and welded joints). Mashynostriyenie, Moskva
58. Shibata M, Ono K (1981) Magnetomechanical acoustic emission—a new method for non-destructive stress measurement. NDT Int 14(5):227–234
59. Lindgren M, Lepisto T (2000) Application of a novel type Barkhausen noise sensor to continuous fatigue monitoring. NDT & E Int 33(6):423–428
60. Song YY et al (2000) The effect of microstructural changes on magnetic Barkhausen noise and magnetomechanical acoustic emission in Mn–Mo–Ni pressure vessel steel. J Appl Phys 87(9):5242–5244
61. Park DG et al (1999) Nondestructive evaluation of irradiation effects in RPV steel using Barkhausen noise and magnetoacoustic emission signals. J Magn Magn Mater 196–197:382–384
62. Schwalbe H-J, Bamfaste G, Franke R-P (1999) Non-destructive and non-invasive observation of friction and wear of human joints and of fracture initiation by acoustic emission. J Eng Med 213(1):41–48
63. Baranov VM, Kudryavtsev EM, Sarychev GA (1997) Modelling of the parameters of acoustic emission under sliding friction of solids. Wear 202(2):125–133
64. Jiaa CL, Dornfeld DA (1990) Experimental studies of sliding friction and wear via acoustic emission signal analysis. Wear 139(2):403–424
65. Druillard TF (1979) Acoustic emission. A bibliography with abstracts. IFI/Plenum, New York
66. Spanner JC (1974) Acoustic emission: technique and applications, vol 12. Intex publ, Co, Evanston, Illinois
67. Williams RV (1980) Acoustic emission. Adam Hilger, Bristol
68. Vakar KB (1980) Akusticheskaya emissiya i ee primenenie dlya nerazrushayuschego kontrolya v yadernoy energetike (Acoustic emission and its application for nondestructive testing in nuclear power generation industry). Atomizdat, Moskva
69. Andreykiv AY, Lysak NV (1989) Metod akusticheskoy emissii v issledovanii prozessov razrusheniya (A method of acoustic emission in investigation of fracture processes). Naukova dumka, Kiev
70. Collacott R (1989) Diagnostika povrezhdeniy (Damages diagnostics) (trans: Babaievskyi PG). Mir, Moskva
71. Stryzhalo VA et al (1990) Prochnost' i akusticheskaya emissiya materialov i elementov konstrukziy (Strength and acoustic emission of materials and structural elements). Naukova Dumka, Kiev
72. Dunegan HL, Tatro CA (1971) Acoustic emission effect during mechanical deformation. In: Bunshan RF (ed) Techniques of metals research, vol 5(2/12). Interscience, New York, pp 273–312
73. Tanaka Kh, Horiuyi Kh, Sakakibara Y (1977) Akusticheskaya emissiya pri plasticheskoy deformazii - metallovedcheskie faktory (Acoustic emission under plastic deformation—material science factors). Kindzoku dzajre 13(2):21–26

74. Wadley HNG, Scruby CB, Speake JH (1980) Acoustic emission for physical examination of metals. Int Met Rev 25(2):41–64
75. Ono K (1994) Trends of recent acoustic emission literature. J Acoust Emis 12(3/4):177–198
76. Skalskiy VR, Demchyna BG, Karpukhin II (2000) Ruynuvannya betoniv i akustichna emisiya (Oglyad). Povidomlennya 1. Statichne navantazhennya i vplyv temperaturnogo polya (Concrete fracture and acoustic emission (A review). Report 1. Static loading and effect of temperature field). Tekhnicheskaia Diagnostika i Nerazrushyuschii kontrol (Technical diagnostics and nondestructive testing) 1:12–23
77. Skalskiy VR, Demchyna BG, Karpukhin II (2000) Ruynuvannya betoniv i akustychna emisiya (Oglyad). Povidomlennya 2. Koroziya zalizobetonu. Aparaturni zasoby. AE – kontrol' ta diagnostyka budivel'nych sporud (Concrete fracture and acoustic emission (A review). Report 2. Corrosion of the reinforced concrete. Equipment. AE examination and diagnostics of building constructions). Ibid 2:9–27
78. Andreykiv OY et al (2000) Rozvytok doslidzhen' prozesiv ruynuvannya iz zastosuvannyam yavischa akustichnoyi emisiyi u FMI Im. G.V. Karpenka NAN Ukrayini (Development of fracture processes researches using the phenomenon of acoustic emission at H.V. Karpenko Physico-Mechanical Institute of the National Academy of Sciences of Ukraine). Preprint, NAN Ukrayini, Fizyko-mechanichnyi institut, 1(2000), L'viv
79. Schnitt-Thomas KG, Stengel W (1983) Möglichkeiten zur Früherkennung von Wasserstoffschädigungen in metallischen Werkstoffen durch Anwendung der Schallemissionanalyse. Werkst Korros 34:7–13
80. Andreykiv AY et al (1990) Zastosuvannya metodu akustychnoyi emisiyi pry doslidzhenni materialiv u vodnevomu ta koroziynomu seredovischach (Application of the method of acoustic emission for investigation of materials in hydrogen and corrosive environments). Fizyko-chimichna mechanika materialiv (Physicochem Mech Mater) 5:26–36
81. Andreykiv OY, Lysak MV, Skalsky VR (1996) A method of accelerated evaluation of Kiscc under stress corrosion cracking. Eng Fract Mech 54(3):387–394
82. Andreykiv AY et al (1992) Metodika opredeleniya K_{Iscc} stali v srede vodoroda s pomosch'yu metoda akusticheskoy emissii (A method of determining K_{Iscc} values of steel in hydrogen using acoustic emission). Tekhnicheskaia Diagnostika i Nerazrushyuschii kontrol (Tech Diagn Nondestr Test) 1:18–26
83. Skalskiy VR (1995) Vliyanie vodoroda na rastreskivanie metallov i kontrol' takich prozessov metodom AE (Influence of hydrogen on metal cracking and testing of such processes by AE method). Ibid 1:52–65
84. Skalskiy VR, Andreykiv OYe, Serhiyenko OM (1999) Ozinka vodnevoyi poshkodzhenosti materialiv za amplitudamy signaliv akustichnoyi emisiyi (Estimation of hydrogen damaging of materials by acoustic emission signal amplitudes). Ibid 1:17–27
85. Hartbower CE, Gerberich WW, Crimmins PP (1967) Mechanisms of slow crack growth in high-strength steels. Aerojet-General Corporation, Sacramento, California. Feb 1967, vol 1, pp 213–245
86. Hartbower CE et al (1972) Use of acoustic emission for the detection of weld and stress-corrosion cracking. Acoustic emission, ASTM STP 505. American Society for Testing and Materials. Philadelphia, Pennsylvania, pp 187–221
87. Jaffrey D (1979) Sources of acoustic emission (AE) in metals. A review. Non Destr Test (Aust) 16(6):16–23

Chapter 3
Analysis of Acoustic Emission Caused by Internal Cracks

An important scientific and technical problem that could be solved using acoustic emission (AE) is the possibility of inspecting nucleation and sub-critical growth of internal crack-like defects in structure elements caused, for example, by hydrogen embrittlement of metals [1–4]. These defects, owing to the loading applied to a structure, can spring up sub-critically—in particular, in local areas near a contour in an existing crack. Such character of destruction is caused by the nature of the stress–strain state near the crack front, as the effect of stress corrosion as well as micro-structure of the material. In many cases, the fracture occurs inside the material and is not observed visually, even in the case of the local sub-critical growth of a crack. However, elastic waves caused by these processes could be received by AET placed on the surface of IO.

For a quantitative estimation of defects by the AE method, it is necessary to establish relationships between AE signals and crack parameters. If we assume that a newly formed crack is considerably less than the geometrical size of a body, then to investigate the given problem it would be necessary to consider a non-stationary dynamic problem of crack formation and its growth in a half-space (Fig. 3.1). Besides, AE inspection is frequently carried out under conditions much different from those of the ambient medium (corrosive environment, high or low temperatures, or when the access to the inspected surface is complicated). In such cases, waveguides are used in AE measurements so one should take into account the effect of a waveguide on AE parameters. Thus, the problem to be solved is establishing a relationship between AE parameters at the waveguide face and the characteristics of a nucleating or growing crack. Assuming that such a general statement is valid, the solution to this problem is rather difficult, so in order to simplify it, we divide it into separate problems (Fig. 3.1): (1) determining AE parameters caused by the formation and sub-critical growth of an internal crack type defect at local areas near its contour in an infinite elastic body; (2) the effect of the body surfaces on AE signals; and (3) the effect of a waveguide on AE signals.

Z. Nazarchuk et al., *Acoustic Emission*, Foundations of Engineering Mechanics,
DOI 10.1007/978-3-319-49350-3_3

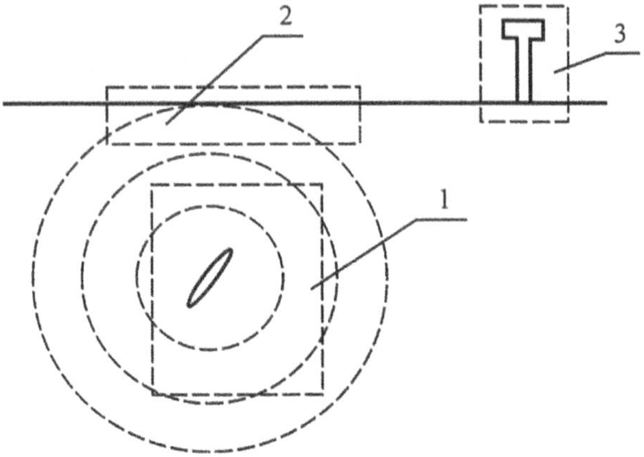

Fig. 3.1 Crack-like defect radiating AE and problems that should be solved for the estimation of parameters of such a defect using AE signals: *1* nucleation and sub-critical crack growth; *2* the effect of a free surface on AE signal parameters; and *3* a waveguide effect on AE signal parameters

3.1 Nucleation and Sub-critical CRACK Growth

3.1.1 Nucleation of a Mode I Penny-Shaped Crack

According to the approach proposed in [5], we replace an arbitrarily shaped crack with a penny-shaped crack in the same area. Suppose that a penny-shaped crack nucleates when the tensile stresses in a certain region of an elastic body reach a certain critical value σ_0 (the integral characteristics of the material breaking strength). The crack formation is accompanied by an instant drop of normal stresses on its surfaces from an initial level σ_0 to zero.

If we consider a system of cylindrical coordinates $Or\theta z$, we see that the origin O coincides with the center of the crack of radius r_0 and the axis Oz is normal for the crack plane (Fig. 3.2). At infinity, tensile stresses σ are applied along Oz axes. At the time $t = 0$, they reach a certain critical value σ_0, resulting in a penny-shaped crack nucleation. Using the known approaches [3, 6, 7], this problem can be reduced to the wave equations

$$\Delta\varphi - \frac{1}{c_1^2}\frac{\partial^2\varphi}{\partial t^2} = 0, \Delta\psi - \frac{\psi}{r} - \frac{1}{c_2^2}\frac{\partial^2\psi}{\partial t^2} = 0 \tag{3.1}$$

with respect to the unknown scalar potentials $\varphi(r,z,t)$ and $\psi(r,z,t)$.

The Eq. (3.1) should satisfy the boundary conditions for half-space $z > 0$

$$\begin{aligned} \sigma_{zz}(r,0,t) &= -\sigma_0 H(t), r \le r_0, \\ u_z(r,0,t) &= 0, r > r_0, \\ \tau_{rz}(r,0,t) &= 0, 0 < r < \infty, \end{aligned} \tag{3.2}$$

and zero initial conditions

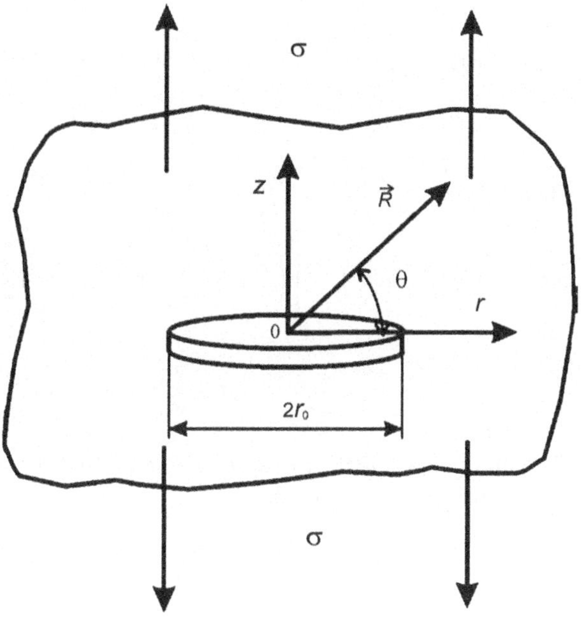

Fig. 3.2 A Mode I penny-shaped crack in an elastic body

$$\varphi(r,z,0) = \psi(r,z,0) = \left.\frac{\partial\varphi}{\partial t}\right|_{t=0} = \left.\frac{\partial\psi}{\partial t}\right|_{t=0} = 0. \tag{3.3}$$

Here, $c_1 = \sqrt{\frac{\lambda+2\mu}{\rho}}$ is the velocity of a longitudinal wave, $c_2 = \sqrt{\frac{\mu}{\rho}}$ is the velocity of a shear wave, λ, μ are Lame's constants, ρ is the material density, and $H(t)$ is the Heaviside function. The relationship between potentials φ and ψ and components of a displacement vector u_r and u_z has the form:

$$u_r = \frac{\partial\varphi}{\partial r} - \frac{\partial\psi}{\partial z} \quad u_z = \frac{\partial\varphi}{\partial z} + \frac{\partial\psi}{\partial r} + \frac{\psi}{r}. \tag{3.4}$$

The boundary problem (3.1–3.3) can be solved using the method of integral transforms. Similar to [7], using the Hankel transform over the spatial coordinate r and Laplace transform over time t, we reduce the problem to the Fredholm integral equation of the second kind:

$$\Lambda(q,s) = \int_0^1 \Lambda(u,s)K(u,q,s)\mathrm{d}u = q \quad 0 \le q \le 1, \tag{3.5}$$

having the kernel $K(u, q, s)$ symmetric with respect to u and q in the form

$$K(u,q,s)=\frac{2}{\pi(1-\varepsilon^2)}\int_0^{\infty} g(\eta/r_0,s)\sin(\eta q)\sin(\eta u)d\eta,\quad 0\le q\le 1,\quad 0\le\eta\le 1,\tag{3.6}$$

where

$$g(\eta/r_0,s)=(1-\varepsilon^2)-\left[\left(1+\frac{1}{2\eta^2n^2}\right)^2\left(1+\frac{1}{\eta^2n^2}\right)^{1/2}\left(1+\frac{\varepsilon^2}{\eta^2n^2}\right)^{-1/2}\right]2\eta^2n^2,$$

$$n=\frac{c_2}{sr_0},\varepsilon=c_2/c_1.$$

The solution of integrated Eq. (3.5) was found numerically and approximated by dependence proposed in [8]:

$$\Lambda(q,s)=\frac{b(q)m(q)}{\sqrt{m^2+p^2}},\tag{3.7}$$

where $b(q)$ and $m(q)$ are the parameters of approximation obtained by the fewest squares method, $p=sr_0/c_2$. Taking into account the dependence (3.7) for the components of a displacement vector, after inverting the Laplace and Hankel transforms using the Parseval and convolution theorems, we obtain the following equations:

$$u_i(\bar{r},\bar{z},T)=AB_i\int_0^1 b(q)m(q)\left[\int_{r-\xi}^{r+\xi}\sum_{j=1}^{2}S_{ij}(\alpha,\bar{z},\tau_j,q)H\left(T-\varepsilon_j\sqrt{\bar{z}^2+\alpha^2}\right)\right.$$
$$\left.\times G_i(\alpha,\bar{r},q)d\alpha\right]dq,\qquad \bar{r}>1,\quad (i=z,r)\tag{3.8}$$

where

$$A=\frac{\sigma_0r_0}{\rho c_1^2\varepsilon^2\pi(1-\varepsilon^2)},\quad \bar{z}=z/r_0,\quad \bar{r}=r/r_0,T=c_2t/r_0,\quad \varepsilon_1=\varepsilon,\quad \varepsilon_2=1,$$

$$\tau_j=T-\varepsilon_j\sqrt{\bar{z}^2+\alpha^2},B_z=\bar{z}\Big/\left(\pi\sqrt{\bar{r}}\right),B_r=\left(\pi\sqrt{\bar{r}}\right)^{-1},$$

$$G_z=\sqrt{\alpha}K(k),\quad G_r=\alpha\sqrt{\alpha}[2E(k)-K(k)],$$

$$k=\sqrt{q^2-(a-\bar{r})^2\Big/(4\alpha\bar{r})},$$

$E(k)$ and $K(k)$ are complete elliptic integrals of the first and second kind respectively,

$$S_{1j}(\alpha,\overline{z},\tau_j,q) = Z_1^j((\alpha,\overline{z},\tau_j,q))J_0(m(q)\tau_j) + Z_2^j((\alpha,\overline{z},\tau_j,q))J_1(m(q)\tau_j) + Z_3^j((\alpha,\overline{z},\tau_j,q))I(m(q)\tau_j)$$

$$S_{2j}(\alpha,\overline{z},\tau_j,q) = R_1^j((\alpha,\overline{z},\tau_j,q))J_0(m(q)\tau_j) + R_2^j((\alpha,\overline{z},\tau_j,q))J_1(m(q)\tau_j) + R_3^j((\alpha,\overline{z},\tau_j,q))I(m(q)\tau_j)$$

$$Z_1^1((\alpha,\overline{z},\tau_j,q)) = \varepsilon(\overline{z}^2+\alpha^2)^{-1}\left[1-2\alpha^2\varepsilon^2(\overline{z}^2+\alpha^2)^{-1}\right] + 3/m(q)^2\tau_1(\overline{z}^2+\alpha^2)^{-5/2}\left[2-5\alpha^2(\overline{z}^2+\alpha^2)^{-1}\right]$$

$$Z_2^1((\alpha,\overline{z},\tau_j,q)) = 3/m(q)\left[T^2-\varepsilon^2(\overline{z}^2+\alpha^2)\right] + (\overline{z}^2+\alpha^2)^{-5/2}\left[2-5\alpha^2(\overline{z}^2+\alpha^2)^{-1}\right]$$

$$Z_3^1((\alpha,\overline{z},\tau_j,q)) = 1/m(q)\left\{(\overline{z}^2+\alpha^2)^{-3/2}\left[1-2\varepsilon^2+3\varepsilon^2\alpha^2(\overline{z}^2+\alpha^2)^{-1}\right] + 3(\overline{z}^2+\alpha^2)^{-5/2} \left[2-5\alpha^2(\overline{z}^2+\alpha^2)^{-1}\right]\left[T^2-1/m(q)^2\right]\right\}$$

$$Z_1^2((\alpha,\overline{z},\tau_j,q)) = 2\alpha^2(\overline{z}^2+\alpha^2)^{-2} + 3\tau_2/m(q)^2(\overline{z}^2+\alpha^2)^{-5/2}\left[5\alpha^2(\overline{z}^2+\alpha^2)^{-1}-2\right]$$

$$Z_2^2((\alpha,\overline{z},\tau_j,q)) = -3/m(q)\left[T^2-(\overline{z}^2+\alpha^2)\right] + (\overline{z}^2+\alpha^2)^{-5/2}\left[5\alpha^2(\overline{z}^2+\alpha^2)^{-1}-2\right]$$

$$Z_3^2((\alpha,\overline{z},\tau_j,q)) = 1/m(q)\left\{(\overline{z}^2+\alpha^2)^{-5/2}\left[5\alpha^2(\overline{z}^2+\alpha^2)^{-1}-2\right]\left[T^2-1/m(q)^2 + (\overline{z}^2+\alpha^2)^{-3/2}\left[2-3\alpha^2\left((\overline{z}^2+\alpha^2)^{-1}\right)\right]\right]\right\}$$

$$R_1^1((\alpha,\overline{z},\tau_j,q)) = \varepsilon(\overline{z}^2+\alpha^2)^{-1}\left[\left[1-2\overline{z}^2\varepsilon^2(\overline{z}^2+\alpha^2)^{-1}\right]^{-1} + (1-2\varepsilon)\right] + 3/m(q)^2\tau_1(\overline{z}^2+\alpha^2)^{-5/2}\left[5\overline{z}^2(\overline{z}^2+\alpha^2)^{-1}-1\right]$$

$$R_2^1((\alpha,\overline{z},\tau_j,q)) = 3/m(q)\left[T^2-\varepsilon^2(\overline{z}^2+\alpha^2)\right] + (\overline{z}^2+\alpha^2)^{-5/2}\left[5\overline{z}^2(\overline{z}^2+\alpha^2)^{-1}-1\right]$$

$$R_3^1((\alpha,\overline{z},\tau_j,q)) = 1/m(q)\left\{(\overline{z}^2+\alpha^2)^{-5/2}\left[5\overline{z}^2(\overline{z}^2+\alpha^2)^{-1}-1\right]\left[T^2-1/m(q)^2\right] + (\overline{z}^2+\alpha^2)^{-3/2}\left[1-\varepsilon^2-3\overline{z}^2\varepsilon^2(\overline{z}^2+\alpha^2)^{-1}\right]\right\}$$

$$R_1^2((\alpha,\bar{z},\tau_j,q)) = 3\tau_2/m(q)^2\left(\bar{z}^2+\alpha^2\right)^{-5/2}\left[1-5\bar{z}^2\left(\bar{z}^2+\alpha^2\right)^{-1}\right]-2\bar{z}^2\left(\bar{z}^2+\alpha^2\right)^{-2}$$
$$R_2^2((\alpha,\bar{z},\tau_j,q)) = 3/m(q)\left[T^2-\left(\bar{z}^2+\alpha^2\right)\right]+\left(\bar{z}^2+\alpha^2\right)^{-5/2}\left[5\bar{z}^2\left(\bar{z}^2+\alpha^2\right)^{-1}-1\right]$$
$$R_3^2((\alpha,\bar{z},\tau_j,q)) = -1/m(q)\left\{3\left[T^2-1/m(q)^2\right]\left(\bar{z}^2+\alpha^2\right)^{-5/2}\left[5\bar{z}^2\left(\bar{z}^2+\alpha^2\right)^{-1}-1\right]\right\}$$

$$I(x) = \int_0^x J_0(\chi)d\chi, J_0(x)$$

and $J_1(x)$ are the zero and first order Bessel functions of the first kind, respectively.

An asymptotic dependence of the displacement vector components can be obtained by leaving in Eq. (3.8) the components, which describes decreasing weaker than $1/R$ ($R^2 = r^2 + z^2$). In a spherical coordinate system originated in the center of a penny-shaped crack, the components of a displacement vector are as follows:

$$u_R = u_r\cos\theta + u_z\sin\theta, u_\theta = u_z\cos\theta - u_r\sin\theta. \tag{3.9}$$

Replacing the internal integrals over α in Eq. (3.8) by their average values as α tends to r, for large R, after simplification we will obtain the approximation

$$u_i(\bar{R},\theta,T) = AB_i(1/\bar{R})\left\{\int_0^1 2qb(q)m(q)J_0(m\tau_i)M_i(k)dq\right\}H(\tau_i)+O(R^{-2}), \tag{3.10}$$

where $i = R,\theta$, $B_1(\theta) = \varepsilon/\pi(1-2\varepsilon^2\cos^2\theta)$, $B_2(\theta) = 1/\pi\sin 2\theta$, $\tau_i = T - \varepsilon_i\bar{R}\bar{R} = R/r_0$, $M_1(k) = K(k)$, $M_2(k) = 2E(k) - K(k)$, $k = q/2R\cos\theta$, and B_1 and B_2 determine the angular dependence of radiation when a crack is modeled by a system of three mutually perpendicular dipoles [9].

From Eq. (3.8), it was found that for distances from a crack greater than $28r_0$, the amplitude values both for longitudinal and shear waves decay as $1/R$, which is predicted in Eq. (3.10). When performing the AE inspection, the distances between AE sources and AET are much larger than the dimensions of AE sources. Therefore, more attention will be paid to the consideration of AE caused by a penny-shaped crack for a large R. The dependencies of a dimensionless value

$$U(\bar{r},\bar{z},T) = \left[u_z^2(\bar{r},\bar{z},T)+u_r^2(\bar{r},\bar{z},T)\right]^{1/2}/A$$

versus dimensionless time T/ε calculated according to Eq. (3.8) for $\bar{R} = 1000$, $\varepsilon = 0.535$ and the Poisson's ratio $\nu = 0.32$ are presented in Fig. 3.3a (longitudinal wave) and in Fig. 3.3b (shear wave). The curves 1 and 2 correspond to angles $\theta = 15^\circ$ and $\theta = 75^\circ$, respectively.

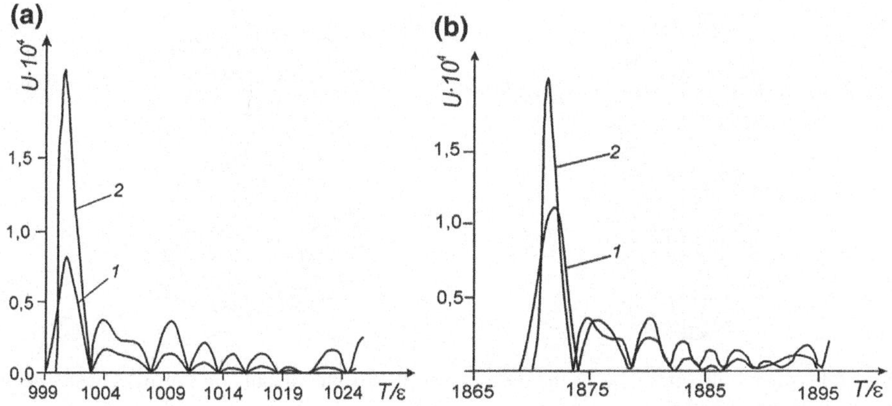

Fig. 3.3 Dependence of the dimensionless module of a displacement vector on dimensionless time T/ε at $\bar{R} = 1000$ for: **a** longitudinal, and **b** transversal waves: Curve *1* corresponds to an angle of observation $\theta = 15°$, Curve *2*—$\theta = 75°$

Similar calculations were carried out for other angles of crack orientation within the range from 0 to $\pi/2$. As a result of the analysis of the maximal values of oscillations for longitudinal and shear waves, the approximation expressions are proposed. They characterize the angular dependencies of peak values of the displacement vector module for $R \gg 1$. For longitudinal waves

$$\Phi_1^{(d)}(\theta) = \frac{1 - 2\varepsilon^2 \cos^2\theta}{\sqrt{1 + \chi_{1Б} \cos^2\theta}} \tag{3.11}$$

and for shear waves

$$\Phi_2^{(d)}(\theta) = \frac{|\sin 2\theta|}{\sqrt{1 + \chi_{2Б} \cos^2\theta}} \tag{3.12}$$

where χ_1 and χ_2 are the parameters of approximation. Their numerical values $\chi_1 = 0.68$ and $\chi_2 = 2.69$ at $\varepsilon = 0.535$ and $\nu = 0.32$ were obtained using the fewest squares method. The angular dependencies of the peak values of $U(\bar{r}, \bar{z}, T)$, calculated according to Eq. (3.8), were compared with Eqs. (3.11) and (3.12). These expressions describe the angular distribution of amplitudes that were calculated with an error not exceeding 4%, and correlate with Eq. (3.10). They are presented in Fig. 3.4a and b for longitudinal and shear waves, respectively.

Taking into account Eqs. (3.11) and (3.12) as well as the character of spatial decay (see Eq. (3.10)) for an estimation of the maximum values of the displacement vector module, the following approximation expressions are obtained:

$$u_{\max}|_{c_i} = \delta_i \frac{\sigma_0 \Phi_i^{(d)}(\theta) r_0^2}{\rho c_1^2 R}, \tag{3.13}$$

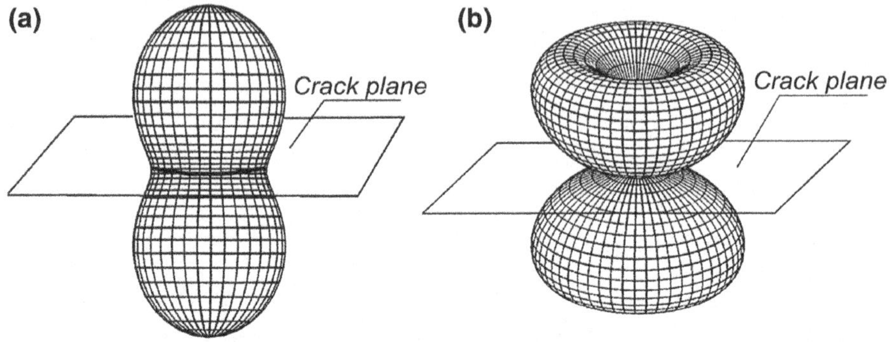

Fig. 3.4 Angular dependencies of the maximum values of the module of a displacement vector for **a** longitudinal, and **b** shear waves

where $i = 1$ corresponds to a longitudinal wave and $i = 2$ to a shear wave, $\delta_1 = 0.452$, $\delta_2 = 0.832$.

In these expressions, R is dimensional. The analysis of the dependence (3.13) and the calculations carried out according to Eq. (3.8) show that for the angles within the range of $15^{\circ} \leq \theta \leq 70^{\circ}$, peak values of the displacement vector for a longitudinal wave, are somewhat lower than for a shear wave (Fig. 3.3). It is necessary to consider this when receiving AE signals and finding the location of the AE source. It was also found that the time interval to reach the first maximum essentially decreases with an increase of the orientation angle θ, and the duration of its rear front changes insignificantly (Fig. 3.3).

During experimental research of AE signals generated by the formation and propagation of cracks in materials, the issue of determining the frequency characteristics of these signals turns out to be quite important. This is explained by the necessity of choosing an AET whose bandwidth corresponds to the spectrum of AE signals radiated by these sources that will further increase the sensitivity of AE storage hardware. Therefore we have carried out an investigation of frequency dependencies for the obtained module of a displacement vector, using the method of fast Fourier transform (FFT) [10].

The results of calculations of frequency dependencies of the displacement vector module are presented in Fig. 3.5a, b ($\bar{R} = 1000$) for longitudinal and transversal waves, respectively. A time sampling was chosen equal to 0.5, and the number of digitizing points $N = 1024$. In these figures, the ordinate axis corresponds to the normalized value of $U_{\Omega} = u(\Omega)/u(\Omega)|_{\max}$ and the abscissa axis corresponds to dimensionless frequency $\Omega = \omega r_0/c_1$ (ω is the angular frequency). It is seen from these figures that the width of the spectrum is greater for $\theta = 75^{\circ}$ than for $\theta = 15^{\circ}$ and correlates with the width of the first maximum of the module of a displacement vector for these angles and R (Fig. 3.3).

The width of the spectrum in Fig. 3.5 is $\Delta\Omega \approx 2.5$. Assuming that the radius of a penny-shaped micro-crack is $r_0 = 5 \times 10^{-6}$ m and $c_1 = 5 \times 10^3$ m/s, we obtain $\Delta f \approx 40$ MHz ($\omega = 2\pi f$).

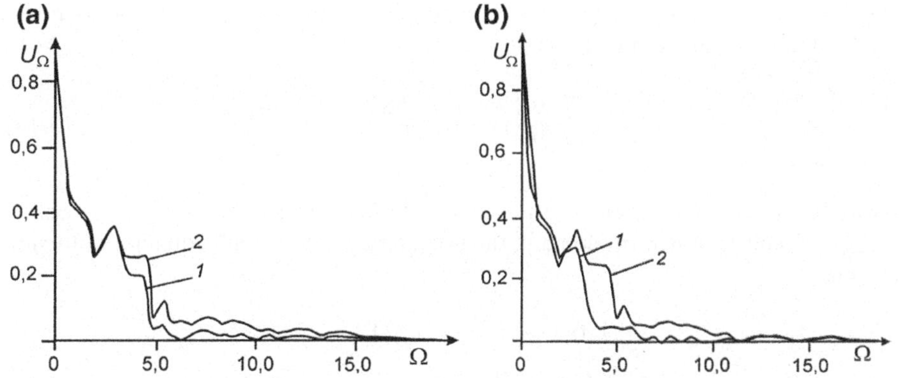

Fig. 3.5 Dependence of dimensionless value of the module of a displacement vector U_Ω on dimensionless frequency Ω at $\bar{R} = 1000$ for: **a** a longitudinal wave, and **b** a shear wave: Curve *1* corresponds to the angle of observation $\theta = 15°$, Curve *2*—$\theta = 75°$

3.1.2 Nucleation of a Mode III Penny-Shaped Crack

This type of loading may be simulated by subjecting the crack surfaces to equal and opposite tractions $\pm\tau_{z\theta}$ that vary linearly with a distance from the center of the crack. When we consider a system of cylindrical coordinates $Or\theta z$, we see that the origin O coincides with the center of the crack of radius r_0 and the axis Oz is normal to the crack plane (Fig. 3.6).

At the time, $t = 0$, $\pm\tau_{z\theta}$ attains a certain critical value τ_0 that results in a penny-shaped crack nucleation. Taking into account the symmetrical conditions with respect to the plane $z = 0$, the problem can be formulated for an upper half-space $z > 0$ with zero initial conditions and boundary conditions [7]

$$\tau_{z\theta}(r,0,t) = -\tau_0(r/r_0)H(t), r < r_0$$

$$u_\theta(r,0,t) = 0, r \geq r_0, \tag{3.14}$$

Fig. 3.6 A Mode III penny-shaped crack in an elastic body

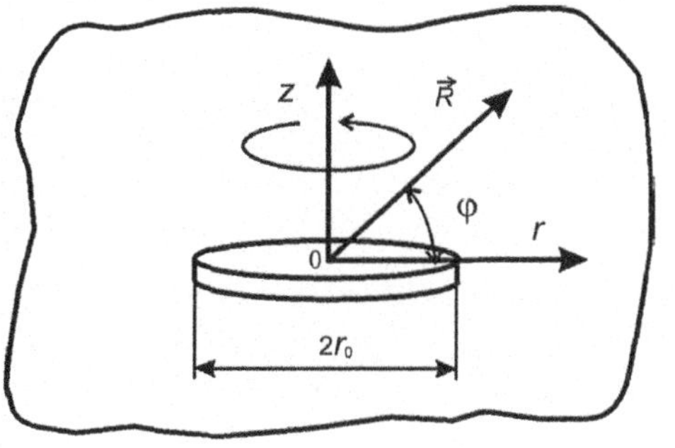

where u_θ (r, z, t) is the only non-zero component of a displacement vector governed by the dynamic equation of torsion:

$$\frac{\partial^2 u_\theta}{\partial r^2} + \frac{1}{r}\frac{\partial u_\theta}{\partial r} - \frac{u_\theta}{r^2} + \frac{\partial^2 u_\theta}{\partial z^2} = \frac{1}{c_2^2}\frac{\partial^2 u_\theta}{\partial t^2}, \tag{3.15}$$

Away from the crack, u_θ (r, z, t) is required to disappear as $(r^2 + z^2)^{1/2} \to \infty$.

In the Laplace transform domain, the boundary condition and equation of torsion become

$$\begin{aligned} &\tau_{z\theta}^*(r, 0, s) = -\tau_0(r/r_0)(1/s), r < r_0, \\ &u_\theta^*(r, 0, s) = 0, r \geq r_0; \\ &\tfrac{\partial^2 \overline{u_\theta}}{\partial r^2} + \tfrac{1}{r}\tfrac{\partial \overline{u_\theta}}{\partial r} - \tfrac{\overline{u_\theta}}{r^2} + \tfrac{\partial^2 \overline{u_\theta}}{\partial z^2} = \tfrac{s}{c_2^2}\overline{u_\theta}. \end{aligned} \tag{3.16}$$

$$\frac{\partial^2 \overline{u_\theta}}{\partial r^2} + \frac{1}{r}\frac{\partial \overline{u_\theta}}{\partial r} - \frac{\overline{u_\theta}}{r^2} + \frac{\partial^2 \overline{u_\theta}}{\partial z^2} = \frac{s}{c_2^2}\overline{u_\theta}. \tag{3.17}$$

Applying a Hankel transform over spatial coordinate r to Eq. (3.17), we obtain an ordinary differential Eq.

$$\frac{\partial^2 \overline{u_\theta}^*}{\partial z^2} - (\xi^2 + s^2/c_2^2)\overline{u_\theta}^*. \tag{3.18}$$

A solution to this problem may be found in a manner similar to the problem discussed above.

As a result, we obtain the following equation:

$$u_\theta(r, z, T) = A\int_0^1 b(q)m(q)[V_1(r, z, T, q) + V_2(r, z, T, q) + V_3(r, z, T, q)]dq \tag{3.19}$$

where

$$A = \frac{4\tau_0 r_0}{\sqrt{2\pi}3\mu},$$

$$V_1(r,z,T,q) = z\int_{r-q}^{r+q}\left[\left[z^2+\alpha^2\right]^{-1/2}\int_{\tau}^{T} J_0(m(q)(T-\tau))\mathrm{d}\tau F(\alpha,r,q)\right]\mathrm{d}\alpha H\left[\sqrt{T^2-z^2}-\alpha\right]$$

$$V_2(r,z,T,q) = -z\int_{r-q}^{r+q}\left[\left[z^2+\alpha^2\right]^{-1/2}\int_{\tau}^{T_2} J_0(m(q)(T-\tau))\mathrm{d}\tau F(\alpha,r,q) H\left[\sqrt{T^2-z^2}-\alpha\right]\right]\mathrm{d}\alpha$$

$$V_3(r,z,T,q) = z\int_{0}^{r+q}\left\{\alpha\left[z^2+\alpha^2\right]^{-3/2}\left[\left[z^2+\alpha^2\right]^{1/2}J_0(m(q)(T-\tau))\right.\right.$$

$$\left.\left.+\int_{\tau}^{T} J_0(m_1(q)(T-\tau))d\tau\right]R(\alpha,r,q)\right\}H\left[\sqrt{T^2-z^2}-\alpha\right]\mathrm{d}\alpha, \tau=\sqrt{z^2+\alpha^2},$$

$$R(\alpha,r,q) = \int_0^\infty \cos(uq)J_1(ru)J_0(\alpha u)\mathrm{d}u, \tag{3.20}$$

$$R(\alpha,r,q) = \begin{cases} 0, & q<\alpha-r \\ \frac{1}{r}\left\{1-\Lambda_0\left[\sin^{-1}\left(\frac{2\alpha}{q+r+\alpha}\right)^{1/2},k\right]+\frac{\alpha-q}{\pi\sqrt{\alpha r}}K(k)\right\}, & |\alpha-r|<q<\alpha+r \\ \frac{1}{r}\left\{1-\Lambda_0\left[\sin^{-1}\left(\frac{q+\alpha-r}{q+r+\alpha}\right)^{1/2},\frac{1}{k}\right]-\frac{1}{\pi k}\sqrt{r/\alpha}K\left(\frac{1}{k}\right)\right\}, & q>\alpha+r \\ \frac{1}{r}, & |q-\alpha|<q+\alpha+r \end{cases}$$

$$\Lambda_0(\phi,k) = \frac{2}{\pi}\left[E(k)F\left(\phi,\sqrt{1-k^2}\right)+K(k)E\left(\phi,\sqrt{1-k^2}\right)-K(k)F\left(\phi,\sqrt{1-k^2}\right)\right],$$

$$k = \frac{\sqrt{q^2-(\alpha-r)^2}}{2\sqrt{\alpha r}}$$

$$F(\alpha,r,q) = \int_0^\infty \sin(uq)J_0(ru)J_0(\alpha u)\mathrm{d}u \tag{3.21}$$

$$F(\alpha,r,q) = \begin{cases} 0, & q<|\alpha-r|, \\ \frac{1}{\pi\sqrt{\alpha r}}K(k), & |\alpha-r|<q<\alpha+r, q,\alpha,r>0 \\ \frac{1}{\pi k\sqrt{\alpha r}}K\left(\frac{1}{k}\right), & q>\alpha+r \end{cases},$$

where $F\left(\phi,\sqrt{1-k^2}\right)$ and $E\left(\phi,\sqrt{1-k^2}\right)$ are the elliptic integrals of the first and second kind, respectively. The integrals in Eqs. (3.17) and (3.18) are found using [11].

The numerical calculation of dimensionless value of $U_\theta(r,z,T) = u_\theta(r,z,T)/A$ for Mode III crack depending on dimensionless time T was carried out according to Eq. (3.19), and the results for $R = 100$ $\left(R = (r^2+z^2)^{1/2}\right)$ are shown in Fig. 3.7. The magnitude of the first maximum that corresponds to the arrival of the shear wave front at the observation point is the highest. The distances between peaks are approximately the same. The calculation of this normalized displacement $U_\Omega = u(\Omega)/u(\Omega)|_{\max}$ depending on dimensionless frequency $\Omega = \omega r_0/c_2$ was made using the fast Fourier transform method [10] and is shown in Fig. 3.8 for the same values of R and φ.

Similar calculations of $U_\theta(r,z,T)$, depending on dimensionless time T, were carried out for various values of R and φ. After the analysis of maximal values of $U_\theta(r,z,T)$, we obtain the following approximation dependence (for $R > 30r_0$):

$$u_{\theta|\max} = 1.138\frac{\tau_0 r_0^2 \Phi_2^{(d)}(\varphi)}{\sqrt{2\pi}3\mu R}, \tag{3.22}$$

where $\Phi_2^{(d)}(\varphi)$ is determined by Eq. (3.20). The approximation error does not exceed 7.5%.

The angular distribution of $u_{\theta|\max}$ value is the same as for Mode I crack shear wave (Fig. 3.4b).

The results of the calculations according to Eqs. (3.13) and (3.22) for two aluminum alloys D18 (chemical composition—Si 0.5%, Fe 0.5%, Cu 2.2…3.0%, Mn 0.2%, Mg 0.5%, Cr 0.1%, Zn 0.1%, other 0.15%) and B95 (chemical composition —Zn 5.0…6.5%, Mg 1.8…2.8%, Cu 1.4,…2.0%, Mn 0.2…0.6%, Cr 0.1…0.25%, Fe 0.25…0.5%, Si 0.1%, Ni 0.1%, Ti 0.05%, others less than 0.1%) used in airplane production, are shown in Fig. 3.7. The following characteristics were used

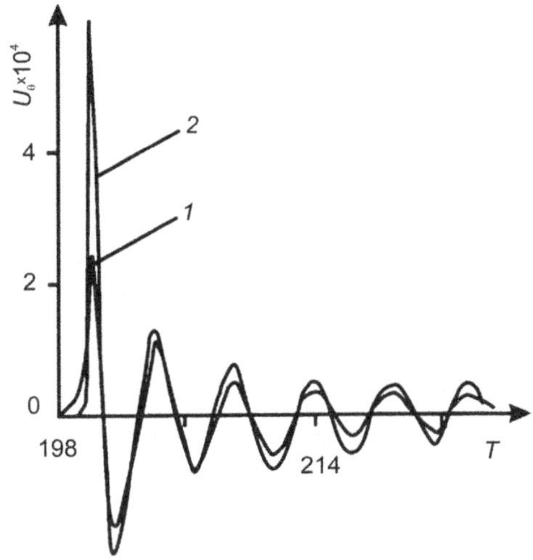

Fig. 3.7 The dependence of $U_\theta(r,z,T)$ on T: *1* $\varphi = 15°$, *2* $\varphi = 45°$, $R = 100$

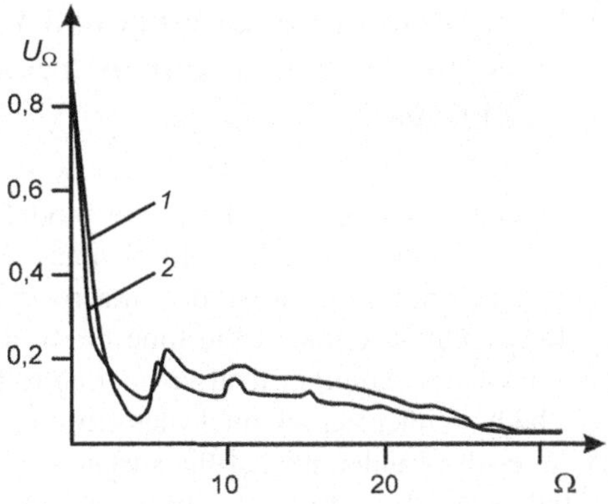

Fig. 3.8 Dependence of U_Ω on Ω: *1* $\varphi = 15°$, *2* $\varphi = 45°$, $R = 100$

for D18 Al alloy: $\sigma_0 = 300$ MPa, $\tau_0 = 200$ MPa, $E = 71000$ MPa, $\nu = 0.32$; for B95 Al alloy: $\sigma_0 = 480$ MPa, $\tau_0 = 32$ MPa, $E = 70000$ MPa, $\nu = 0.33$ and $\rho = 2700$ kgm^{-3} $R = 0.5$ m $\pi r_0^2 = 10^{-6}$ m^2 for both materials [12].

Supposing that $u_{\max}|_{c_i}$ and $u_\theta|_{\max}$ are directly proportional to the amplitude *V* of the electric signal at the sensor output, we can rewrite Eqs. (3.13) and (3.22) as follows:

$$V = \beta_i \frac{\sigma_0 \Phi_i^{(d)}(\theta) r_0^2}{\rho c_1^2 R}, V = \beta_\theta \frac{\tau_0 r_0^2 \Phi_2^{(d)}(\varphi)}{\sqrt{2\pi 3 \mu R}}. \quad (3.23)$$

For Mode I and Mode III cracks, respectively, β_i and β_θ are the proportionality factors for these cracks. They depend on AE device characteristics (amplification factor, frequency response), AE sensor characteristics, etc. Therefore, β_i and β_θ should be determined experimentally.

Using both proposed models of crack nucleation, the analytical dependencies between parameters of a crack and parameters of AE signals are obtained. These dependencies have shown that the amplitude of an AE signal emitted by the forming of a penny-shaped crack under sudden tensile or twisting loading is proportional to the crack surface area (r_0^2) and decays as 1/*R*. The angular distribution of amplitudes of AE signals can be used to determine the crack space orientation by a system of AE sensors. The obtained dependencies for amplitudes of AE signals can be used to develop the techniques of AE testing of nucleation and propagation of isolated cracks of the normal and twisting fracture modes.

3.2 Modelling the Sub-critical Crack Growth at Local Areas of Its Contour as a Source of Acoustic Emission Signals

In order to develop an appropriate model for radiation of AE signals caused by sub-critical crack growth at local areas of its front, we can consider an elastic half-space with a planar Mode I macro-crack bound by a smooth contour *L*.

Let us suppose that at the time $t = 0$ in local area, where stresses (or deformations) reach a certain critical value, due to the application of external tensile forces to a body, a micro-crack nucleates close to a contour of this macro-crack (Fig. 3.9). As a result of unloading of the surfaces of this newly formed micro-crack from an initial level down to zero, elastic waves are radiated. They reach the inspected object surfaces and can be received by AET (Fig. 3.1).

To simplify the problem, we replace this micro-crack with a penny-shaped Mode I crack of an equal area. If we also suppose that the radius of this penny-shaped micro-crack is much smaller than the radius of the macro-crack contour curvature, in this case, instead of the above-mentioned problem, we may consider the problem of a sudden nucleation of a penny-shaped micro-crack near the edge of a semi-infinite macro-crack.

If we estimate the resulting components of the dynamic displacement field, we can now consider a dynamic problem of the growth of a semi-infinite through-crack in a homogeneous isotropic elastic body. The following angular dependencies of $\Phi_i^h(\beta)$ were obtained in [13] for this problem. They correspond to longitudinal and transverse waves that are valid for the wave front region.

$$\Phi_i^h(\beta) = V_i(\beta)C_i(\beta)P_i(\beta), \ i = 1, 2 \tag{3.24}$$

Here, the functions

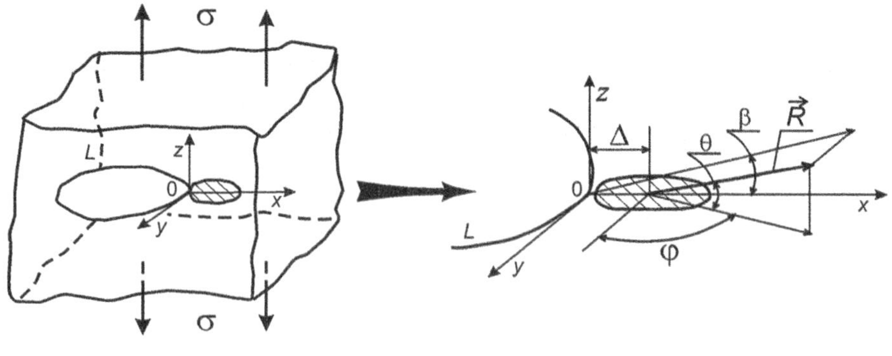

Fig. 3.9 The local growth of an internal crack

$$V_i(\beta) = \frac{1}{1 - c_c/c_i \cos\beta} \tag{3.25}$$

describe the change in the angular dependence of radiation caused by the crack edge propagation with velocity c_c for a component of longitudinal (i = 1) and shear (i = 2) waves. Functions

$$C_i(\beta) = \frac{\sqrt{1 + c_1/c_i \cos\beta}}{(1 + c_R/c_i \cos\beta)K_-(-s_i \cos\beta)} \tag{3.26}$$

where

$$K_\pm(\eta) = \exp\left\{-\frac{1}{\pi}\int_{s_1}^{s_2} \frac{g(z)dz}{z \pm \eta}\right\}, s_i = 1/c_i, \tag{3.27}$$

$$g(z) = \tan^{-1}\left\{\frac{4z^2\sqrt{z^2 - s_1^2}\sqrt{s_2^2 - z^2}}{(s^2 - 2z^2)^2}\right\} \tag{3.28}$$

describe the effect of the free surface of a semi-infinite crack on the angular distribution of radiation. Functions

$$P_1(\beta) = 1 - 2(c_2/c_1)^2\cos^2\beta; P_2(\beta) = \sin 2\beta \tag{3.29}$$

determine the angular dependencies of maximum values of the displacement vector module at the fronts of longitudinal and shear waves radiated by the edge of this crack; c_R is the Raleigh wave velocity.

The comparison of expression (3.29) with the dependencies (3.11) and (3.12) for the case of a penny-shaped crack nucleation shows their similarity. Therefore, if the radiation of elastic waves is caused by the nucleation of a penny-shaped micro-crack close to the contour of a macro-crack, then the effect of the free surfaces of a macro-crack on the displacement field caused by the formation of this penny-shaped crack can be estimated using the dependence (3.26). Therefore, we express a radiation field at a distance much greater than the micro-crack radius as a product of displacement components for an instant nucleation of a penny-shaped crack in an elastic body (see Eq. (3.9)) by functions $C_i(\beta)$ in Eq. (3.16) accounting for the effect of a free surface.

In a spherical coordinate system $OR\theta\varphi$, whose origin coincides with the center of a penny-shaped crack, the components of a displacement vector at distances $R \gg r_0$ can be written as follows:

$$u_R(\mathbf{R},t) = C_1(\beta)u_R^{(d)}(\mathbf{R},t),$$
$$u_\theta(\mathbf{R},t) = C_2(\beta)u_\theta^{(d)}(\mathbf{R},t), \tag{3.30}$$

where functions $u_R^{(d)}(\mathbf{R},t)$ and $u_\theta^{(d)}(\mathbf{R},t)$ are determined by Eqs. (3.8) and (3.9),

$$\cos\beta = \frac{\cos\theta\,\sin\varphi + \delta}{\sqrt{(\cos\theta\,\sin\varphi + \delta)^2 + \sin^2\theta}} \tag{3.31}$$

$\delta = \Delta/R$, Δ is the distance between the center of a penny-shaped crack and the edge of a semi-infinite crack. For large R, $\delta \to 0$. The components u_R and u_θ obtained in Eq. (3.30) have the same time dependence as appropriate components $u_R^{(d)}$ and $u_\theta^{(d)}$. However, the directional characteristic of radiation in this case will differ from those obtained for a penny-shaped crack.

The numerical results calculated according to the dependencies (3.30) make it possible to get the directional characteristic of radiation for the case of a penny-shaped crack nucleation in the vicinity of the macro-crack front. In Fig. 3.10, this directional characteristic of the maximum values of components $u_R(\mathbf{R},t)$ is presented for longitudinal wave and angle $\varphi = \pi/2$. The points mark experimental data obtained in [14]. We can see in this figure that the angular dependence obtained differs mostly from the angular dependence for an isolated penny-shaped crack (Fig. 3.4a) at θ close to π. In the case of $|\theta| \le \pi/2$, the angular dependencies practically coincide. Therefore, if an AET is located in this region of angles θ, the effect of the free surface on AE signals is insignificant.

It follows from the dependencies (3.30) that the magnitude of displacements at the front of a radiated longitudinal wave is proportional to r_0^2

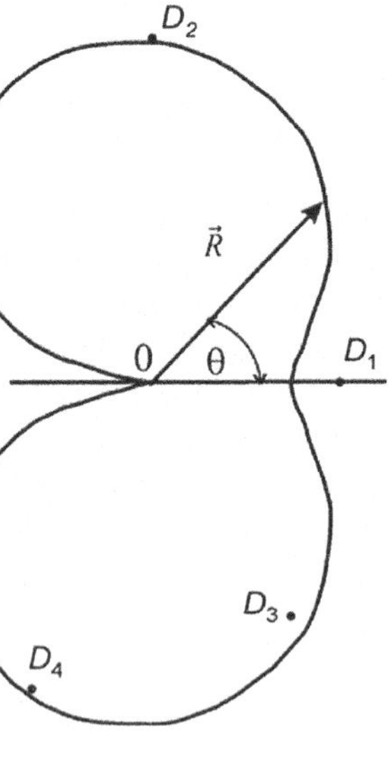

Fig. 3.10 Angular distribution of a maximum of the displacement vector module for longitudinal wave during nucleation of a penny-shaped crack near an internal crack front

$$u_{\max}|_{c_1} = \frac{\delta_1 \sigma_0 r_0^2}{\rho c_1^2 R} \Phi_1^{(d)}(\theta) C_1(\beta) \tag{3.32}$$

i.e., proportional to the area of a nucleated defect.

One can go back to considering the problem on a micro-defect of the S area nucleation in the vicinity of the internal macro-crack contour. Assuming that S is equal to the area of a micro-crack and that the amplitude A of the AE signal is proportional to the magnitude of displacement at the front of a longitudinal wave, we find that

$$A = \frac{\lambda S}{R} \Phi_1^{(d)}(\theta) C_1(\beta), \tag{3.33}$$

where $\lambda = \lambda_0 \lambda_1$ and λ_0 is a proportionality factor between electrical signals in the output of AET and the maximum values of displacements at the front of a longitudinal wave, $\lambda_1 = \delta_1 \sigma_0 / (\pi \rho c_1^2)$. If during the local growth of an internal crack, such micro-defects are formed in the vicinity of its contour N, then the total area of the internal macro-crack growth will be

$$\Delta S = \sum_{k=1}^{N} b A_k, \tag{3.34}$$

where b is the proportionality factor.

Thus, as follows from dependencies (3.34), the amplitudes of AE signals caused by the internal flat crack growth in local areas of its front are proportional to the total growth area. A relationship of this sort, between A and S, was confirmed experimentally [15, 16].

3.3 The Effect of Body Boundaries on AE Signals Caused by the Growth of an Internal Defect

The recording of AE signals is carried out by an AET located on the surface of the IO. For this reason, an important issue in AE diagnostics is the study of the effect of the free surface on AE parameters. To investigate the change in the displacement vector component caused by the boundaries of a body, we will consider the problem of determining the surface motion due to a sudden nucleation of the internal penny-shaped crack in a body.

Such a problem can be solved by using the Helmholtz potentials method [17, 18]. In this approach, the solution to the problem could be reduced to the numerical-analytical solution of the system of singular integral equations with respect to the unknown Fourier transforms over the time of displacement vector jumps on the crack surfaces. The necessity of using the numerical methods, both for solving the singular integrated equations and for inverting Fourier transforms, complicates the obtaining of the results and their analysis, in particular, finding the dependencies between the maxima of surface motion and orientation of a penny-shaped crack. Therefore, in order to estimate the surface displacement caused by a sudden nucleation in a half-space of such a crack, we will use the approximation approaches used in [19].

The geometry of the problem is shown in Fig. 3.11. The plane of a crack forms an angle Φ with normal to the boundary of a half-space. In a toroidal coordinate system $OR\alpha\psi$, the angle ψ defines a running point on the crack contour. For each plane ψ = constant, coordinates (r', α) form a polar coordinate system whose origin coincides with this point. The plane $\psi = 0$ is a plane of symmetry, and d is the distance between the nearest point of the contour of a penny-shaped crack and the boundary of a half-space. The $Oxyz$ Cartesian coordinate system will be also used. Its origin is the point of intersection of the crack plane, the symmetry plane, and the boundary of a half-space.

For this model, using the results obtained in [19] in the near zone, the overall (incident plus reflected) displacement on the surface caused by an incident longitudinal wave emitted by the crack in the plane of symmetry $\psi = 0$ is given by

$$\mathbf{U}_1(R,\alpha) = \left[R_x^{(1)}(\alpha)\mathbf{i} + R_y^{(1)}(\alpha)\mathbf{j}\right] u_R^{(d)}(R,\theta) \tag{3.35}$$

where

$$\begin{aligned} R_x^{(1)}(\alpha) &= \frac{2\varepsilon^{-1}\sin 2(\alpha+\Phi)\sin\delta}{\varepsilon^2\cos^2 2\delta - \sin 2\delta\,\sin 2(\alpha+\Phi)}, \\ R_y^{(1)}(\alpha) &= \frac{2\varepsilon^{-2}\cos(\alpha+\Phi)\cos 2\delta}{\varepsilon^2\cos^2 2\delta - \sin 2\delta\,\sin 2(\alpha+\Phi)}, \end{aligned} \tag{3.36}$$

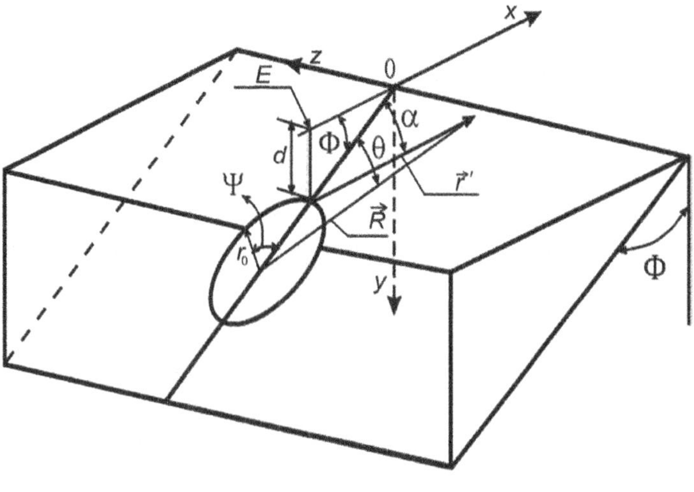

Fig. 3.11 A penny-shaped crack in an elastic half-space

$u_R^{(d)}(R,\theta) = u_R^{(d)}(R,\theta,t)$ or $u_R^{(d)}(R,\theta,\omega)$ are components of a displacement vector for the problem of nucleation of a penny-shaped crack in the infinite body both for the time and frequency domains, respectively, determined by Eqs. (3.8) and (3.9), $\boldsymbol{i}$ and $\boldsymbol{j}$ are the unit vectors, R is the distance from the crack center to the observation point located on the boundary of the half-space, $R = (r_0 + d\cos\alpha/\cos(\alpha+\Phi))/\cos\theta$, θ is the angle between the plane of the crack and the direction vector that originates in the center of the defect to the observation point, and the angle δ satisfies equation

$$\cos\delta = -\varepsilon\sin(\alpha+\Phi), \tag{3.37}$$

the motion on the free surface caused by an incident transverse wave for angles $\alpha+\Phi$, such that $|\sin(\alpha+\Phi)| < \varepsilon$, is given by

$$\mathbf{U}_2(R,\alpha) = \left[R_x^{(2)}(\alpha)\mathbf{i} + R_y^{(2)}(\alpha)\mathbf{j}\right]u_\theta^{(d)}(R,\theta), \tag{3.38}$$

where

$$\begin{aligned} R_x^{(2)}(\alpha) &= \frac{2\varepsilon^{-2}\cos(\alpha+\Phi)\cos 2(\alpha+\Phi)}{\varepsilon^{-2}\cos^2 2(\alpha+\Phi) - \sin 2\gamma\,\sin 2(\alpha+\Phi)}, \\ R_y^{(2)}(\alpha) &= \frac{-2\sin 2\gamma\,\cos(\alpha+\Phi)}{\varepsilon^{-2}\cos^2 2(\alpha+\Phi) - \sin 2\gamma\,\sin 2(\alpha+\Phi)}, \end{aligned} \tag{3.39}$$

$u_\theta^{(d)}(R,\theta)$ is equal to $u_\theta^{(d)}(R,\theta,t)$ or $u_\theta^{(d)}(R,\theta,\omega)$, which are the components of a displacement vector for the problem of a penny-shaped crack nucleation. They are determined by (3.8) and (3.9), both for the time and frequency domains, respectively. The angle γ should satisfy the Equation

$$\varepsilon\cos\gamma = -\sin(\alpha+\Phi). \tag{3.40}$$

Equations (3.35) and (3.38) are correct for times corresponding to the near-front wave region and condition $d \gg r_0$. Taking into account the Eqs. (3.13), (3.14) and (3.36), (3.39) for the maximum value of the module of the displacement vector caused by incident longitudinal and shear waves, we obtain the following equation

$$u^{(k)}_{\max|c_i} = \frac{\delta_i\sigma_0\Phi_i^{(d)}(\theta)r_0^2}{\rho c_1^2 R}R_{(k)}^{(i)}(\alpha), \quad (i,k = \overline{1,2}), \tag{3.41}$$

where functions $\Phi_1^{(d)}(\theta)$ and $\Phi_2^{(d)}(\theta)$ are defined by Eqs. (3.11) and (3.12) for longitudinal and shear waves, respectively. Note that for $d \gg r_0$, the difference between the angles θ and α is insignificant.

In Figs. 3.12 and 3.13, dimensionless maximum values of the displacement vector module on the surface of the half-space $U_{\max|c_i} = \left[u^{(x)2}_{\max|c_i} + u^{(y)2}_{\max|c_i}\right]^{1/2}/A$ for longitudinal and shear waves, respectively, are presented with respect to the

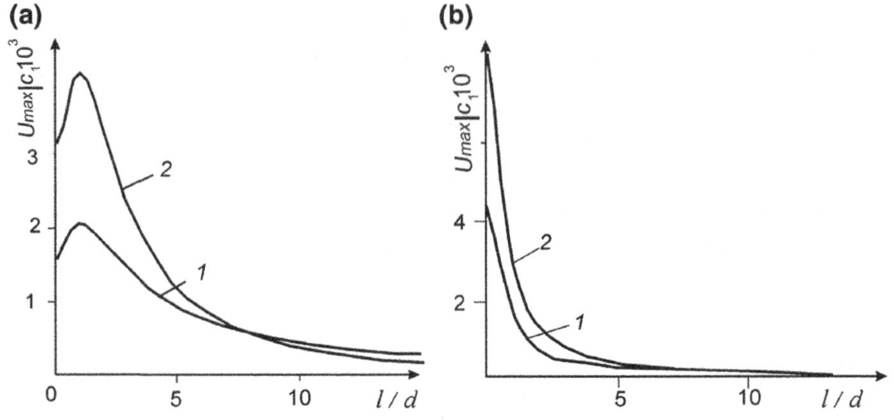

Fig. 3.12 Dependence of the dimensionless value of $U_{\max|c_1}$ on the half-space surface on the dimensionless distance *l/d* at $d/r_0 = 200$ for a longitudinal wave at the angles of a crack orientation **a** Φ = 0°, and **b** Φ = 75°; Curve *1* corresponds to the surface motion caused only by an incident wave, and Curve *2* corresponds to the surface motion caused by an overall wave

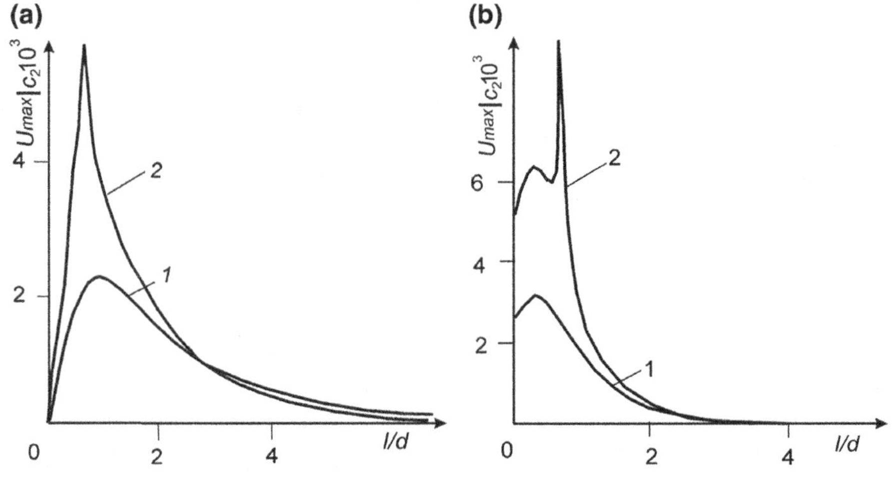

Fig. 3.13 Dependence of the dimensionless value $U_{\max|c_2}$ on the half-space surface on the dimensionless distance *l/d* at $d/r_0 = 200$ for a shear wave at the angles of a crack orientation **a** Φ = 0° and **b** Φ = 75°; Curve *1* corresponds to the surface motion caused only by an incident wave, and Curve *2* corresponds to the surface motion caused by an overall wave

dimensionless distance *l/d* between the epicenter (point *E* in Fig. 3.8) and the observation point located on the boundary of the half-space, $l = dtg\Phi + x$.

The calculations were carried out for crack orientation Φ = 0° (Figs. 3.9a, 3.10a) and Φ = 75° (Figs. 3.12b and 3.13b), $d/r_0 = 200$. In these figures, 1 denotes the maximum values of the displacement vector module for an incident wave, and 2 denotes the maximum values of the surface displacement, both calculated by Eq. (3.41). As shown in Fig. 3.12 for the case of longitudinal wave, the maximum values of surface motion exceed the appropriate ones for an incident wave by nearly a factor of two.

For distances *l/d* greater than about seven depths of the defect location, the difference between the maxima of motion caused by an incident wave and an overall motion on the surface is insignificant. In the case of a shear wave, this difference becomes small at the distance from the epicenter, approximately equal to four depths of the defect location.

3.4 The Waveguide Effect on the Change of the Parameters of AE Signals

While using waveguides in AE measurements, the following should be taken into account: A waveguide has eigenfrequencies that depend on its geometrical sizes. AET also has a resonance frequency band; therefore, in order to obtain the fewest losses during the transfer of AE signals, it is important to match these frequencies to the components of the measuring system. A cylindrical waveguide is most frequently used in AE measurements since it is the simplest to manufacture and can easily be attached to the specimen or structure element.

The finding of frequency eigenvalues for a cylindrical waveguide of arbitrary sizes is a difficult problem [20, 21]. Besides, the use of the solution obtained in these papers for engineering calculations seems to be rather problematic, so the approximation approach for selecting the sizes of cylindrical waveguides is proposed in this study. This approach is based on the method of limiting interpolation [22], whose efficiency is proved for many problems of the fracture mechanics and AE. As a result, the following approximation expression for calculation of frequency eigenvalues for a cylindrical waveguide of arbitrary length and diameter [23] is obtained:

$$a_{p,q} = a_{p,q}^{(d)} a_{p,q}^{(d_*)} a_{p,q}^{(s)} \left\{ \left[\left(a_{p,q}^{(d_*)} - a_{p,q}^{(d)} \right) a_{p,q}^{(s)} \right]^m + \left[a_{p,q}^{(d)} + a_{p,q}^{(d_*)} \right]^m \right\}^{-1/m} \tag{3.42}$$

where $a_{p,q}^{(d)}$ and $a_{p,q}^{(s)}$ are the functions appropriate to frequency eigenvalues for a disk and a rod, respectively, $a_{p,q}^{(d_*)}$ is the value of $a_{p,q}^{(d)}$ at $l/a \gg 1$, l is the length of the cylinder, a is its radius (Fig. 3.14). Equation (3.42) precisely takes into account the limiting cases of interpolation (disk and long rod). Parameter m can be found from the exact solution of the problem in some intermediate points.

One can now consider the limiting cases of interpolation. For a thin disk, there will be no stresses in the plane perpendicular to the Oz axis (Fig. 3.14). The equation for finding frequency eigenvalues for this case can be written as [24]

Fig. 3.14 The cylindrical waveguide

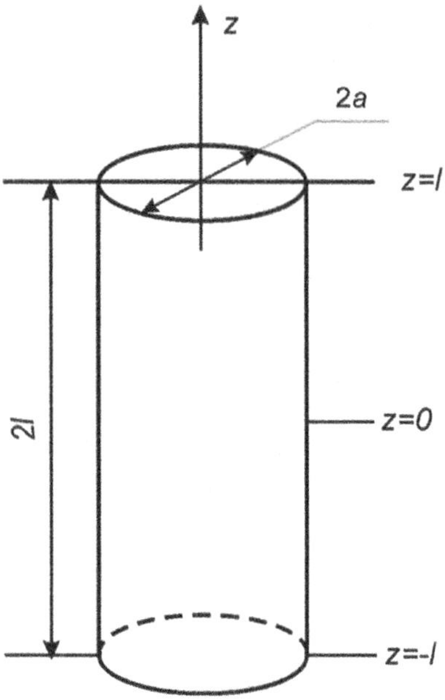

$$\frac{th\beta l}{th\alpha l} = \frac{4\alpha\beta g^2}{\left(\beta^2 + g^2\right)^2}, \tag{3.43}$$

where α and β are phase constants for tension and shear along the cylinder axis, g is a phase constant for a radial direction. These constants are related by the dependence

$$g^2 = \alpha^2 + h^2 = \beta^2 + k^2, \tag{3.44}$$

where $h = \omega c_1$, $k = \omega c_2$ and g is defined by the following equation [24]:

$$\frac{d^2 J_0(ga)}{d(ga)^2} = \frac{g^2 + \beta^2}{2g^2} J_0(ga). \tag{3.45}$$

Frequency eigenvalues of oscillation for a thin disk $f_{p,q}$ can be found by solving the system of Eqs. (3.43), (3.45)

$$f_{p,q} = a_{p,q}/(2\pi l c_2), \tag{3.46}$$

where $a_{p,q} = (kl)_{p,q}$, p and q are integers, $(al)_q$ and $(ga)_p$ are q-th and p-th roots of Eqs. (3.44) and (3.46), respectively. The equation for determining frequency eigenvalues of a rod has the form of

$$(x-1)^2\varphi(h'a) + (1-ex)[x-\varphi(k'a)] = 0, \tag{3.47}$$

where $\varphi(y) = \frac{yJ_0(y)}{J_1(y)}$, $x = c^2/c_0^2(1+\nu)$, $e = \frac{1-2\nu}{1-\nu}$, $c_0 = \sqrt{E/\rho}$ is the velocity of sound in a rod, and E is Young's modulus, c is a phase velocity.

Phase constant g for a long rod could be divided into two terms h' and k' corresponding to tension and rotation components. A phase constant γ of wave propagation in an axial direction is unique. The relationship between these constants is

$$h' = h^2 - \gamma^2; k' = k^2 - \gamma^2, \tag{3.48}$$

where h and k are given by Eq. (3.44). If the rod has a finite length, the following condition should hold

$$\gamma_q = \frac{\pi q}{2l}, \tag{3.49}$$

where q is a positive odd number. The eigenvalues of $a_{p,q}$ can be found from the solution of a system of Eqs. (3.47) and (3.49):

$$a_{p,q} = \pi q\sqrt{x_p/2}. \tag{3.50}$$

Thus, Eqs. (3.42), (3.46) and (3.50) enable us to estimate the frequency eigenvalues of a cylindrical waveguide of arbitrary geometrical sizes. As an example, the value of $a_{1,1}$ at $m \approx 4.6$ was calculated (Fig. 3.15).

The theoretical results obtained were used in choosing the optimum sizes of the waveguide to estimate the threshold value of K_{Iscc}. According to the design requirements for the test equipment, it was important that the waveguide be 132 mm long. The frequency band of the resonance-type AET was 200–350 kHz.

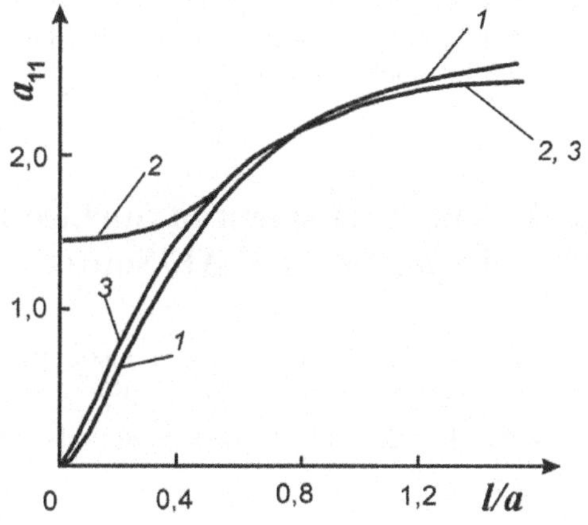

Fig. 3.15 Dependencies of the value of $a_{1,1}$ on l/a calculated for a disk (Curve *1*), rod (Curve *2*) and by approximation expression (32) (Curve *3*)

The waveguide diameter should be chosen so that the frequency eigenvalues lie within the bandwidth of AET. The calculations by approximation Eq. (3.42) have shown that the given condition is satisfactory if $p = 3$, $q = 3$ and $p = 1$, $q = 13$. According to the dependence (3.46), $f_{p,q}$ will be greater than 200 kHz if $a_{p,q} \geq 27,2$. Herein follows the condition $l/a \geq 10$: As the length of a rod is equal to 132 mm, the waveguide diameter of 13 mm was chosen. (Note that while $l/a \geq 10$, the value of frequency eigenvalues calculated by approximation expression (3.42) and by dependence (3.50) are identical.)

The normal displacement W at the end face of the waveguide for frequencies close to eigenvalues can be estimated by the expression given in [24]. Then, using the dependence (3.41) we obtain the following estimation for the maximum normal displacement at the end face of the waveguide caused by a crack nucleation in the elastic half-space:

$$U_{\max}|_{c_i} \propto \frac{\delta_i \Phi_i^{(d)}(\theta) r_0^2}{\rho c_1^2 R} R_{(y)}^{(i)}(\alpha) W, \quad i = 1, 2. \tag{3.51}$$

Moreover, using dependencies (3.33) and (3.51), the amplitudes of AE signals can be expressed as

$$A \propto \frac{\lambda S \Phi_i^{(d)}(\theta)}{R} R_{(y)}^{(i)}(\alpha) W, \quad i = 1, 2. \tag{3.52}$$

The investigation of the AE changes caused by the waveguide of a chosen dimension was carried out within the experimental determination of the threshold value of K_{Iscc} [25]. The research has shown a slight loss in amplitude due to the waveguide, which confirms the validity of the method of limiting interpolation in solving these types of problems.

Using the proposed models of nucleation and growth of a crack, as well as the research of the body boundary and the waveguide effect on the change of the parameters of AE signals, the analytical dependencies between parameters of a crack and parameters of AE signals are obtained. These dependencies can be used in developing the appropriate techniques of AE testing and the equipment as applied to crack detection.

3.5 The Assessment of Surface Displacements Caused by an Internal AE Source

If we consider the Rayleigh acoustic waves caused by an internal source pulse irradiation [26, 27], we know that the surface of the tested body additionally changes the parameters of AE signals generated by a crack. Such changes are ambiguous and depend on the body size and geometry. The results seen in the literature mainly refer to the research on the epicenter of irradiation [28, 29]. We can also consider some typical cases that arise during AE testing.

A peculiar feature of the AE recording on large objects is the presence of Rayleigh waves in addition to longitudinal and transversal ones. Depending on the transducer location with respect to the epicenter source, waves of a certain type will prevail. Therefore, at first it is necessary to set the limits of the prevailing waves and then find the equation for calculating the displacement vector components.

To compare Rayleigh waves with other waves, we can consider the formation of an internal source in a homogeneous half-space [27]. For simplification, we assume that at the initial moment of time, at a distance z_0 from the free surface, a point source of irradiation begins to act (Fig. 3.16). Then we should introduce the system of cylindrical coordinates $0r\theta z$ with the center (point O) located in the epicenter on a half-space boundary and with the Oz axis perpendicular to it.

Boundary conditions on a half-space surface:

$$\sigma_z = \tau_{rz} = 0 \text{ at } z = 0. \tag{3.53}$$

The problem formulated is an axisymmetric one. Wave potentials φ and ϕ satisfy the system of Eq. (2.54) with an additional term in the right-hand part of the first of the Eqs. (2.54) $\varphi_0\delta(r)\delta(z - z_0)H(t)/(2\pi r)$, and the initial conditions are zero

$$\begin{aligned} &\nabla^2\varphi - c_1^{-1}\partial^2\varphi/\partial t^2 = \varphi_0\delta(r)\delta(z - z_0)H(t)/(2\pi r); \\ &\nabla^2\phi - \phi/r - c_2^{-1}\partial^2\phi/\partial t^2 = 0, \end{aligned} \tag{3.54}$$

where $\nabla^2 = \partial^2/\partial r^2 + r^{-1}\partial/\partial r + \partial^2/\partial z^2$; φ_0 is the source intensity, $\delta(\ldots)$is the Dirac delta function.

Applying the Hankel integral transform to the wave equations with respect to coordinate r, and the integral Laplace transform with respect to time t, we get the components of a displacement vector in the Laplace transform domain.

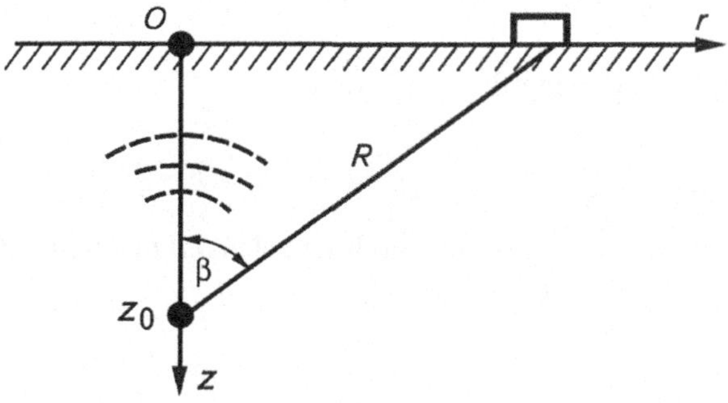

Fig. 3.16 Schematic presentation of the AE source in a half-space

$$\bar{u}_r = \int_0^\infty \alpha^2[B\gamma_2 \exp(-\gamma_2 z) - A \exp(-\gamma_2 z) - \varphi_0 \exp(-\gamma_1 |z - z_0|)] J_1(\alpha r) d\alpha,$$

$$\bar{u}_z = \int_0^\infty \alpha[-A\gamma_1 \exp(-\gamma_1 z) + B \exp(-\gamma_1 |z - z_0|) + \varphi_0 \exp| - \gamma_1 z|/s J_0(\alpha r)] d\alpha,$$

$$(3.55)$$

where

$$\gamma_1 = \left(\alpha^2 + s^2/c_1^2\right)^{1/2}, (l = 1, 2).$$

The unknown constants A and B are found from the system of two algebraic equations obtained from the dependences (3.56) that connect the wave potentials with the components of a displacement vector

$$u_r = -\partial\varphi/\partial r - \partial\phi/\partial z; u_z = \partial\varphi/\partial z + \partial\phi/\partial x + \phi/r \tag{3.56}$$

and boundary conditions (3.54). Substituting these constants A and B in (3.55), we find displacements of the free surface in the Laplace transform domain

$$u_r = -\partial\varphi/\partial r - \partial\phi/\partial z; u_z = \partial\varphi/\partial z + \partial\phi/\partial x + \phi/r \tag{3.57}$$

where

$$R(\alpha, s) = \left[\alpha^2 + s^2/\left(2c_2^2\right)\right]^2 - \alpha^2\gamma_1\gamma_2$$

is the Rayleigh function.

Writing the Bessel function $J_0(\ldots)$ in terms of the Hankel function of the first kind $H_0(\ldots)$ as

$$J_0(x) = 0.5\left[H_0^{(1)}(x) - H_0^{(1)}(-x)\right]$$

[30] and using the expression

$$H_0^{(1)}(-iz) = 2K_0(z)/\pi i,$$

where $K_0(\ldots)$ is the modified Hankel function, after changing the integration variable $\alpha = \eta/i$,we get

$$\bar{u}_z(r,0,s) = 2\varphi_0 s/\pi i c_2^2 \int\limits_{-i\infty}^{i\infty} \eta \exp(-\gamma_1 z_0)\left(\frac{s^2}{2c_2^2} - -\eta^2\right)\frac{K_0(\eta r)}{R(\alpha,s)}d\eta. \quad (3.58)$$

We find the Laplace inverse transform using the Cagniard method [31]. For this purpose, modifying the expression (3.58) by introducing a new variable of integration $p = s\eta$ and taking into account the relationship $K(z^*) = [K(z)]^*$, we obtain

$$\bar{u}_z(r,0,s) = \frac{8\varphi_0 s}{\pi c_2^2}\mathrm{Im}\left(\int\limits_0^{i\infty} \frac{p(c_2^{-2} - 2p^2)\exp(-s\bar{\gamma}_1 z_0)K_0(spr)}{R(p)}dp\right). \quad (3.59)$$

We can now introduce a real variable of integration $\tau = pr + \bar{\gamma}_1 z_0$. Then, the relationship

$$p(\tau) = \begin{cases} \left[r\tau + iz_0\left(\tau^2 - R_0^2/c_1^2\right)^{1/2}\right]/R_0^2, \tau \geq R_0/c_1, \; R_0 = \left(z_0^2 + r^2\right)^{1/2}; \\ \left[r\tau - iz_0\left(R_0^2/c_1^2 - \tau^2\right)^{1/2}\right]/R_0^2, \tau < R_0/c_1 \end{cases} \quad (3.60)$$

describes a contour in a complex plane that consists of the segment of a real axis

$$[-z_0 c_1/R_0, \; rc_1/R_0] \text{for } \tau < R_0/c_1$$

and a hyperbola for $\tau \geq R_0/c_1$. In order to apply the Jordan lemma, consider a closed contour with an imaginary semi-axis $[0, i\infty]$, a line (3.60) and a part of a circle of a dimensionless radius that encloses an imaginary half-space with this contour. In the region that binds a closed contour, the integrand does not have singularities and branch points. The function $K_0(x)$ for large values of x has the asymptote

$$\pi \exp(-x)\left[1 + o(x^{-1})\right]/(2x)^{1/2}.$$

Therefore, based on the Jordan lemma, the imaginary axis integration in the expression (3.59) can be replaced by the contour of integration (3.60) over τ that changes from 0 to ∞. After finding the original of the Laplace transform of the expression $\exp(-s\gamma_1 z_0)K_0(spr)$ and its differentiation, we find the relationship to calculate the displacements on the half-space surface:

$$u_z(r,0,t) = \frac{-8\varphi_0}{\pi c_2^2}\frac{\mathrm{d}}{\mathrm{d}t}\mathrm{Im}\left(\int\limits_{R_0/c_1}^{t} \frac{\Phi_1(r,t;\tau)}{\Phi_2(r,t;\tau)}\mathrm{d}\tau\right), \quad t > R_0/c_1, \quad (3.61)$$

where

$$\Phi_1(r,t;\tau) = ip\bar{\gamma}_1\left(\bar{\gamma}_2^2 - p^2\right),$$
$$\Phi_2(r,t;\tau) = R(\tau)(t-\tau)^{1/2}(t-\tau-2pr)\left(\tau^2 - R_0^2/c_1^2\right)^{1/2}.$$

Derivation by time t and normalization of variables

$$\tilde{r} = r/z_0, \tilde{t} = tc_1/z_0, \tilde{\tau} = \tau c_1/z_0, \tilde{R} = R_0/z_0, \tilde{u}_z = -u_z z_0^2/\left(8\pi\varphi_0\varepsilon^2\right), \tilde{\gamma}_1 = c_1\bar{\gamma}_1, \tilde{\gamma}_2 = c_1\bar{\gamma}_2,$$

yield a final dependence for calculation of the displacement vector components on the half-space boundary [27]

$$\begin{aligned} u_z(r,0,t) = & \left(\frac{1}{t-R_0}\int\limits_{R_o}^{t}\frac{F_1(r,t;\tau)}{F_2(r,t;\tau)}\mathrm{d}\tau\right)H(t-R_0) \\ & + \left(-\int\limits_{R_o}^{t}\frac{\mathrm{Re}F_3(r,t;\tau)}{F_2(r,t;\tau)}\mathrm{d}\tau\right)\delta(t-R_0), \end{aligned} \tag{3.62}$$

where

$$\begin{aligned} F_1(r,t;\tau) = {} & 0.5\Big[\mathrm{Re}F_3(r,t;\tau)(\tau+R_0)^{-1/2} - (t-R_0)\mathrm{Re}(F_3(r,t;\tau)/(t-\tau-2pr)) \\ & \times(\tau+R_0)^{1/2}\Big] - \mathrm{Im}\Big\{\left[\left(\gamma_2^2-p^2\right)\left(\gamma_1^2-\gamma_2^2\right) - 4p^2\gamma_1^2\right]/\left[R(p)(t-\tau-2pr)^{1/2}\right]\Big\} \\ & \times(\tau-R_0)/(\tau+R_0) + \mathrm{Im}\{4p\gamma_1F_3(t,\tau)\left[2\gamma_1\gamma_2 - (\gamma_1/\gamma_2+\gamma_2/\gamma_1)p^2 - \right. \\ & \left. - 2(\gamma_2-p^2)\right]/R(p)\}(\tau-R_0)^{1/2}/(\tau+R_0) - \mathrm{Re}\{F_3(r,t;\tau) \\ & \left[2ir\gamma_1 - \left(\tau^2. \quad + R_0^2\right)^{1/2}\right]/[2(t-\tau+2pr)]\}(\tau-R_0)^{1/2}/(\tau+R_0); \end{aligned}$$

$$F_2(r,t;\tau) = (t-\tau)^{1/2}\left(\tau^2 + R_0^2\right)^{1/2};$$
$$F_3(r,t;\tau) = p\gamma_1\left(\gamma_2^2 - p^2\right)/\left[R(p)(t-\tau-2pr)^{1/2}\right].$$

Note that in the relationship (3.62) the "~" sign above the normalized values is omitted for simplification.

From correlation (3.62), it yields that when the longitudinal wave arrives, we have a delta-pulse, because a source has been chosen as the uniform extension center. As expected, the decay of amplitudes of the Rayleigh wave with the growth of the distance between the epicenter and the observation point is weaker than for longitudinal or transversal waves in space generated by an internal source. At small distances from the epicenter, the Rayleigh wave is only slight, or even absent (Fig. 3.17a). With the increase of $\tilde{r}$, the form of an elastic oscillation changes at the

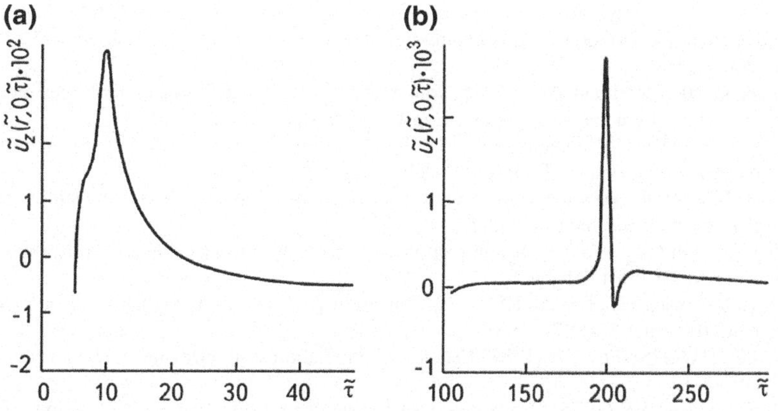

Fig. 3.17 Variation of the dimensionless displacement $\tilde{u}_z = -u_z z_0^2/(8\pi\varphi_0\varepsilon^2)$ with time $\tilde{\tau} = \tau c_1/z_0$ at distances $\tilde{r} = r/z_0 = 1$ (**a**) and $\tilde{r} = 25$ (**b**)

moment that the Rayleigh wave arrives, and the back and forward fronts of the pulse become more distinct (Fig. 3.17b).

Using the approach proposed in [32], we can see that the Rayleigh wave becomes evident if the following inequality is satisfied

$$|p(r/c_R) - 1/c_R| \leq 1/c_R - 1/c_2. \tag{3.63}$$

Substituting into (3.62) the value from (3.60) instead of p at $t \geq R_0/c_1$, we get [27]

$$tg\beta \geq [2/(c_{2R} - 1)]^{1/2}, c_{2R} = c_2/c_R. \tag{3.64}$$

The value of c_{2R} is found by putting the function $R(p)$ equal to zero and choosing the largest solution among those obtained [33]. For example, for metallic materials at $\nu = 0.3$, the Rayleigh wave will be observed at distances of $r \geq 5z_0$. This conclusion agrees with the calculations carried out directly from the Eq. (3.62).

Note that when locating the defects or evaluating their parameters, it is essential to take into account the type of wave that is received. Putting the type of a wave into calculations incorrectly can yield considerable errors.

References

1. Hartbower CE, Gerberich WW, Liebowitz H (1968) Investigation of crack-growth stress wave relationships. Eng Fracture Mech 1(2):291–308
2. Gerberich WW, Hartbower CE (1969) Monitoring crack growth of hydrogen embrittlement and stress corrosion cracking by acoustic emission. Fundamental aspects of stress corrosion cracking. NACE Houston, Houston, pp 420–438

3. Andreykiv AYe, Lysak MV (1989) Metod akusticheskoy emissii v issledovanii prozessov razrusheniya (A method of acoustic emission in investigation of fracture processes). Naukova Dumka, Kiev
4. Dunegan HL, Tetelman AS (1971) Nondestructive characterization of hydrogen embrittlement cracking by acoustic emission. Eng Fracture Mech 2(4):387–402
5. Lysak MV (1994) Acoustic emission during jumps in subcritical growth of cracks in three-dimensional bodies. Ibid 47:873–879
6. Lysak MV (1996) Development of the theory of acoustic emission by propagating cracks in terms of fracture mechanics. Ibid 55:443–452
7. Chen EP, Sih GC (1977) Transient response of cracks to impact loads. Elastodynamic Crack Prob Ser Mech Fract 4:1–58
8. Sih GC, Embley GT, Ravera RS (1972) Impact response of a finite crack in plane extension. Int J Solids Struct 8(7):977–993
9. Wadley HWG, Scruby CB (1983) Elastic wave radiation from cleavage extension. Int J Fract 23(2):111–128
10. Ahmed N, Rao KR (1975) Orthogonal transformations for digital signal processing. Springer, New-York, Heidelberg, Berlin
11. Prudnikov AP, Brychkov YA, Marichev OI (1983) Integraly i ryady. Spezial'nye funkzii (Integrals and Series. Special Functions). Nauka, Moskva
12. Ostash OP et al (2007) Miznist' i dovgovichnist' aviaziynych materialiv ta elementiv konstrukziy (Strength and Durability of Airplane Materials and Structural Elements). In: Panasyuk VV (ed) Mechanika ruynuvannya ta miznist' materialiv (Fracture Mechanics and Strength of Materials), vol 9. Lviv
13. Achenbach JD, Harris JG (1979) Acoustic emission from a brief crack propagation event. J Appl Mech 46(1):107–112
14. Scruby CB, Wadley HNG, Rusbridge KL (1983) Origin in acoustic emission in Al-Zn-Mg alloys. Mater Sci Eng 59(2):169–183
15. Nesmashnyi YV et al (1984) O svyazi amplitudy signala AE s prirascheniem ploschadi treschiny (On the relation of the amplitude of AE signals with increment of the crack area). In: Sbornik tezisov i dokladov I Vsesoyuznoy konferenzii "Akusticheskaya emissiya materialov i konstrukziy" (Proc. All-Union Conf. "Acoustic Emission Mater. and Struct.", 11 – 13 Sept. 1984, Rostov-upon the-Don), vol 1. Rostov-na-Donu, 1984
16. Gerberich WW, Alteridge DG, Lessar JF (1975) Acoustic emission investigation of microscopic ductile fracture. Met Trans A 6A(2):797–801
17. Mykhaskiv VV, Stankevych VZ, Khai MV (1993) Granichnye integral'nye uravneniya trechmernych zadach ob ustanovivshichsya kolebaniyach poluprostranstva s ploskimi treschinami (The boundary integral equations for three-dimensional problems on the steady-state vibration of a half-space with flat cracks). Izvestiya AN Rossii, Mechanika tverdogo tela 6:44–53
18. Khay MV (1993) Dvumernye integral'nye uravneniya tipa n'yutonovskogo potenziala i ich prilozheniya (Two-dimensional integral equations of a Newton type potential and their application). Naukova Dumka, Kiev
19. Harris JG, Pott J (1984) Surface motion excited by acoustic emission from a buried crack. Trans ASME J Appl Mech 51(1):77–83
20. Hrinchenko YV (1978) Ravnovesie i ustanovivshiesya kolebaniya uprugich tel konechnych razmerov (Balance and steady-state vibration of elastic bodies of the finite dimension). Naukova Dumka, Kiev
21. Imenitova YV, Chernyshov VV, Shegai VV (1976) O raschete svobodnych kolebaniy uprugich zilindrov konechnoy dliny (On calculation of free vibration for elastic cylinders of finite length). Izvestiya AN SSSR 226(2):315–317
22. Andreykiv AY (1982) Prostranstvennye zadachi teorii treschin (Spatial problems in the crack theory). Naukova Dumka, Kiev
23. Lysak MV et al (1994) Doslidzhennya vplivu chvilevodu na zminu parametriv signaliv akustichnoyi emisiyi (Investigation of waveguide influence on change of parameters of

acoustic emission signals). Fizyko-chimichna mechanika materialiv (Physicochemical Mech Mater) 3:64–71
24. Kikuchi Y (1969) Ultrasonic transducers. Corona Publishing Company Ltd, Tokio
25. Andreykiv AYe et al (1992) Metodika opredeleniya K_{Iscc} stali v srede vodoroda s pomosch'yu metoda akusticheskoy emissii (A method of determining KIscc values of steel in hydrogen using acoustic emission). Tekhnicheskaia diagnostika i derazrushyuschii kontrol 1:18–26
26. Serhiyenko OM, Lysak MV Akustychna emisiya pry dokrytychnomu rosti trischiny (Acoustic emission during a subcritical crack growth). In: Mechanika ruynuvannya ta miznist' materialiv (Fracture Mechanics and Strength of Materials), vol 5. Lviv, pp 695–714
27. Andreykiv OY et al (1993) Doslidzhennya vkladu chvil' Releya v akustichne pole, scho vinikaye pri rozvitku vnutrishn'ogo defektu (Investigation of Rayleigh waves contribution into acoustic field that arises during internal defect propagation). Fizyko-chimichna mechanika materialiv (Physicochemical mechanics of materials) 2:12–19
28. Shibata M (1984) A theoretical evaluation of acoustic emission signals—the rise-time effect of dynamic forces. Mater Eval 42(1):107–120
29. Sinclair JE (1979) Epicentre solution for point multipole sources in an elastic half-space. J Phys D Appl Phys 12(8):1309–1315
30. Abramowitz M, Stegun I (1979) Dovidnyk matematichnich funkziy (Handbook of mathematical functions). Nauka, Moskva
31. Cagniard I (1962) Reflection of progressive seismic waves. McGraw Hill, New York
32. Aki A, Richards PD (1983) Kolichestvennaya seysmologiya. Teoriya i metody (Qualitative seismology. Theory and methods), vol 1. Mir, Moskva
33. Seimov BM (1976) Dinamicheskie kontaktnye zadachi (Dynamic contact problems). Naukova Dumka, Kiev

Chapter 4
Some Methodological Foundations for Selecting and Processing AE Signals

Alongside the doubtless success of AE diagnostics of materials and structural elements, there is a number of methodical difficulties in applying this phenomenon. First of all, there is the problem of ensuring the reliability of AE results and comparing them with similar ones obtained in various laboratories and research centers. Such a situation can be explained by the lack of general methodical requirements concerning the AE testing of materials containing crack-like defects. To partly close this gap, in this chapter we generalize the approaches to AE estimation of crack growth resistance characteristics of materials, AE testing of products and structures, and others. Specific features of the AE investigations are related to the study of the AE and physico-mechanical processes. On the one hand, they should meet the requirements of fracture mechanics, and on the other [1], take into account the specificity of applying the AE method [2–4]. The greatest attention is given to the least investigated problems, such as the criterion for the selection of the AES generated by cracks among the signals emitted by other sources, the selection of the most informative AES parameters for crack examination, the selection of the AET operating frequency, and its location; some of these problems have been studied theoretically [5, 6]. Here we also focus on some other methodical aspects of the problem.

4.1 Some General Methodical Guidelines on the Use of the AE Method in the Mechanical Testing of Materials with Cracks

Note that there are no universal recommendations for AE investigations because they are performed in dissimilar conditions under the action of such factors as temperature, aggressive environment, loading, etc. However, it is possible to make

Z. Nazarchuk et al., *Acoustic Emission*, Foundations of Engineering Mechanics,
DOI 10.1007/978-3-319-49350-3_4

a certain systematization of the approaches by generalizing the known theoretical results and experiments' data.

In crack growth resistance testing of materials, the chart presented in Fig. 4.1 is most often used. The force *P*, created by a loading device 2, passes to a specimen 1 via a strain gauge. During propagation of macro-cracks in it, AET *4* and *5* receive the elastic waves, which are then passed on as an electric signal to a preamplifier *6* and afterwards to the block of filters *7*. The AES selected within a certain frequency band enter the coincidence block *8*, where they are subjected to spatial selection. The principle behind it consists of the following: the subsequent passage of the useful AE signal from AET *5* occurs only in the case when AES from AET *4* comes simultaneously to the coincidence unit *8*. Besides the AES filtration, this enables an additional elimination of the effect of mechanical noises on the results of AE research.

Next, the coincidence unit *8* opens the AE path (only in the case of the AES coincidence with AET *4*) and useful AES from AET *5* come to the block of data acquisition *9*, from which they can be selected for further processing and analysis.

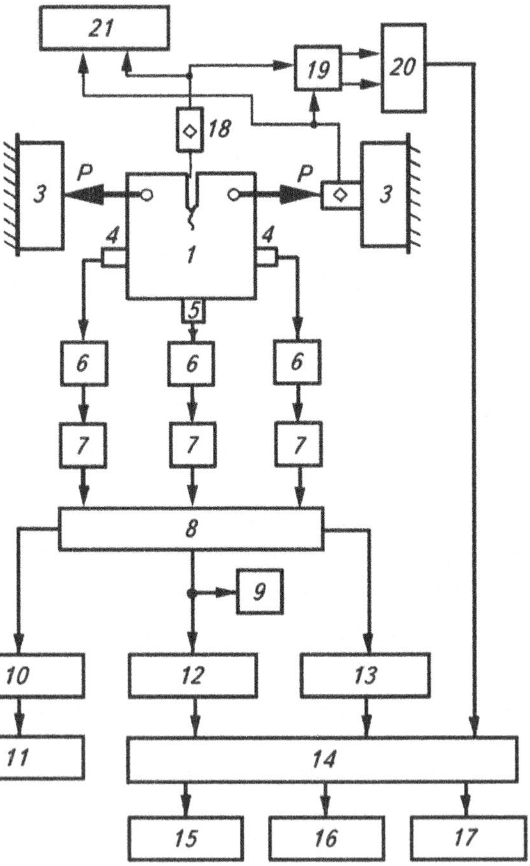

Fig. 4.1 Experimental testing setup

The AES can be simultaneously processed in an analog form by the simplest production run AE device *10* connected to the fast-acting recorder *11* of the required parameters. To obtain more complete information on a fracture, a fast-acting analog-to-digital converter (ADC) of AES *12* or spectrum analyzers *13* connected to personal computers *14* are used. Digital electric signals of loading forces can be supplied through the corresponding interfaces from a strain gauge *2* and crack opening displacement from the primary transducer of displacements *18*, which are preamplified by a two-channel measuring amplifier *19* and are converted into digits by a respective ADC *20*. A fracture diagram in the coordinates "*P* load—crack opening displacement *v*" can be recorded separately on a graph plotter *21*. Other necessary information is reproduced by peripheral devices *15* (monitor), *16* (digital graph plotter), and *17* (printer).

This setup may be either more complicated or simpler, depending on the experiments' tasks and on the available facilities and equipment. For example, channels for AE recording can be added in order to evaluate the emission source coordinates; a waveguide of the AE signals [7–9] may be applied for testing at temperatures different from the ambient or in an aggressive environment; and spectrum analyzers may be used to investigate the shape or the fine structure of the AE signals, etc.

Specimens and equipment are prepared for the experiments almost in the same way as for conventional crack growth resistance tests [1, 10–12]. Taking into account the above-mentioned, we present some basic ideas of this procedure. The primary AES transducer should have good acoustic contact with a specimen. For this purpose, the surface of a specimen is carefully treated in the location of mounting the AET. To improve the acoustic contact, appropriate auxiliary materials are used as a layer between AET and IO, and special techniques for fixing the AET upon the specimen are used [13, 14].

When preparing the equipment, at first the arrangement of a loading device is studied to predict the probable sources of external noises caused by the machine drives, friction in joints, and so on. Then the background hindrances are insulated from a specimen by anti-friction gaskets, appropriate joints, etc.

Preparation of specimens. The sizes of specimens and a method of their preparation are chosen, taking into consideration the specified tasks of an experiment: the type of a test, the loading mode, chamber size and geometry for making a required service environment and its parameters, type of the investigated material and its strength properties, structural features of loading devices, etc. Therefore, there are no general universally regulated methods of specimen manufacture and preparation for experimental researches. However, national standards determine some recommendations that consider the factors listed above. Thus, for instance, for a sheet metal that is 1–10 mm thick, it is recommended to use plane specimens, and for 20–100 mm thick, to use compact prismatic specimens subjected to eccentric tension [1]. In the 10–200 mm thickness range, prismatic specimens are used. For investigation purposes, cylindrical specimens are manufactured from round cross-section materials of 12–40 mm in diameter, and from bars of a square or rectangular cross-section up to 40 mm thick. Geometry and sizes of the specimens cut out of the structural shapes, for example a T-section, channel and corner bars, pipes, etc., are chosen by considering the highest concentration of stresses and strains in the structural element of a certain type of rolling. In most cases for low- and medium-strength metals with a relative elongation of $\delta_5 \geq 15\%$ in three-point bending tests, the specimens that are simple in production and do not require large loadings during testing are used. In view of the strength properties of materials, the most universal are eccentric tension specimens.

When preparing specimens, the problem of choosing a specimen thickness or diameter arises. It is usually chosen considering the elasticity modulus E and the yield strength of a material σ_y, except for magnesium alloy specimens. Values of specimen thickness or diameter recommended for production are proposed in codes [1, 12, 15].

After choosing the specimen type, the specimen is mechanically processed to the required sizes, geometry and cleanness of the heat-treated surface. If such processing is complicated, a specimen is preliminarily manufactured with a tolerance not less than 0.5 mm, which is afterwards removed by polishing. Depending on the specimen type (milling, polishing, sawing through, etc.), electrochemical or mechanical methods are used to induce notches-concentrators of stresses. A fatigue crack is induced by the load that does not exceed 0.5 σ_y for stress ratio $R = 0.1\ldots 0.2$. For a cylindrical specimen under bending, $R = -1$.

It should be stressed that in preparing the specimens for AE research, it is important to prepare the place where the AET will be mounted. Most often it is prepared by mechanical or electrical polishing after grinding the specimen surface.

During specimen testing using the AE method at temperatures that differ from ambient, it is necessary to keep the requirements [1, 12, 15] of the corresponding codes dealing with test peculiarities. In cases of the AET mounted directly on the specimen, the waveguides of the AE signals are used, and AETs are affixed to them [16, 17].

Testing equipment. Application of the AE method to the research of strength characteristics of materials is related to certain requirements ascribed to testing machines and loading equipment. The basic requirement is a noiseless operation of the loading devices, since dynamic processes in a deformed material generate AE signals of insignificant power. If the minimum level of noises of such devices is not ensured, the information contained in AE signals can be lost in the background. Apart from noiseless operation, the loading machine should provide necessary forces and a loading law as well.

In the production-type testing machines, the hydraulic effects in servo-valves as well as operation of drives and friction in moving joints are potential sources of noise. For these reasons, many machines are not suitable for AE investigations in certain frequency ranges without an appropriate modernization. To this end, proceeding from the structural features of machines, the sources of noise should be revealed and removed. Thus, in designing the testing machines, the usage of hydraulic pumps with alternate motion should be avoided. To decrease the impact force phenomena under a pulsating loading, hydraulic accumulators are used, with their recharging pumps located outside the machine frame. In mechanical loading devices, anti-friction gaskets and Teflon coatings are used to decrease the background noise caused by the friction of machine parts and units. Sometimes during loading, certain machine elements come into contact, thus generating low-frequency background noises in the range of 0.2–200 kHz. This greatly complicates the decoding of AE data, since these frequencies are often included in the operating range of an AE device. In order to provide isolation from such noises, special sound insulators are used that prevent the elastic pulses from reaching a specimen, or guide bearings are used to reduce the friction between the equipment units as much as possible [18]. An important factor causing a decrease of non-informative AE signals is the correct mounting of a specimen in the machine grips and the absence of plastic yielding of the material of a specimen in places where it gets into contact with the supports and grips. To improve the protection against the elastic background vibrations and electromagnetic waves, a machine or its operating part is placed in a special unit, and mineral fiber pads are placed between the connecting parts of a machine. These methods of protection against noises can be used for testing machines in AE investigations of structural materials.

However, even after modernization, many production-type machines are of little use for precise AE research. Therefore, machines with special loading devices have been designed for these investigations. They do not practically generate background noises [18]. There are also testing machines and equipment in which the noise reduction is accomplished by using a hand drive, vessels filled with water, thermal expansion and weight, etc., as a loading device. However, these devices have a few drawbacks; mechanical machines with a hand drive are low-powered and their application is limited. The devices in which the loading is created by the vessel weight gradually filled with water are used more frequently. The rest of the original devices used in the AE investigations do have a drawback: only simple loading laws are realized.

4.2 Technical Aspects of Preparation for AE Tests

Depending on the goal of the research, the available AE equipment [19–21] and the method of specimen testing, it is necessary to: (1) select the amplitude-frequency response of AET; (2) set the AE device operating frequency range that is outside the frequency band of the noise of loading devices; (3) set the threshold level and amplification factor of the AE path; and (4) calibrate the primary AE transducers by simulators. In Fig. 4.2, the simplest set-up for AE investigation of the crack growth resistance of structural steel specimens under three point bending is presented.

Recommendations for selecting the AET type, the method of its mounting on IO, and the provision of a reliable acoustic contact between the surfaces of a transducer and the investigated material were discussed earlier [22–24] and will be briefly discussed hereinafter. Therefore, here we discuss only the selection of the frequency band, the threshold level, and the amplification factor. The Kaiser effect is used for this in laboratory conditions under a repeated-static loading of the specimen without

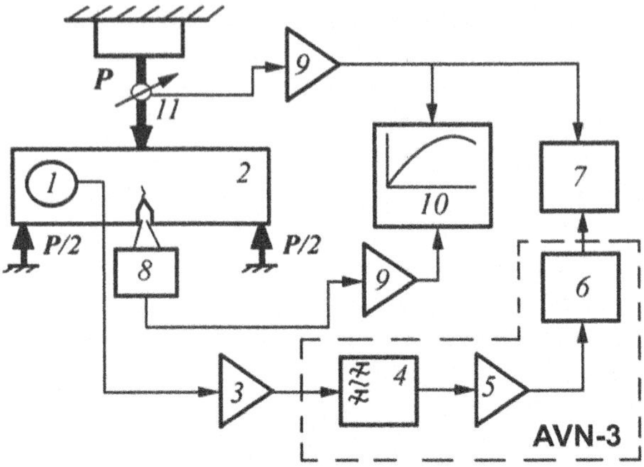

Fig. 4.2 Tensile testing of specimens with application of the AE phenomenon: *1* is the specimen; *2* is the primary transducer of displacement; *3* is the AET; *4* is the strain gauge; *5* is the loading device; *6* is the preamplifier; *7* is the block of filters of low and high frequencies; *8* is the power amplifier; *9* is blocks of the AES processing; *10* is the recording device; *11* is the amplifier of direct current; and *12* is the graph plotter

a crack by forces that exceed the critical values for a cracked specimen when the sub-critical crack growth begins. The specimen is loaded up to the values that exceed the critical load, and then it is unloaded and loaded again. Under a repeated loading, a lower frequency is selected, the threshold level is increased, and the amplification factor of a measuring path is optimized until the background noises vanish. In this case, the selection of the frequency band of noise filtration causes changes in the average amount of the AE pulses, which are acceptable [25].

During AE research, especially in the operating conditions, it is sometimes impossible to completely eliminate the background noises, whether mechanical, electric, or electromagnetic. Such noises can be part of the recorded AE data and can substantially complicate their decoding and interpretation, and in some cases can even lead to false conclusions. In this case, additional measures for the AE path protection are to be taken. For this purpose, in [26] a new effective method has been developed for the AE examination of cracks separating the AE signals from mechanical, electric, and electromagnetic noises. First, the remote zones of a specimen and the adjoining equipment are detected, where the highest values of the AE signal parameters from a crack and noises, respectively, are expected, and thus the difference between the arrival time of the signals from a noise zone is less than the AE event. The AETs with approximately identical amplitude-frequency response are mounted at these zones. The signals from these AETs arrive at two parallel similarly sensitive channels. The AES selection modes are set so that AE signals with almost similar amplitudes pass through the channels (or with a higher amplitude on the second channel than on the first because its transducer is closer to the noise area). During crack growth, when there are no noises, the AE signal passes through the first channel and is absent in the second (or much less than in the first one), which enables its selection on the AE diagram. Thus, the AES parameters caused by the crack growth are obtained by eliminating AES that are recorded simultaneously by all AE transducers with practically identical or somewhat higher amplitudes originated from the area of noise action (the second recording channel).

Before testing, it is helpful to preload a shunted specimen in a testing machine under a loading that exceeds the critical loading by 10...15% [27] in order to grind grips and reduce noise in them, i.e., application of the Kaiser effect. In testing, the AE data and load, displacements, etc., should be recorded synchronously and continuously. Depending on the purpose and experimental conditions, the recommendations and the type of loading can change.

To compare the results of the AE studies in various laboratories, it is necessary to know (1) the amplitude-frequency response of AET, its sensitivity and sensitivity of the whole AE system; (2) the operating frequency range of the system, the selected threshold level, the specimen material, and the geometry; (3) the AET location on the specimen and the method of its mounting; (4) the type of loading device, the method of holding a specimen in grips; and (5) the location of the loading device elements and the specimens, and so on.

4.3 Selection of Informative Parameters of AE Signals

As already mentioned, modern AE systems permit the recording and processing of many of the AES parameters: amplitude, pulse duration, frequency spectrum, number of pulses and events, change of these parameters in time, and others. However, the amplitude, count rate, and cumulative count of the AE signals are used most frequently. Using these parameters, the moment of the crack start and propagation were determined, enabling the evaluation of the corresponding stress intensity factors (SIF). Here, we will briefly discuss the key aspects of the problem.

Many researchers consider the possibility of establishing the moment of the crack start using the AE method. For this purpose, it is recommended to record the AES [28] appearance or a certain level attained by them. However, the first AE signals arise almost at the beginning of loading and they have small amplitudes [29], while the AES amplitudes and the cumulative count of AE depend on the device sensitivity, threshold level, and the selected frequency range [30]. All these prevent the researchers from obtaining reliable results.

During plastic deformation, the long-term and low amplitude AES are recorded, while during brittle fracture, short pulses with large amplitudes are recorded [31, 32]. The number of AE pulses is an integral characteristic depending on the signal amplitude and its duration. When the AES duration increases, the number of pulses recorded at a constant amplitude, in particular, using a resonance type of AET, also increases. Therefore, one can expect the appearance of AE signals with small amplitudes and a comparatively large cumulative count of AES. The initiation and propagation of micro- and macro-cracks will generate the AE signals with large amplitudes and a relatively small cumulative count.

These features are effectively used to detect the crack start and the AES identification from a propagating crack when there are only two mechanisms of the AE generation, i.e., plastic deformation and a macro-crack growth. However, in some materials, additional AE sources appear, e.g., fracture of brittle inclusions or micro-crack formation. These processes can generate the AE signals that are often comparable by an amplitude with the AE signals caused by a macro-crack propagation that complicates the interpretation of AE diagrams and the evaluation of the crack length increment. In this case, the quantitative analysis of some amplitudes of AE signals is low-effective for evaluation of the crack start in materials that generate AES of considerable amplitudes under loading before the sub-critical crack propagation stage. Then it is essential to work out a criterion based on other AES parameters, such as the signal wave form, spectral features, etc., which would be effective for this situation.

4.4 Simulation of AE Sources

The methods of AE diagnostics provide a potential possibility of access and examination of the entire structures and buildings, for which it is very important to ensure a high level of integrity of elements, such as aircraft, bridges, pressure vessels, pipelines, etc. The sequence of events causing the AES that could be detected—namely, the processes of AE source formation, elastic wave propagation, transformation and processing of AE signals—is presented schematically in Fig. 4.3 [33]. It follows that in order to clearly determine the AE characteristics and thus identify the results obtained, it is very important to have the model AE sources for the AET verification and calibration of sensitivity of the whole AE path. In particular, the absence of such sources has caused an imperfect testing of materials and poor AE examination and diagnostics of structures for a long time.

A model source should possess the following characteristics [33]: repeatability, high reproducibility, and a well-established mechanism of a source action. These characteristics should be close to the real AE sources and relatively simple to be applied.

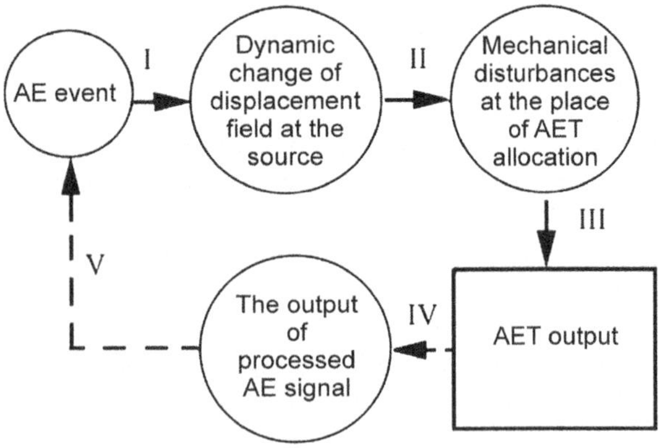

Fig. 4.3 Logical sequence of the AE event analysis: *I* is characteristic of the AE source; *II* is wave propagation in a solid; *III* is AET amplitude frequency response; *IV* is the AES processing; *V* is interpretation of AE data

Papers [34, 35] describe one of the first models of the AE sources, i.e., a source that is simulated by the failure of a glass capillary and satisfies, to a certain extent, the formulated requirements; this method of simulation was improved later [36]. A glass capillary was replaced by high-quality graphite lead, one end of which is pressed to the tested object with a constantly increasing pressurizing force until it fails. Thus, during fracture, the force of a stepwise function type is generated, and its value is chosen by adjusting the diameter and hardness of the lead. This simple method is accepted by the standardization groups of AE research in the U.S., Europe, and Japan as a standard method of AE verification and calibration of sensitivity of the channels of the AES selection and processing. The device with the lead provides a smooth loading and high reproduction of the shape of AE signals ensuring the process of absolute calibration.

Since the role of the AE source simulators in solving the problems of metrology prescriptions of the AE research facilities used for non-destructive testing of materials, products, and structures is very important, their development becomes very topical. Practical use of the AE source simulators shows that they also enable solving a number of additional problems related to optimizing the calibration of the AE facilities. Thus, simulators can be used for (1) checking the serviceability and adjustment of the channel sensitivity of AE devices; (2) recording the amplitude-frequency response of primary transducers and their directional diagrams; (3) evaluating the decay of AE signals and their propagation velocities in real constructions; (4) minimizing the measurement errors by the AET rational location and selection of its orientation; and (5) selecting the optimum frequency range and the threshold level of the AE signals, etc. [29].

The analysis of the published data shows that insufficient attention has been paid to the creation of the AE source simulators. Most of the researchers use different simulators with unknown technical characteristics; moreover, the principles of simulation have a number of drawbacks. The varying conditions of an acoustic contact between the AET and specimens result in low reproducibility of the AE signals, which are simulated and transmitted into the object tested. The mechanical resonance elements in the simulators lead to a misfit between the shape and the parameters of the simulated and real AE signals.

Some requirements for the AE source simulator result from the previously obtained characteristics of real signals: short duration, insignificant energy of a single pulse, high intensity, etc. (see Chap. 2). The amplitude of the mechanical displacement in the AE pulse can attain $10^{-7}\ldots10^{-14}$ m; a step-wise change in stresses in a source can take place within the small area of a square millimeter. It is also necessary to take into account the operating requirements of the AE application for the non-destructive testing of objects in special conditions. For instance, the quality control of welding during the formation of a weld requires the AE signals to be measured and simulated at high temperatures, and the presence of a corrosive environment or irradiation imposes special requirements for simulator stability or non-contact simulation. To decrease the AE research errors, it is necessary to provide a high reproducibility (stability) of signals induced by a simulator.

In practice, the simulators of the AE sources are based on the techniques of ultrasonic vibration excitation of a solid by a mechanical impact, reverse piezoelectric effect, magnetostriction transducer, spark discharge, or a light pulse. The analysis of these methods shows that none of them fully satisfies all the requirements [29].

When a solid—for instance, a ball—strikes the object under investigation, the mechanical vibrations that have ultrasonic components are excited in the object. However, to decrease the duration of a contact, it is necessary to increase the collision velocity that causes the growth of the energy of the excited vibrations. Calculations show that for the case of a steel ball with a diameter of 5.0 mm, the duration of a contact less than 100 μs occurs at a very-high collision velocity, yielding the energy of vibrations introduced in IO, which is not comparable with the AE.

The methods of vibration excitation by emitting AET based on a reverse piezoelectric effect consist of the following: AET eigenfrequency is chosen to lie within the frequency range of the AE excitation electric pulse. Then, AET emits a sequence of pulses, each of them being a decaying oscillation with a frequency equal to its resonance frequency. To avoid this phenomenon, it is necessary to increase the AET resonance eigenfrequency by applying very thin piezoelectric plates (for high frequencies). This complicates the sonic transmitter construction and deteriorates its reliability. One of the drawbacks of this method is a direct contact with the investigated object, a strong dependence of excitation efficiency on the quality, and the stability of a contact. The application of piezoelectric ceramics is limited by its relatively low values of the Curie temperature.

A magnetostriction transducer of the AE signals is widely used in experiments as well [37]. It is excited by the current pulses of 0.5 A, and duration of 3 μs with a frequency within 10 and 60 Hz. A waveguide in the form of a thin wire with basic resonance frequency of about 100 kHz is attached to the transducer that ends with a thin quenched point. Pulses excited in a waveguide in the form of longitudinal waves pass into the IO through the point pressed to it. The advantage of a simulator is the point area of transmitting the AE signals into IO and the possibility to simulate the AES at high temperatures. The disadvantage of the method is the necessity for an acoustic contact that reduces the reproducibility of the AE signals that are simulated and transmitted into the IO, and the presence of resonance elements. Therefore, a pulse duration increases, and it is impossible to adjust its shape. Besides, magnetostriction transducers effectively operate only at a frequency of up to 1 MHz, because the coefficient of transformation of the electric energy into the acoustic energy decreases proportionally to the frequency squared.

The electric discharge between the electrode and the surface of the IO is used as the AES simulator, which permits the formation of acoustic pulses of high power and short duration. According to data from the Dunegan/Endevco Company [38], a spark method is used for piezoelectric AET calibration. No information is given on the character and parameters of the AE pulses. Although the method actually provides the point introduction of vibrations, in the authors' opinion [29] it is still characterized by a low reproducibility of the elastic AE waves.

A method of acoustic wave excitation by a laser light pulse has its own advantages. Some modern solid lasers make it possible to get light bunches of power up to $10^{12}\ldots10^{16}$ W/cm^2 with the simultaneous focusing of radiation in the spot of a diameter of 10…100 μm, creating rather powerful acoustic pulses. This method does not need an acoustic contact; therefore, it can be used for high-temperature, hard-to-reach objects. A light spot can be of virtually any configuration, such as point, narrow band, ring, etc. It is easy to inspect the amplitude of simulation pulses and their high stability. A drawback of this method is the short duration of a pulse in a single-pulse mode of irradiation and insufficient power of widely used lasers [29].

The fitting criterion of the simulation action of the AE source and a propagating defect is the identity of the elastic fields in close vicinity, where elastic waves are transmitted by a simulator. The fitting is established only by the parameters, which carry the information on the defect parameters. Instead of the above-mentioned criterion, it is possible to use the correspondence of the force or the kinematics effect of the simulator in the area of the vibrations input with the force or kinematical conditions on the simulated defect surface, assuming that the latter is on the input surface.

Some other methods of AES excitation in solids and liquids are also known [25, 28, 39, 40]. They use various physical principles of elastic wave generation and provide both discrete and continuous AES. These methods have both advantages and disadvantages and can be used in a laboratory and in production and field conditions of AE testing of materials and structures. However, progress in the research of crack growth resistance in structural materials, testing and diagnostics of products and constructions by the AE signals, brings up the problem of creating new, more perfect approaches to the simulation of AE signals in order to increase the reliability of the results, and consequently to provide the reliable operation of products and constructions.

Simulation of the AE source during evaluation of the static crack growth resistance of materials. Calibration of the instrumental AE path was done to compare the experimental results obtained by various authors in various research conditions. The methods of calibration of the AE channels turn out to be the most correct if the source of the AE simulation and AET are located directly on the IO. This is explained by the fact that the change of the place or the method of the AET mounting can substantially change the character of the response of the system "AET-IO" and thus affect the results of measurements.

In crack growth resistance testing of structural materials, it is important to provide a maximal sensitivity of the AE channel by calibration in order to reveal the early stages of sub-critical crack growth, and especially the moment of its start. It is known [5] that on the Mode I crack at the moment of its jump on the juvenile surface, an instantaneous decrease of stresses from the initial level to zero occurs; this is accompanied by the emission of elastic vibrations. Such single jumps of a crack very often provide a very small increment in the new surface area (for instance, for U8 steel it can be $(1.3\ldots1.5) \times 10^{-9}$ m^2). As a result, the AE have low amplitudes of vibrations that propagate in a three-dimensional solid. Thus it follows that we should not neglect the AE wave decay during precise AE measurements. It means that to simulate this decay in a solid, it is necessary to approach, as closely as possible, the real conditions of the elastic AE wave initiation and propagation in the material investigated.

It is shown [41] that the AE sources during crack propagation in a brittle body are mainly concentrated (about 70%) in the internal part of the specimen at a distance of 2 mm at both sides from the crack propagation plane. In this study, it is also shown that during a single crack jump in the body, the sequence of the arrival of AE elastic waves to AET is as follows: First, a longitudinal, then a transversal wave, then spatial waves reflected from the specimen surface, and, finally, surface waves arrive at the transducer. The AE vibrations that arose at the specimen surface continue with a greater amplitude than the vibrations excited inside the specimen (see also paper [42]). Therefore, the results of AE research will be more reliable if we model the AE signals from the internal part of a specimen near the probable plane of the crack growth with the fewest amplitudes and then perform calibration of the AE tools and AET by these signals. This is the purpose of the proposed model for the AE source simulation.

Thus, a glass capillary was used for the AE source simulation that maximally corresponds to a single crack jump. Glass, as the material for a brittle simulator, was chosen due to its homogeneity and isotropy. This permits, due to breaking the capillary, generating the same types of elastic waves with approximately identical characteristics of the AE signals. Figure 4.4 shows a capillary location in a compact [1] specimen *1* of the investigated material with an induced fatigue crack of a required length. A cylindrical hole, which nearly reaches the fatigue crack front, is made in the central part of the concentrator. A glass capillary *2* is inserted into the hole and clamped by a corresponding filler *3*. After preparing the specimen for testing, the AET is mounted on the specimen, and the AE signals from AET are transmitted electronically to the corresponding units of AE equipment. The AE path parameters, such as the amplification factor, threshold level of the AE signals, operating frequency band, etc., are selected by breaking the capillary with the AES recording. In this case, the optimum place for the AET location on the specimen surface is chosen [43].

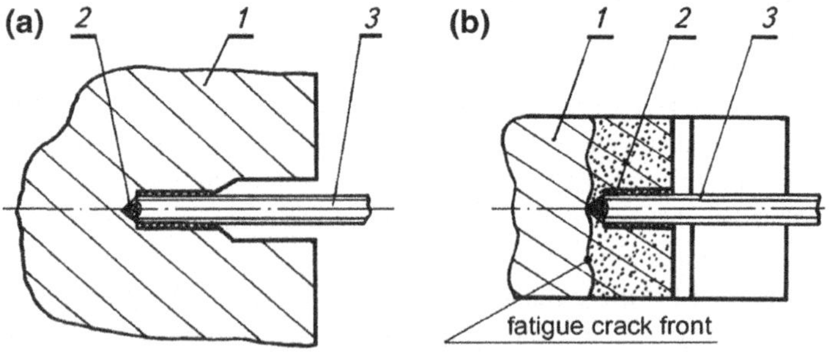

Fig. 4.4 Chart of a glass capillary clamping in the specimen body, vertical (**a**) and horizontal (**b**) axial cross-sections: *1* is a specimen, *2* is a capillary, *3* is a filler

Thus, when determining the static crack growth resistance of structural materials by the AE signals, the proposed method of the AE simulation eliminates the disadvantages of the AE source model proposed in [4]. The proposed method, in contrast to the one shown above, provides (1) a rigid clamping of the capillary end inside the specimen body that excludes its friction under loading; and (2) excitation of spatial waves that maximally represent the AE nature during a single crack jump with the minimum area of a fresh surface in the region of its most probable formation inside the material. The formation of the surface elastic Rayleigh and Lamb waves [44] is possible only if spatial waves are reflected from the specimen surfaces, thus permitting the reduction of their effect on the AE research results to the minimum by special measures [32].

Simulation of the AE source on the large-scale objects examined under field conditions. The application of the AE methods for non-destructive testing of industrial objects requires a very careful calibration of sensitivity of the measuring channels. This is an especially urgent problem, when the extent of the effect of the object geometry on the received AE signals, environmental effect, resonance of the object or the system "object-AET" are not well known. Quite often, when determining the AE source location on the object under investigation, some AETs have different sensitivities and respond ambiguously to some types of elastic waves, which cause an incorrect determination of the AE source and its orientation.

To interpret the results of measurements correctly and to compare them with the known verified or rated data and determine the degree of safe operation, it is necessary to simulate the AE sources of a high degree of repeatability with the approximation of crack-like defects propagation to the real conditions. Thus, the model of the AE source will be the more effective, the easier and simpler the simulation of the reproduction of AE signals in the conditions of detrimental effect of background noises that are plentiful in the industrial and field conditions of AE diagnostics.

It is known from the laboratory investigations [45] that pressure vessels and other constructions made of structural steel during sub-critical crack growth can generate the AE signals with the following gradation of their amplitude values: (1) weak AE signals caused by plastic deformation; (2) discrete AE signals with middle amplitudes caused by the fracture and/or separation of sulphide and silicide inclusions; and (3) the AE signals caused by the growth of a crack with wide-ranging amplitudes, depending on the local microstructure. Plastic deformation and fracture or separation of inclusions can arise during the formation of a plastic zone at the defect tip. Taking this into account, the simulation of AE signals consists of using the elastic waves emitted during crack propagation in the specimens made of the tested object material. By choosing the proper loading, and by measuring the crack length increment, it is possible to simulate various areas of fracture surfaces during an actual crack increment. This permits maximally approaching the real mechanisms of plastic zone formation and the stages of sub-critical growth of crack-like defects in the investigated object. Accordingly, having adjusted the AE equipment using the simulation of the AE sources, effective diagnostics and testing of objects by the AE signals can be performed [46].

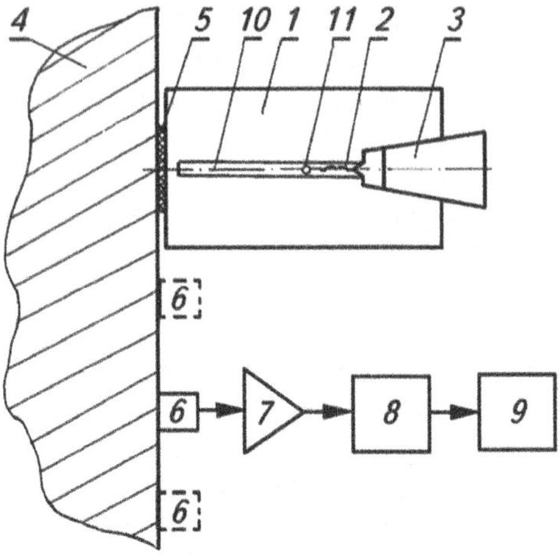

Fig. 4.5 Simulation of the AE source on IO permitting optimization of the AET location (schematically)

Figure 4.5 illustrates a scheme of the simulation process. Specimen *1* of the IO material in the form of a T-cantilever beam [47] with the preliminarily induced fatigue crack *2* is loaded to some calculated value *P*. Then, the loading is fixed on the specimen by a hard insertion *3* [48] and the specimen is mounted on the object *4* through a contact layer *5*. Specimen *1* is rigidly attached to the inspected object *4* by a threaded connection or gluing or soldering, etc., depending on the test conditions. At some distance from the specimen *1*, AET *6* is mounted, which is connected with a preamplifier *7* by a block of filtration and processing of the AE signals *8*, and with the recording device *9*.

The value of the loading force *P* should be calculated to prevent the propagation of the existing fatigue crack under a static loading by a rigid wedge. Catalysts are used to provoke its growth, i.e., the Rebinder effect [49]. When this happens, a crack begins to grow and move in the direction of the groove *10* to the hole *11*, which stops its further propagation, and plastic deformations appear at the hole due to the redistribution of mechanical stresses.

A repeated simulation on the same specimen permits selecting the service conditions of the AE facility, calibrating the AE sensitivity, and choosing the places of their optimum mounting with corrections for finding the coordinates and orientation of the AE sources. Thus, such an AE source simulation simplifies its excitation and maximally approximates the levels of signals to the real ones for different mechanisms of AE generation during propagation of the crack-like defects in the investigated object.

4.5 Simulation of AE Events at the AET Output

The methods for evaluating the AE path sensitivity of the facilities by excitation of the elastic AE waves, which arise and propagate in a solid and are received by AET, are described above. In some cases, in order to adjust the AE equipment units, it seems promising to use the simulation of stochastic AES, which reproduce the sequence of the AE events that are expected at the AET output. The principle of the design and operation of such a device is described in [50].

The method of stochastic decimation of the Bernoulli sequence is realized in the simulator, which reproduces a stationary quasi-Poisson sequence of pulses with a given value of intensity of events or an exponentially decaying intensity of events. The initial intensity value m_0 and the parameter, time constant, T_0 of its decaying in time, are preset. The relative difference of dispersions of intervals between pulses at similar intensity of the flow of the AE events is a criterion of dissimilarity of a generating quasi-Poisson flow from the actual Poisson flow. The amplitude A_0 of the pulses can be regulated in a dynamic range of 10 dB, and their form at this stage is described by a cosine function with frequency ω exponentially decaying in time with the frequency ω. The parameter γ determines the exponential decay of this function, i.e., an informative component of the electric signal is generated in the form

$$A(t) = \sum_{i=0}^{\infty} A_0 \exp[-\gamma(t - t_i)] \cos[\omega(t - t_i)]\chi(t - t_i), \tag{4.1}$$

with distribution of the number of pulses in time

$$P(N) = \frac{\lambda^N}{N!} e^{-\lambda}, \tag{4.2}$$

where

$$\lambda = \int_{t_0}^{t_0+T} m(t)\mathrm{d}t, \quad m(t) = m_0$$

during generation of a stationary pulse flow of the AE events and

$$m(t) = m_0 \exp\left(-\frac{t - t_0}{T_0}\right)$$

during generation of a pulse flow with exponentially decaying intensity. Indicator function

$$\chi(t - t_i) = 1 \quad \text{if} \quad t \geq t_i, \quad \text{and} \quad \chi(t - t_i) = 0 \quad \text{if} \quad t < t_i.$$

The change in the value of the probability $p(t_j - t_0)$ of a pulse appearance for the simulator timing period according to the exponential law

$$p\left(t_j - t_0\right) = p_{\max} \exp\left(-\frac{t_j - t_0}{T_0}\right), \tag{4.3}$$

where $p_{\max}$ is the maximum value of probability of pulse appearance in a timing period, permits managing the value of the parameter T_0. The change of the clock period or clock frequency permits controlling the intensity m_0 of a stationary flow of the AE events.

A simulator (Fig. 4.6) consists of clock-pulse generator *1*, phase-pulse frequency dividers *2* and *3*, *S*-input 4, and three input elements of logical "and" 5, generator of the pseudo-random uniformly distributed binary numbers (GPUDBN) *6*, generator of digital exponent *7*, shaper of output pulses *8* and comparator of binary numbers *9*.

During each operation of a clock-pulse generator *1*, the next change of the pseudo-random uniformly distributed number at the outputs of the GPUDBN *6* occurs and l bits of this number are transmitted to the first inputs of a comparator circuit *9*, and other s positions are transmitted to the inputs of the logical element *4*. Pulse flow from the output of clock-pulse generator *1* through the frequency divider *2* with the division factor k is connected to the first input of logical element *5*, and through divider *3* with the division factor k_2 to the control input of the generator of digital exponent *7*. Under the control of the output pulses from divider *3*, the sequential output of values of l bit digital exponential function is transmitted from generator *7* to the second group of inputs of l bit comparator circuit *9*. Binary numbers are compared so that the potential of logical unit appears at the comparator circuit output only if the pseudo-random binary number is lower than the value of a digital function. The former of output pulses *8* sets the pulse amplitude, configuration, and polarity. In this case, configuration of the pulses is set by a pulse current impact excitation of an oscillating *RLC*-contour. To generate a stationary quasi-Poisson flow of pulses, the third input of element *5,* instead of the comparator circuit *9* output, is connected with the potential of a logical unit. Then, the value of intensity

$$m_0 = \frac{f_0}{2^s k}, \tag{4.4}$$

can be set by a corresponding switching of division factor k of frequency f_0.

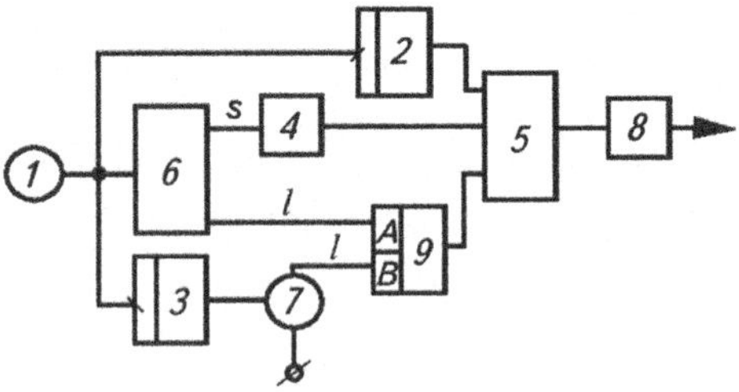

Fig. 4.6 Functional schemes of the simulator of AE events

The assigned exponential law of non-stationarity of the pulse flow intensity is introduced by setting a non-stationary digital function *Y*(*t*). Since the sequential output of the function *Y*(*t*) values is realized by the pulses of divider *3*, then by switching its division coefficient k_2, it is possible to adjust the rate of the change of the digital function, i.e., to compress or extend the specified functional dependence of the pulse flow intensity change in time.

The functional circuit of the digital exponent generator is shown in Fig. 4.7. It consists of $l + \alpha$-bit registers *1* and a subtraction circuit *2*. Its operation is organized so that from $(l + \alpha)$-bit number $Y(i - 1)$, the number $Y(i - 1)2^{-\alpha}$ is subtracted, i.e., the same number shifted by α digits to the right (in the direction of the least significant bits). Then, during the next recording of the result of subtraction in register *1* under control of the output pulse from divider 2 (Fig. 4.7), the number $Y_j = Y_{j-1} - Y_{j-1}\,2^{-\alpha}$ will be recorded into register *1*, which is connected to the corresponding inputs of the comparator circuit. Thus, generator *7* produces an *l*-bit digital exponent.

$$Y(t) = Y_0 e^{-\frac{t}{T_0}} \tag{4.5}$$

The simulator of AE events is characterized by the following technical data:

- Intensity of the AE events flow $m_0 = (1, 2, 5)\cdot 10^{a}\ \mathrm{s}^{-1}$ ($a = -1; 0; 1$);
- Basic relative error of setting the intensity $\delta_{n_0} < 1\%$;
- Time constant $T_0 = 0.08 \cdot 2^{b}$ s ($q = 0, 1, 2, 3, 4, 5, 6$);
- Relative error of setting the time constant T_0 of intensity decay $\delta_{T_0} < 1\%$;

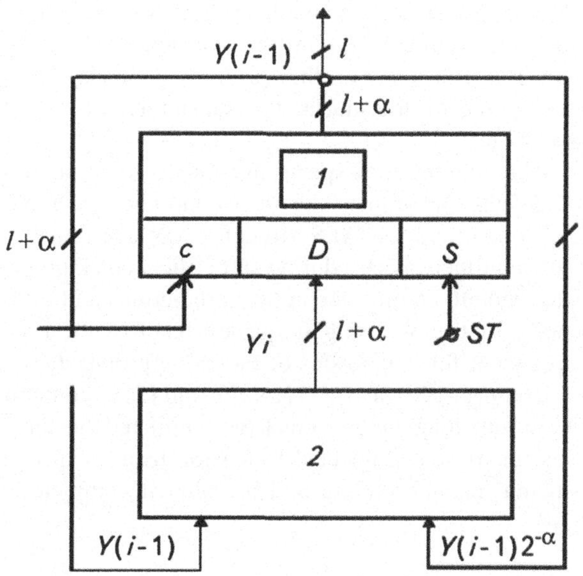

Fig. 4.7 Functional circuit of the digital exponent generator

- Difference of a generated flow from actual Poisson pulse sequence in terms of the criterion of relative difference of interval dispersions is less than or equal to 6%; and
- Initial pulses configuration of the simulator $F(t - t_i) = A_0\cos[\omega(t - t_1)]\exp[-\gamma(t - t_i)]$, $t \geq t_i$, where A_0 is regulated within the range of (0.2…2) V, ω = (0.3; 0.5; 1) MHz, γ = (10^3; 10^4; 10^5) s^{-1}.

4.6 Spectrum of the AE Signals During Macro-crack Growth

The analysis of the AES spectra can provide important information on the stages of sub-critical crack propagation. On this basis, it is possible to build the criteria of assessing the state of structural elements or other products that contain crack-like defects [32, 51].

The theoretical investigations of the AES structure (including spectrum), have shown that the narrowing of the AES frequency band with the crack growth depends on SIF [5, 6]. According to calculations [52, 53], a frequency band can be of several megahertz, depending on the size of the defects. However, the operation of AE devices in the megahertz band is complicated for several technical reasons that can be often avoided: Experimental studies indicate that a frequency range lower than 4 MHz [54, 55] is sufficient for a crack inspection. Thus, the AES spectrum is in the range of up to 1 MHz, and an amplitude maximum of the frequency spectrum is at the frequency of less than 400 kHz [56]. Then, the results of the tests will only slightly depend on the type of material, AE transducer, specimen sizes, and the method of AES processing. However, it is known that the AES spectrum contains information not only on physical processes that generate AE; it also depends on the specimen geometry and size, as well as the characteristics of the AE transducers that substantially change the AES generated by a defect [40, 57–62].

Although the AES spectrum contains important information on the development of defects, sometimes the data contradict each other. Therefore a method was developed, and the AES structure was additionally investigated based on it—in particular their spectra during sub-critical crack propagation. In this way, an attempt was made to confirm theoretical predictions and technical recommendations on the choice of the operating frequency band of the primary AES transducer and AE equipment for diagnostics of the crack propagation stages and mechanisms.

Compact 1201-T 6 × 75 × 79 mm [51] aluminum alloy specimens were tested. A 35 mm long fatigue crack was initiated on the lateral edge of the specimen. Specimens were subjected to tension using a special SVR-5 device that provided smooth manual loading and the almost complete absence of background noises [18].

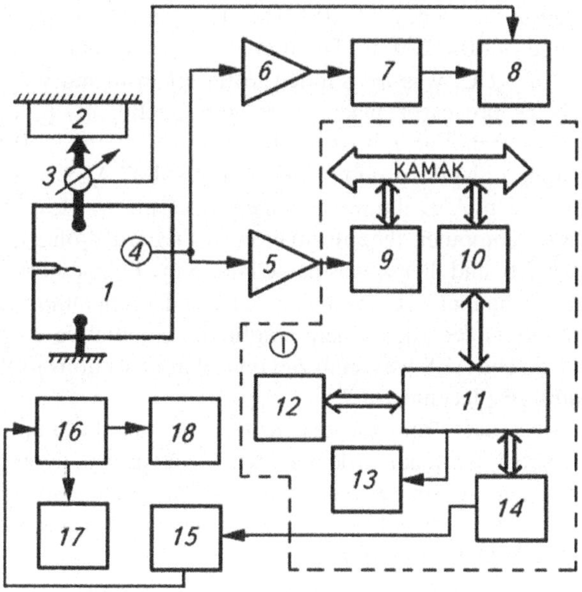

Fig. 4.8 A setup of the experimental research: *1* is a specimen; *2* is a loading device; *3* is a dynamometer; *4* is an AET; *5* is a preliminary amplifier; *6* is a measuring amplifier; *7* is an AE equipment AVN-3; *8* is a fast-acting recorder; *9* is an ADC; *10* is a dynamic set of the modules; *11* is a "MERA" computer; *12* is a processor of "Elecronika MT-70" equipment; *13* is a display; *14* is an E-256K quasi-disk; *15* is a translator-display; *16* is a computer; *17* is a monitor; and *18* is a printer

Available AE research equipment of the AF-15, AVN-3, and other types made it possible to get only the traditional parameters of AES (amplitude, power, duration, cumulative count, and count rate of the AES), but did not permit analyzing the fine AES structure, in particular their spectral distribution. Therefore, a specially designed informative measuring complex was used in the experiments [51] (Fig. 4.8).

The AET was used as an AE transducer. It is supplied with a set of AF–15 equipment with a pass band of 0.2…2.0 MHz and equal amplitude frequency response in the frequency range of 175…500 kHz. The AET was fixed by the clamp to the specimen through a contact layer of a lubricant. A block of two concatenated preamplifiers from the set of the AF-15 device served as a preliminary amplifier. The total amplification factor of the block was 70 dB, and a pass band was from 0.1 to 1.5 MHz.

The testing, selection, and processing of the AES were performed as follows: Under specimen loading, the crack grew, radiating elastic waves. These waves reached the lateral surface of the specimen and were transformed by AET into electrical signals, and then passed on for pre-amplification. From the amplification unit, the signals passed to the fast-acting F-4226 ADC with the main memory of 1 Kb permitting the storage of information that was transformed into a digital code with the digitization frequency of 20 MHz. The period of internal timing was chosen in the range of 50–3200 ns. The digital sound, by sampling 1 Kb from the internal memory of ADC, was transmitted to the main memory of the fast-acting "Electronika MT-70" processor, where AES were functionally processed. The data array obtained was then transmitted to the memory of the "MERA" computer whereupon, through a display group of the "DYNAMO" AES modules, they were visualized on the screen. A total time of signal processing was less or equal to 1.5 s. Using the "MERA" computer, the information was recorded on a quasi-disk with a capacity of of 256 Kb, and then it was passed on to the PC memory via a translator. As an indicator, a channel was used that consisted of preamplifiers, an AVN-3 device, and an N-338/4 recorder, where amplitude, cumulative count of AES, and load were recorded. The AES was analyzed on different sections of the curve "Load P − crack opening displacement δ."

To obtain correct data from the experiment, at first the amplitude-frequency response of the "specimen-loading device-AET" system was investigated. For this purpose, a short-range AE pulse was used that was previously obtained for the linear section of the P-δ curve [51]. The spectrum of such a system had a wide frequency range of 2 MHz (Fig. 4.9).

For AES, as shown in Fig. 4.10a, the maximum irradiation is at the 630 kHz frequency. Taking into account the loading diagram, it is possible to assume that the signals recorded at the initial stages of a fracture diagram were caused by plastic deformations and micro-cracking at the crack tip.

Under further loading, a sub-critical crack growth affects the spectral characteristics of AES (Fig. 4.10b). Immediately after the crack start detected by an indicator channel, the frequency band of AE signals narrows jump-like to a frequency of 1 MHz and the spectrum maxima shifts towards the lower frequencies. For instance, the width of the frequency range (Fig. 4.10c) at the level of 0.7 of their maximal value, recorded at the middle stage of sub-critical crack growth, is 140…960 kHz. The narrowest frequency band at the point of deviation from linearity on the P-δ diagram (Fig. 4.10d) is only 90…530 kHz, and its maximum is at a frequency of 370 kHz. These AES correspond to the final stage of sub-critical

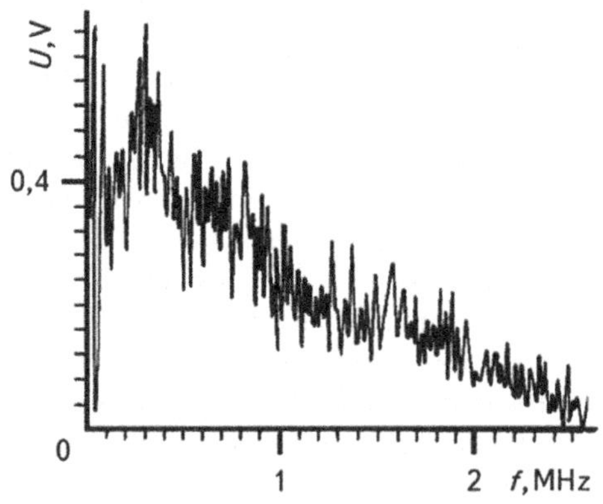

Fig. 4.9 AFC of the system "loading device-specimen—AET"

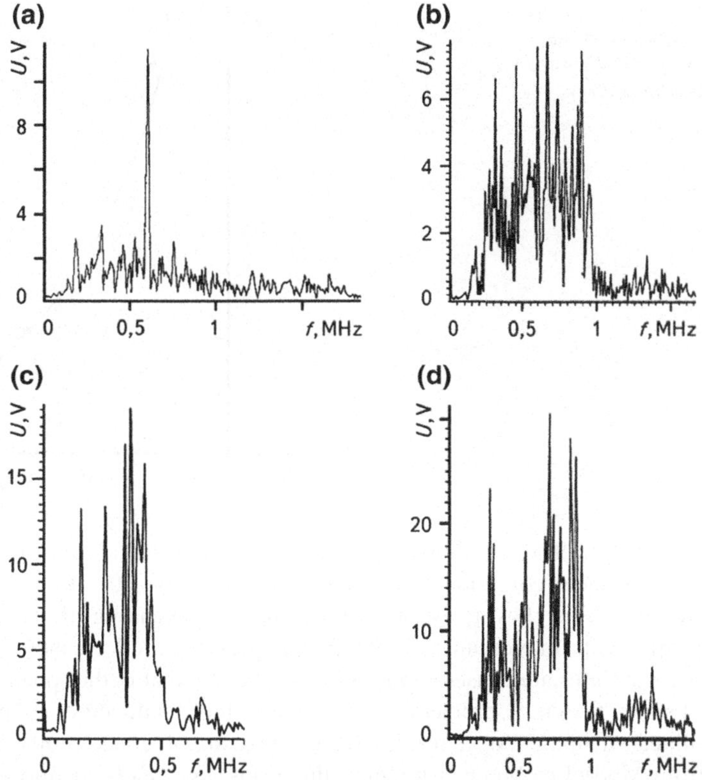

Fig. 4.10 The spectrum of AE signals at various loading stages: **a** on the linear section of the *P*-*δ* diagram; **b** at the initial stage of sub-critical crack growth; **c** at the middle stage; **d** at the final stage

crack growth. Thus, there are two important conclusions from the above-mentioned: Firstly, a jump-like narrowing of the frequency band can be used as a criterion for identifying the AES caused by the crack growth. Secondly, the narrow-band AE transducers having the resonance frequency band within 100… 600 kHz should be chosen. The low frequency limiting of the pass band is caused by significant noises of the loading devices at low frequencies.

As was theoretically predicted [5, 32], the width of the frequency band of AES at the stage of subcritical Mode I crack growth is inversely proportional to the SIF square

$$\Delta f = \beta K_I^2, \tag{4.6}$$

where $\beta = 0.25\ \alpha c_2/a\pi$; α is the proportionality factor; a is the material constant.

Theoretical predictions [63] (Fig. 4.11, solid line) concur with experimental data (dots) for the value of β = 878.6 (MPa)^{2}m/s. Thus, the width of the frequency band

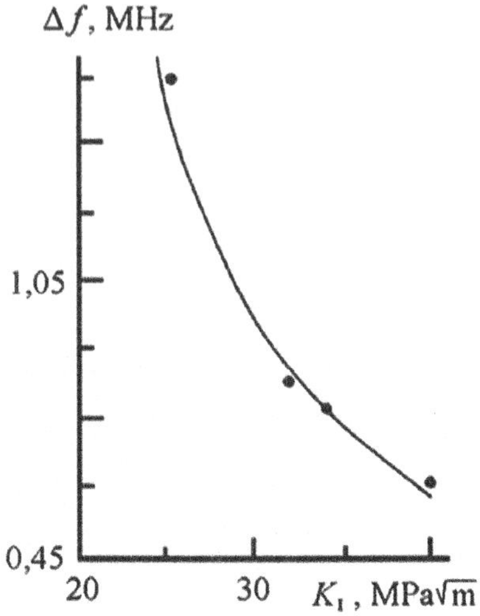

Fig. 4.11 Dependence of the frequency bandwidth of the AE signals Δf on SIF K_I at the stage of sub-critical crack growth

can be used for the SIF determination as well as for the calculation of a structural element strength under condition of its safe exploitation $K_I < 0.7\ K_{IC}$.

After crossing the highest point of the P-δ diagram, spreading of the frequency band, the shift of its maximum toward high frequencies, and an increase of the maxima by 2…3 times are caused—obviously by the fact that in the specimen at the overcritical stage, the crack propagates not in the initially induced crack plane, but in a curvilinear manner. As a result, the mechanism of crack propagation has changed from Mode I to Mode II fracture; this leads, as already mentioned, to the frequency band extension.

During the Mode II crack growth, it should be taken into consideration that the AE signals can be generated not only by overcritical crack propagation, but also by the friction of its faces. The AES with a wide frequency band and with maxima both at high and low frequencies, were detected up to the complete failure of the specimen. The width of the frequency band becomes the same as at the initial stages of sub-critical Mode I crack growth, but with the maxima larger by 2…3 times. The signals with a narrow frequency band were recorded among the AES with a wide frequency band, which means that at the overcritical stage, the crack develops due to various mechanisms. The indicator channel (with the use of the AVN-3 device) recorded the AES with large amplitudes.

The results were obtained using a certain investigative procedure (Fig. 4.8). When there is no indicator channel connected to the same AE transducer as the basic channel, the results can quantitatively differ to some extent. However, the main regularities established in this way should remain unchanged.

Figure 4.12 illustrates the frequency spectra obtained while evaluating the durability and static crack growth resistance of various materials.

Thus, with the start of sub-critical crack growth, the frequency band of AES narrows jump-like, which can be used as a criterion for the identification of AES caused by a crack. At the stage of sub-critical crack growth, as predicted

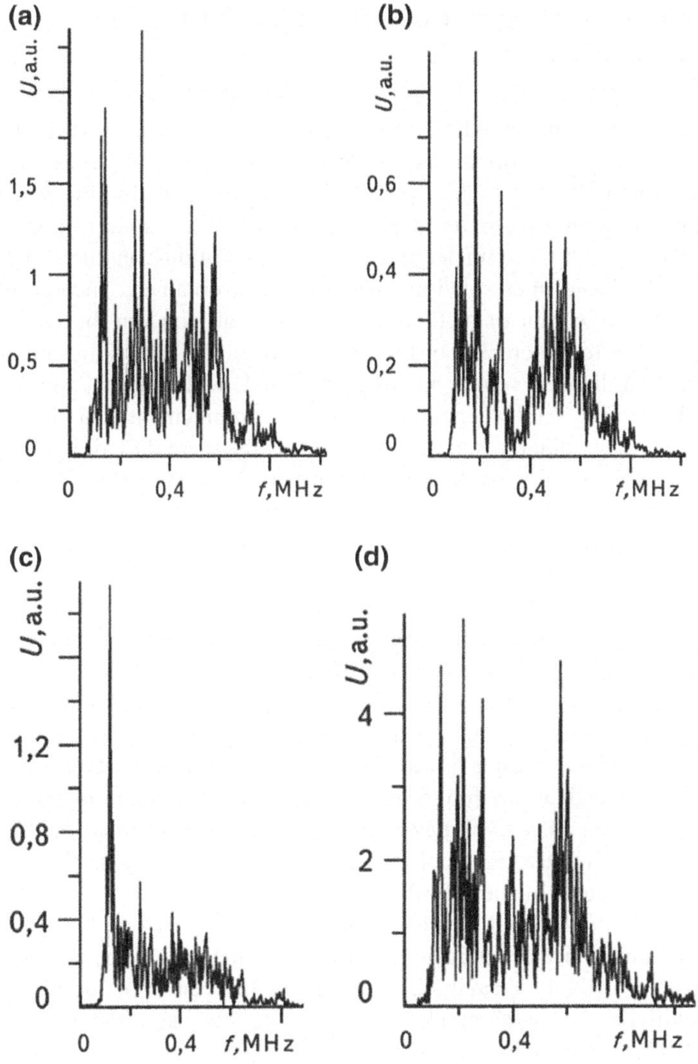

Fig. 4.12 The AES spectra at the initial (**a**, **b**) and final (**c**, **d**) stages of fracture of structural material specimens: **a** is the smooth cylindrical specimen of a diameter of 4 mm, steel 45, as delivered; **c** and **d** are prismatic precracked specimens of sizes 10 × 20 × 120 mm, 38XN3MFA steel, three-point bending

theoretically, the width of the frequency band is inversely proportional to the square of the stress intensity factor in the range of up to 1 MHz, and at final stages up to 600 kHz. With a crack transition to the overcritical stage, according to the Mode I–Mode II fracture, a frequency band extends and becomes approximately the same as at the initial stages of sub-critical crack growth. However, the frequency maxima differ by 2…3 times.

4.7 Directional Diagram of AE Radiation During Macro-crack Growth

The directional diagram of AE (DDAE), which is an angular distribution of the AES amplitudes, is very important for AE testing. Every AE source (crack, dislocation, cavity, etc.) has its inherent DDAE, which is related to the mechanism of the defect initiation or growth, its shape, and its type. When analyzing the DDAE, it is possible to develop certain technical recommendations on the most effective selection of AES and interpretation of the AE data obtained. Such recommendations, from the viewpoint of methodology, first of all concern the location of the AET with respect to the crack. For this, it is necessary to take into account the fact that a crack as a directed source has its own DDAE with maxima and minima [5]. On the other hand, AET, depending on their arrangement, can have different receiving directional diagrams and, consequently, there are directions of the highest and the lowest sensitivity of AES reception and detection [64]. Therefore, coordinating the radiation and reception diagrams by choosing the directions of mutual coincidence of their maxima, it is possible to increase the efficiency of AES detection and processing without taking additional measures.

At the stage of AE data interpretation using DDAE, the mechanism of crack formation, its sizes and orientation, defect type, and truth or falseness of the recorded AES can be established [65]. Below, after a short theoretical substantiation, there follows a description of an experimental method for determining the DDAE caused by a crack using serial AE devices.

The modern understanding of the acoustic radiation directivity is mainly based on theoretical investigations [65–68]. We will generalize some of them that deal with DDAE. First of all, we will pay attention to the investigations, where analytical dependencies for the DDAE calculation are obtained, simplifying the analysis and giving a visual pattern of the DDAE.

During the formation and development of a cavity, only one longitudinal wave emanates. Its directivity depends on the cavity's shape and on the mechanism of its formation or growth. If a cavity is a sphere in the center of radial tension, the radiation will be the same in all directions. In this case, the function of AE directivity, which describes DDAE, is $\Phi(\theta) = 1$ [29], where $OR\theta z$ is a spherical coordinate system. If a sphere-like cavity is in a solid under uni-axial tension, then the directivity of acoustic radiation appears, which is the angular dependence with respect to the plane of the crack location, and for a longitudinal wave, according to [68], $\Phi(\theta) = 1 - g\cos^2\theta$, where $g = 2(c_2/c_1)^2$. The same DDAE has a defect, which is a system of three mutually perpendicular dipoles or a penny-shaped Mode I crack [6, 63].

The theory of elastic waves radiation by a newly formed or sub-critical crack was presented in detail earlier [5, 6]; we will relate only to the results concerning DDAE. Formation of a Mode I or a Mode II crack in an infinite body excites not one but two types of waves—longitudinal and transversal. Consequently, if during AE testing only a longitudinal wave is detected, one can conclude that the defect is a cavity rather than a crack. Identifying the cracks, one has to take into account that a Mode II crack generates one wave, which is transversal but not longitudinal.

DDAE is individual for every type of wave and depends on the mechanism of the crack formation (normal rupture, longitudinal, or transversal shear). For example, in the far-zone, the DDAE radiation is described for a Mode I crack by the following functions [69]:

- longitudinal wave

$$\Phi(\theta) = 1 - g\cos^2\theta, \tag{4.7}$$

- transversal wave

$$\Phi(\theta) = \sin^2\theta, \tag{4.8}$$

where angle $\theta = 0$ in the polar coordinates $Or\theta$ corresponds to the plane of the crack location. Note that in dependencies (4.7, 4.8), only the function of AE directivity is presented, and the proportionality factors, used in expressions for calculation of amplitudes, are omitted.

In the case of a crack jump of a half-disk shape, the function of AE directivity of a longitudinal wave is [63]

$$\Phi(\eta,\theta) = (1 - 2\varepsilon^2\cos^2\eta)(1 + \cos\eta_1)/[(1 + c_{R1}\cos\eta_1)D(-\cos\eta_1), \tag{4.9}$$

where

$$\cos\eta_1 = \cos\eta\left[\cos^2\eta + \sin^2\eta/\sin^2\theta\right]^{-1/2}; \quad \varepsilon = c_2/c_1; \quad c_{R1} = c_R/c_1;$$

$D(-\cos\eta_1)$ is integral, which is found numerically [70]; $OR\eta\theta$ are spherical coordinates. For a transversal wave in a half-space at the crack tip, this function can be written with sufficient accuracy as (4.8).

Comparison of dependencies (4.7)–(4.9) shows that DDAE depends on the mechanism of the defect initiation and growth, its shape, and its type. We can compare the theoretical results obtained according to formula (4.9) and by known experimental data [65, 71]. Note that in this study, the purposefully made wide-band and capacitance AE transducers, as well as the AE tools, were used, which permitted recording only a longitudinal wave in the band of several megahertz. Usually, resonance transducers are used for the AE tests. They do not permit separating longitudinal and transversal waves. In such a case, the recorded data represent a superposition of longitudinal and transversal waves, which is why it is imperative to investigate the DDAE that is the result of these two types of waves.

In paper [69], compact and disk-like 1201-T aluminum alloy specimens of of 120 mm in diameter and 12 mm thick, with a preliminarily induced fatigue edge crack, were tested for eccentric tension. Since theoretical results were obtained for an infinite body in order to compare them with the experimental ones, it is necessary to decrease the reflection of waves from the specimen surfaces as much as possible. Therefore, in order to accelerate the decay of the reflected signals, the specimens were glued with the layers of rubber, which on one side had cuts for attaching the AE transducer.

AET were installed at various angles to the crack plane (0°, 30°, 45°, 60°, 90°), and the amplitudes of AES were recorded by five parallel channels. Each of the channels, except for the AE transducer, had a preliminary AE amplifier, AF-15 device, storage oscilloscope C8-12, and photographic recorder. The measurement conditions were: the pass band of the AE transducer was within 0.2…2.0 MHz with approximately identical amplitude-frequency response in the whole pass band, the amplification factor of the measuring path was 40 dB with the pass band within 0.2…1.0 MHz, and the transmission factor of filters was equal to 1.

Before being measured, the channels were calibrated receiving AES from the breakage of the graphite lead at the crack tip. Having adjusted the sensitivity of each channel, various levels of output signals were obtained, with the difference not exceeding 6%.

While testing the compact and disk specimens, the P-δ diagrams and AE diagrams were synchronously recorded (Fig. 4.13). By the cumulative count of the AE, N the moment of the crack start at the linear section of the P-δ diagram was determined. Three characteristic points corresponding to different stages of sub-critical crack growth were chosen for the analysis: Point I is the initial stage, Point II the final, and Point III the transitional, i.e., from sub-critical to over-critical propagation. Test results of compact and disk specimens show the similarity of the AES shapes recorded by each channel.

Angular distribution of the AES amplitudes (normalized to the amplitude that corresponds to an angle 0°) was calculated in the polar coordinate system (Fig. 4.14), assuming for an aluminum $\varepsilon = 0.266$ [66] and choosing the function of AE directivity from a condition $A = \max(A_1, A_2)$, where A_1 and A_2 are amplitudes

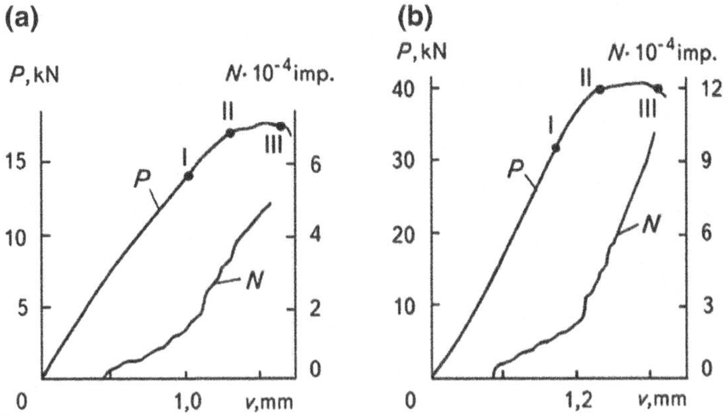

Fig. 4.13 Diagrams "load P − crack opening v" and the dependence of the AES cumulative count N change for compact (**a**) and disk specimens (**b**)

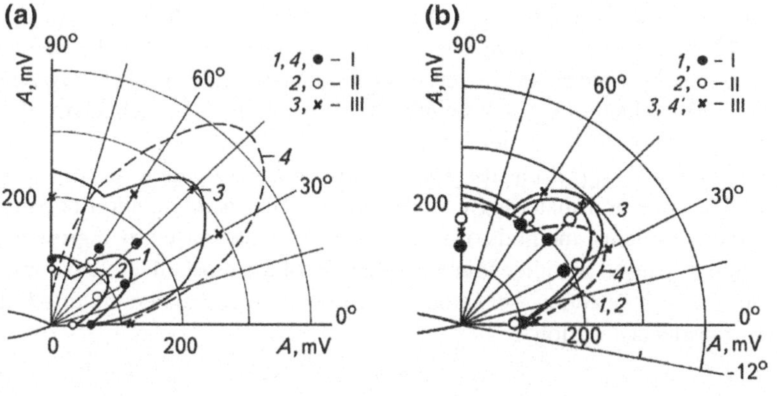

Fig. 4.14 Theoretical (*solid lines*) and experimental (*symbols*) DDAE for compact (**a**) and disk (**b**) specimens for different load values: curves *1*, *2*, *3* correspond to points I, II, III, respectively; *4* is DDAE for a dislocation rupture; *4′*—for a crack turned by an angle of 12°

of longitudinal and transversal waves, respectively, that were recorded by the channel. The DDAE type of each specimen is the same, i.e., AES were excited by the same defect—a sub-critically growing crack.

The agreement of theoretical and experimental data is better for a compact specimen. The ratio of the AES amplitudes, for example, for the angles 0, 45, and 90° is 1:3.5:2, respectively, and by theoretical calculations for a crack it is 1:2.8:2.1. If we assume that AES were caused by the plastic zone growth, the agreement is considerably worse. In fact, simulation of the plastic zone development by a dislocation rupture (constant jump of displacements on the fracture surface) according to Mode II fracture mechanism [72] at an angle of 45° to the plane of the crack location gives the ratio of the relative amplitudes 1:7.8:1 (dotted line in Fig. 4.14a). Theoretical results will substantially differ from experimental data if we consider that the basic source of the recorded AES is the plastic zone growth. Thus, one more criterion of identifying the AES caused by a crack and a plastic zone is the comparison of their DDAE.

For a disk specimen, theoretical and experimental results will differ somewhat, if we assume that the sub-critical crack grows in the plane of the preliminarily induced fatigue crack. Therefore, the angle defining the plane of crack propagation is recalculated. The analysis of the fracture surface showed that in a specimen, the crack propagates at an angle of 5°, while on a lateral surface at 30°. The best agreement of theoretical and experimental data is observed if the crack propagates at a 12° angle to the fatigue crack. Taking this into account, the DDAE is reconstructed (dotted line in Fig. 4.14b). Then, the ratio of theoretical and experimental values of amplitudes for angles 0, 45° and 90° is 1:2.2:1 and 1:2.4:1.5, respectively, which improves their correlation. Theoretical data also correctly coincide with the experimental results under cyclic loading of the 30XGSNA steel specimens [73].

The investigation of DDAE allows drawing a few conclusions and giving recommendations. Thus, the DDAE shape is affected by the mechanism of the defect initiation or growth, its geometry, and its type. The proper selection of the optimum place of AE transducer mounting (by adjusting the defect DDAE with the receiving diagram of the AE transducer) increases the efficiency of AE tests. Using DDAE, it is possible to evaluate the crack length and orientation, the mechanism of its formation, and identify AES generated by the crack growth and other sources. As a result, a technical approach for the experimental evaluation of crack growth resistance of structural materials was developed. The main attention is paid to the least studied problems: methods of selection of the AES caused by a growing crack among signals caused by other sources, selection of the most effective parameters of AES for the crack detection, selection of the operating parameters of an AE device, and AE transducer location.

4.8 Estimation of AE Signals Caused by Propagation of Internal Crack-like Defects

A dynamic non-stationary problem on the instantaneous formation of a penny-shaped Mode I crack in the elastic homogeneous isotropic medium is investigated in [5, 6, 74]. It is assumed that the initiation of such a penny-shaped crack takes place when tensile stresses in some region inside the material attain a certain critical value σ_0. The appearance of the crack is modelled by an instantaneous fall of tensile stresses at the faces of the defects from initial level σ_0 to zero.

Based on the investigations of the first maxima of a displacement vector module for longitudinal and transversal waves at the observation angle θ ($\theta = \mathrm{arctg}(z/r)$) in the range from 0 to $\pi/2$, and at distances from the observation point to a source larger than 30 crack radii, the following approximation expressions were proposed:

For a longitudinal wave

$$\Phi(\theta) = 1 - 2\varepsilon^2 \cos^2\theta \left(1 + \beta_1 \cos^2\theta\right)^{-1/2}, \tag{4.10}$$

and for a transversal wave

$$\Phi(\theta) = \left|\sin^2\theta\right| \left(1 + \beta_2 \cos^2\theta\right)^{-1/2}, \tag{4.11}$$

where $\beta_1 = 0.68$, $\beta_2 = 2.69$, and $\varepsilon = c_2/c_1$, which determine the angular distribution of radiation with an error less than 4%. Taking into account the dependencies (4.10) and (4.11) the following approximation fprmulas for evaluation of the maximal values of the displacement vector modulus were proposed:

$$\left|\vec{U}\right|_{\max|c_i} = \alpha_i \sigma_0 \Phi_i(\theta) r_0^2 \left(\rho c_1^2 R\right)^{-1} \quad (i = 1, 2), \tag{4.12}$$

where r_0 is the radius of a penny-shaped crack, R is the distance to the observation point, $i = 1$ corresponds to the longitudinal, and $i = 2$ to the transversal waves, $\alpha_1 = 0.452$, $\alpha_2 = 0.832$.

Dependencies (4.12) permit evaluating the size and orientation of a penny-shaped crack by the amplitude values of AES recorded by AET. For this purpose, at first the defect location should be found using the method of spatial location with four AET. Then, the procedure for finding the crack orientation is reduced to minimizing the following function:

$$F(\theta_0, \varphi_0) = \sum_{j=2}^{4} \left[\Gamma_j - \Phi_i(\alpha_j)/\Phi_i(\alpha_1)\right]^2, \tag{4.13}$$

where $\Gamma_k = \frac{A_k R_k}{A_1 R_1}$, $A_k(k = \overline{1,4})$ is the AE signal amplitude recorded by the k-th AET, which is considered to be proportional to the maximum of the displacement vector module determined by Eq. (4.13), R_k is the distance from the crack center to the k-th AET, the angles θ_0 and φ_0 are assigned in the spherical coordinate system $OR\theta\varphi$ with the origin that coincides with the crack center, α_j is the direction of the unit normal to the crack, functions $\Phi_i(\alpha_k)$ are determined by Eqs. (4.11) and (4.12), for longitudinal ($i = 1$) and transversal ($i = 2$) waves, respectively, $\alpha_k = \pi/2 - \theta_k$. Varying θ_0 in the range $[0, \pi/2]$, and φ_0—from 0 to 2π, for Γ_k being known from the experiment, and the values of functions $\Phi_i(\alpha_k)$ calculated according to (4.10) or (4.11) for longitudinal and transversal waves, respectively, we find a minimum $F(\theta_0, \phi_0)$ and the respective angles θ_0^* and ϕ_0^*, which determine the defect orientation that was sought. Using the dependence (4.4) for the known R_k, A_k, σ_0 and the crack orientation, we can calculate its radius.

To numerically verify the correctness and stability of the method proposed for the determination of a defect orientation by the AE signals, an error of ±20% was added to the values of Γ_k, which in general are determined experimentally, in comparison with their exact values found from Eq. (4.12), when we consider a certain case of mutual location of the crack and the AET system. According to Eq. (4.13) the error in determining the orientation angles θ_0^* and ϕ_0^* did not exceed 7%.

Investigating the AE radiation caused by sub-critical internal crack propagation in a material—in particular, in local regions near its front—is an important problem of AE diagnostics. To develop a corresponding model of the AE due to a small crack extension, an elastic body with a plane Mode I macro-crack bound by a smooth contour L was considered. A crack-like defect is formed under loading at some point in time, which is assumed to be initial in a small region near the crack contour where tensile stresses attain a critical value. Unloading its faces from an initial level to zero causes the emission of elastic waves that are recorded by AET. To simplify the problem, the formation of a penny-shaped Mode I crack of the same area at the macro-crack front is considered instead of a micro-defect. It is assumed that the radius to its contour is considerably smaller than the curvature radius of a plane macro-crack. Then, instead of the initial problem, there is a problem on the displacement field caused by the sudden formation of a penny-shaped Mode I crack near a semi-infinite crack front.

An approximate solution to this problem in the spherical coordinate system $OR\theta\varphi$, whose center coincides with the center of a penny-shaped crack, is given as a product of the components of a displacement vector obtained from the solution to the problem on a penny-shaped crack formation [5, 6], and functions the $C_i(\alpha)$ that determine the effect of a semi-infinite crack-free surface on the angular distribution of emission in the form [70]:

$$\cos\alpha = (\cos\theta\sin\varphi + \delta)\left[(\cos\theta\sin\varphi + \delta)^2 + \sin^2\theta\right]^{-1/2}, \quad \delta = \Delta/R, \tag{4.14}$$

where c_R is the Rayleigh wave velocity, $K_-(\cdot)$ are the known functions,

$$\cos\alpha = (\cos\theta\sin\varphi + \delta)\left[(\cos\theta\sin\varphi + \delta)^2 + \sin^2\theta\right]^{-1/2}, \quad \delta = \Delta/R,$$

is the distance between the center of a penny-shaped crack and the tip of a semi-infinite crack.

Determination of orientation and size of a penny-shaped crack that formed near a semi-infinite crack front is evaluated by dependencies (4.13) taking into account the relationship (4.14). For a transversal wave, the angular dependence of radiation, due to the presence of a head wave in the region of angles α close to $\cos^{-1}(c_2/c_1)$, has a sharp maximum, which substantially complicates the determination of the defect orientation. Therefore, the advantage in AE investigations should be given to the reception of a longitudinal wave that is also used for the localization of the AES source by a triangulation method. The analysis of the displacement fields shows that the maximal values of displacement are proportional to the area of a new defect that is proved by experimental results [75, 76]. The angular distribution of radiation for a longitudinal wave in a plane $\varphi = 0$ shows that the AES amplitudes A caused by a micro-defect formation in front of the MC, can be determined by the following dependence:

$$A = \lambda S R^{-1} \Phi_1^d(\theta) C_1(\alpha), \tag{4.15}$$

where $\Phi_1^d(\theta)$ is evaluated from expression (4.10), and λ is the proportionality factor between an electric signal at the AET output and maximal displacement at the longitudinal wave front.

If during a sub-critical crack growth, M micro-cracks were formed, the total increment of a crack area will be

$$\Delta S = \sum_{k=1}^{M} s_k = \sum_{k=1}^{M} a_k A_k, \tag{4.16}$$

where

$$a_k = R_k / \left[\lambda \Phi_1^d(\theta_k) C_1(\alpha_k) \right].$$

The increment of a through-crack with a rectilinear front, when the formed defects go through the whole cross-section of a specimen (the cross-section thickness being equal to h), can be approximately evaluated by the formula

$$\Delta a = b/h \sum_{k=1}^{M} A_k, \tag{4.17}$$

where b is the constant to be evaluated from the experiment.

Investigating the free surface effect on AES caused by a penny-shaped crack, the surface displacements were sought as a product of the appropriate displacements obtained from the solution to a problem on the instantaneous formation of a penny-shaped Mode I crack in an elastic space, and the reflection coefficients. They were found in [77] from the approximate solution to the problem on a penny-shaped crack propagation in an elastic half-space within the framework of a ray theory in a wave front approximation. These displacements nearly coincide with the exact ones for the zone close to the epicenter on the condition that the crack radius is much smaller than the distance from its center to the half-space boundary. Using the performed calculations and the dependence (4.6), the following approximate formulas have been obtained for the maximal surface displacement values:

$$U^{(k)}_{\max|c_i} = \alpha_i \Phi_i(\theta) r_0^2 \left(\rho c_1^2 R \right)^{-1} R_1^{(k)}(\theta), \quad k = x, y; \tag{4.18}$$

where $R_i^{(k)}(\theta)$ are the reflection coefficients for longitudinal (i = 1) and transversal (i = 2) waves, respectively. Calculations made according to (4.18), have shown that for a longitudinal wave (Fig. 4.15a, b) the maximum values of the surface displacement in the epicenter almost twice exceed the respective values for an infinite body. For a transversal wave (Fig. 4.16c, d) at small distances from the epicenter, the character of the change in the surface displacement maximum is more complicated in comparison with a longitudinal wave. By combining formulas (4.15) and (4.18), we get a relationship:

$$A = \lambda S R^{-1} \Phi_1^d(\theta) C_1(\alpha) R_1^{(k)}(\theta), \tag{4.19}$$

that determines the AES amplitudes at the half-space surface caused by an incident longitudinal wave during sub-critical internal crack growth.

Dependencies (4.14) and (4.19) can be used to evaluate the sizes and orientation of the local fracture region by AES detected by the system of three AET that are located in the symmetry plane at a half-space boundary.

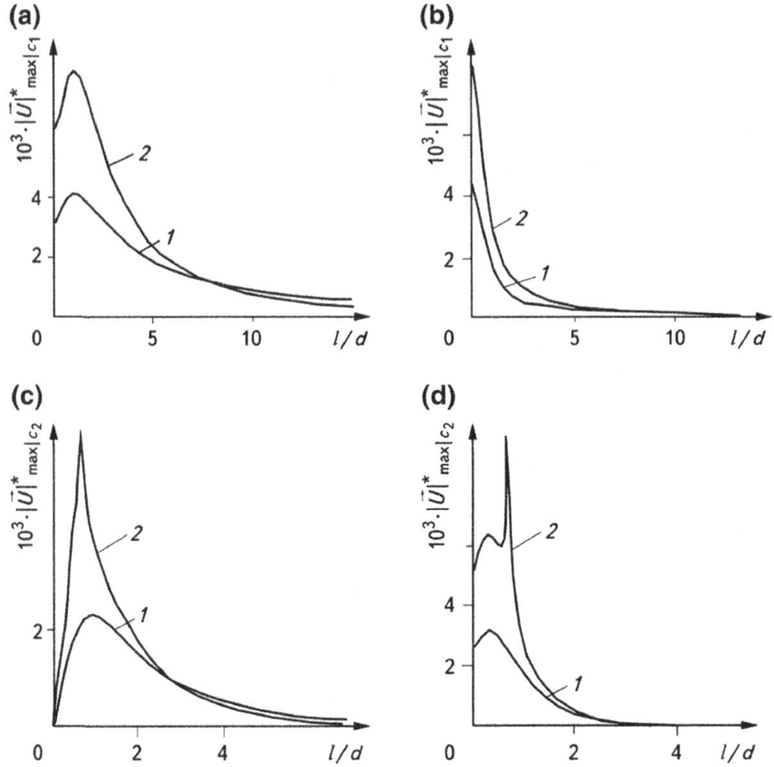

Fig. 4.15 Dependence of dimensionless value of the maximum of the displacement vector module at the surface of a half-space on dimensionless distance *l*/*d* for a longitudinal wave, $d/r_0 = 200$ for a longitudinal (**a**, **b**), and transversal (**c**, **d**) waves, the orientation angles $\eta = 0^{\circ}$ (**a**, **c**) and $\eta = 75^{\circ}$ (**b**, **d**): *1* is the incident wave, *2* is the total wave

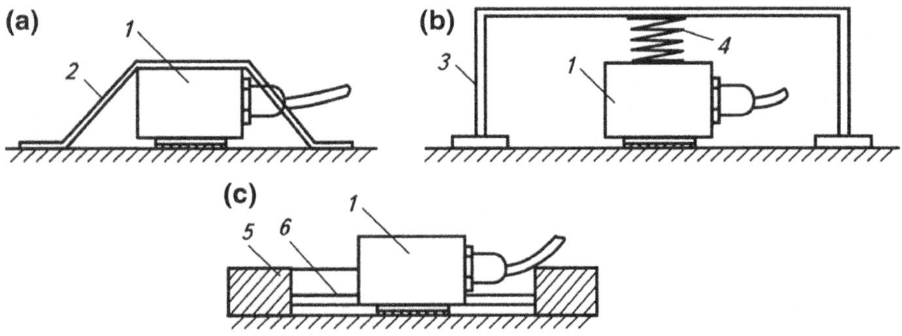

Fig. 4.16 Methods of the AET mounting on IO by adhesive tape (**a**), a spring (**b**), and a magnet (**c**): *1* is AET, *2* is adhesive tape, *3* is a spring, *4* is a carriage spring, *5* is a magnet, and *6* is a diaphragm

4.9 Methods of the AET Mounting at IO

Modern IO are characterized by a complicated geometrical shape, various physical properties of structural elements, service conditions, and the like. Therefore, the methods of the IO testing specify rigorous requirements to provide a reliable acoustic contact between IO and AET at the place of its mounting. In order to improve the contact between them, an intermediate acoustic transparent layer is used.

A thin intermediate layer between AET and solid IO can change the mechanical resonance frequency of the system and consequently decrease the resistance to wave propagation at the resonance frequency. This corresponds to an increase of mechanical quality factor, and to the narrowing of the band pass width of the system “AET-IO”. The presence of a thin contact of acoustic transparent layer leads to an increase of the transmission factor by several times in comparison with a similar factor without the layer [78].

A wide range of techniques for mounting AET on IO are well known: methods using adhesive tapes (Fig. 4.16a), magnets (Fig. 4.16b), pressed springs with carriage springs (Fig. 4.16c), rubber vacuum chuck devices, etc. However, they cannot always be attached to IO due to their complicated shapes, low quality of surface treatment, physical peculiarities, etc. Besides, these methods do not always provide a reliable acoustic contact during product testing and do not ensure its stability in time. In addition, the devices for the AET mounting on IO with the application of magnets are as a rule rather bulky and inconvenient to use. Other facilities, in particular, fastening by an adhesive tape or gluing during deformation of the IO, frequently turn out to be the additional sources of noise and do not exclude the AET shift along the investigated object surface, although certain progress in the quality of the fastening is observed [14, 79].

Paper [80] describes the original device for the AET mounting on IO, which, being free from numerous disadvantages in traditional methods of mounting the AET, is still structurally complicated and bulky, and thus can be used only in some cases. The device (Fig. 4.17) consists of the AET *1*, connected to the AE

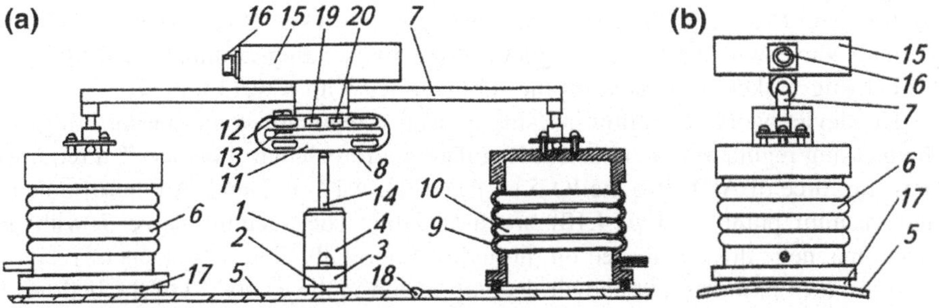

Fig. 4.17 Schematic of the device for the AES reception with a mechanism of mounting the AET on IO surface (**a**); side-view for testing the component with a curvilinear surface (**b**)

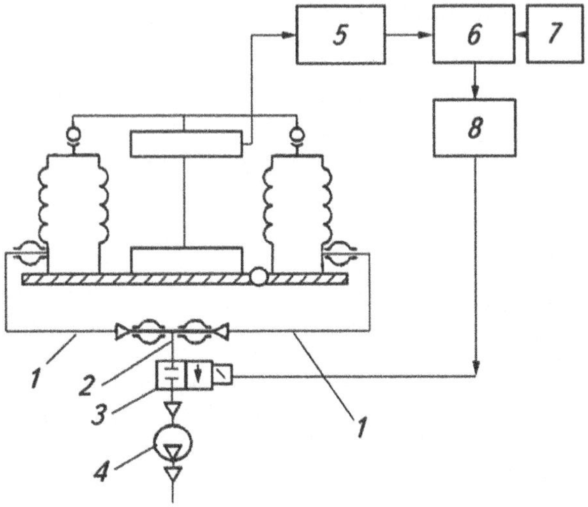

Fig. 4.18 Functional circuit of the AET holding mechanism control

waveguide 2 through the electro-acoustic transducer *3*, screening case *4* and the mechanism of the AET fastening to the IO surface 5. The mechanism for holding the AET *1* is manufactured in the form of vacuum chucks *6*, their number being determined by the IO geometry. The chucks are pivotally connected by beam *7*, which act through the pressing regulator *8* on the AET *1*. Vacuum chucks *6* are manufactured as bellows *9*, the cavities *10* of which are connected by tubes *1* (Fig. 4.18) to the vacuum pump *4*. Between tubes *1*, with the common nipple *2* and the vacuum pump *4*, there is a vacuum valve *3* with electromagnetic drive. Pressing regulator *8* (Fig. 4.17) is manufactured as an elastic element *11* with the strain sensors *12* and *13* fastened to it. The first one is located on the elastic element *11* from the side of the beam 7, and the second from the opposite side. The other side of elastic element *11* through the regulative element *14* interacts with AET *1*. Strain sensors *12* and *13* are plugged into an electric circuit that consists of a measuring amplifier *5* (Fig. 4.18), comparator *6*, force-control device *7*, relay *8* and vacuum valve *3* with electromagnetic drive, and a vacuum pump. Measuring amplifier *5*, comparator *6*, force-control device *7* and relay *8* are assembled into a common control unit *15* (Fig. 4.17), which through power socket *16*, is connected to the vacuum valve with an electromagnetic drive for sending commands. Bellows *9* have a ring gasket *17* located on the IO *5* surface with a weld joint *18*.

The device works as follows: Using a balance regulator, a measuring bridge of the pressing regulator *8* is adjusted. Regulating a force-control device *7*, a required pressing force of AET *1* to the IO *5* surface (Fig. 4.17) is found. A control unit *15* and vacuum pump *4* (Fig. 4.18) are turned on. The vacuum valve *3* with an electromagnetic drive is turned on through the normally closed contacts of relay *8*. As a result of pumping out and air decompression in the cavity *10* of the bellows *9* (Fig. 4.17), the latter get compressed. Through the beam *7*, the compressing force is transmitted to the pressing regulator *8*, regulating bar *14* and AET *1*. As a result, deformation of the elastic element *11* occurs; it is transmitted to strain sensors *12* and *13*. The deformation of strain sensors results in the change of their electrical resistance value, which causes unbalancing of the bridge circuit; the unbalancing is increased by measuring amplifiers *5* (Fig. 4.18) and the increased voltage passes to comparator input 6. The comparator combines voltage of the amplifier and reference voltage from the force-control device *7*. When the voltage at the amplifier input *5* reaches the reference value, a positive voltage appears at the comparator output turning on the relay 8, which consequently turns off the power circuit of a vacuum valve *3* with an electromagnetic drive. As a result, a spring rod of the valve *3* closes the valve and stops pumping out the air from the cavities *10* of the bellows *9* (Fig. 4.17) by tubes *1* (Fig. 4.18).

An increase of air pressure in cavities *10* of bellows *9* (due to outside leakage) causes a reduction of pressing forces of the elastic element *11*, elastic deformation of resistors *12* and *13*, unbalance of the measuring bridge, turn-off of the comparator *6*, and turn-on of the vacuum valve *3* (Fig. 4.18) with the electromagnetic drive. The valve of the latter is turned on until a necessary force of pressing elastic element *11* is insured, and hence the AET *1* to the surface of IO *5* (Fig. 4.17). To remove the AET *1* from IO *5*, a vacuum pump 4 (Fig. 4.18) and control unit *15* are turned off and the air is let in by tubes into cavities *10* of bellows *9*.

The authors of paper [80] recommend using the device for AE diagnostics of welded joints in non-revolving joints of a large-diameter pipeline, storage, and petrochemical pressure vessels for the aviation industry, both during their use and their production. Such materials as austenitic steels, which are heat-resistant, heat-proof, cryogenic, etc., and titanium- and aluminum-based alloys, can be used for pipelines and vessels. This device is less effective in the NDT and technical diagnostics of cylindrical products with a small radius of curvature and with a considerable roughness of the IO surface. Thus, the correct use of the methods of the AET fastening to IO substantially increases the reliability of the AE test results that are obtained.

4.10 Selection of Useful AES During AE Tests

The experience of laboratory AE investigations and NDT of production objects [2, 9, 20, 24, 25, 29, 45, 51, 81] has shown that the most frequent difficulties in utilizing the AE phenomenon are related to the noises generated by the working devices, equipment, etc., and that it is hard to avoid their effect on the experiments' results. In most cases, various AE pseudo-signals can be observed: mechanical noises, such as friction in joints, vibrations, impact interaction of separate units, etc.; hydraulic noises conditioned by flow and leakage, boiling of working liquids; operation of power units etc., and electric interferences (internal thermal noises of devices, commutation of power machines and equipment, interferences of thyristor power control devices, and so on). Therefore, there is always a problem of reducing the IO noises not only related to the application of the measures described above, (i.e., improving the loading machines and devices, calibration of sensitivity of the AE path for AES selection and processing, correct preparation of the operating place and the AET mounting, calibration of their sensitivity, application of parallel channels [82], etc.) but also in connection with the utilization of other technical approaches, the main of which are described below. Their practical realization is impossible without the previous detailed study of the parameters of noises generated in IO in real AE research conditions.

4.10.1 Selection of a Working Frequency Band of AE Facilities

Figure 4.19 schematically represents the AE elastic wave transformation during its propagation in IO from the AE source to the AET output which, having its own geometrical sizes, additionally (except for IO (Fig. 4.20b) modulates an elastic wave due to its amplitude-frequency response (Fig. 4.20b)).

In production conditions, a mechanical noise (MN) is the most widespread type of hindrance and they can have the greatest effect on the results of AE investigations. MN means elastic vibrations of the IO surface, which are not connected with the AE source action in the IO. Numerous experiments prove [25, 29, 32, 83, 84] that amplitude-frequency characteristics are conditioned by various mechanisms of their generation and can differ substantially (see Figs. 4.20, 4.21, 4.22, 4.23 and 4.24).

As seen in Fig. 4.20, a significant decrease in the MN effect during aerodynamic pipe operation can be expected at a frequency higher than 0.3 MHz [29], and the noise background of the working power reactor "Dresden 1" [83] is very weak, at a frequency higher than 0.5 MHz (Fig. 4.21). The MN of roller bearings depends on the available lubricant (Fig. 4.22) and has two typical frequency domains: in the range of 0.05…0.2 and 1.0….1.2 kHz [84], while the noise background of cyclic testing machines is limited by the frequency band of 0.002…0.3 MHz (Fig. 4.23) [29]. Mechanical loading devices have somewhat better noise characteristics in comparison with similar hydraulic machines for static loading (Fig. 4.24). The background noise, which accompanies the leakage of liquids and gases, has its own amplitude-frequency characteristics (see Figs. 4.25 and 4.26) [85].

Thus, when conducting the AE research, a number of questions arise in relation to the choice of the operating frequency band of AE equipment, taking into consideration the action of background noise. It is very important to determine the frequency spectrum of various kinds of noises generated by the equipment operating in its own or adjacent housing. Having generalized what was presented above, and the other known data from experiments [86, 87], it is possible to state that frequency characteristics of noise of various mechanical and hydraulic loading devices and vibration machines, as well as electric interferences, have particular frequency bands. The range of the frequency spectrum in MEN is 20…80 kHz, for

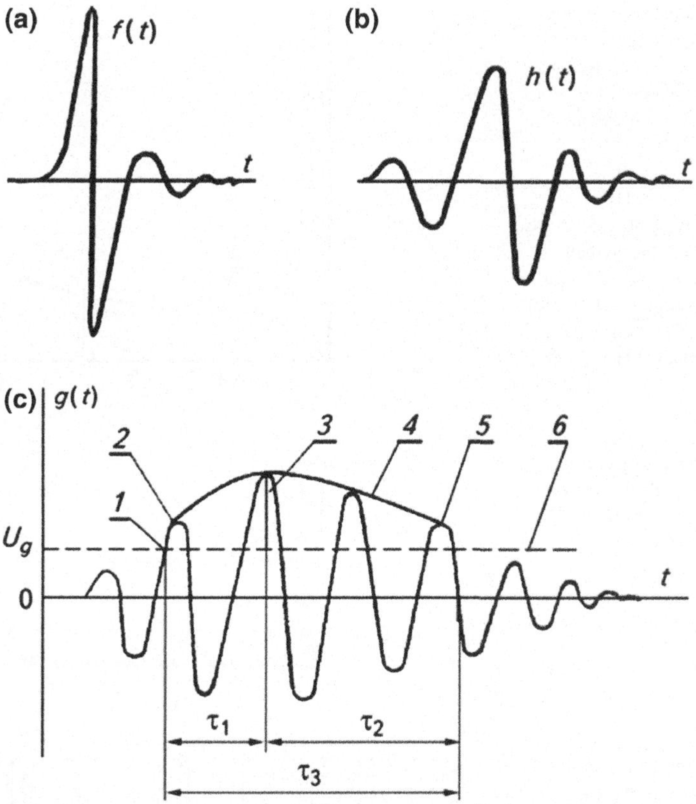

Fig. 4.19 Functions of the frequency distribution of the AE (**a**) event, response of the solid surface (**b**), and the AET (**c**): *1* is the moment of the AES arrival, *2* is the first pulse, *3* is the amplitude of an envelope curve, *4* is the envelope curve, *5* is the last pulse, τ_1 is the rise time of the wave front of the envelope curve, τ_2 is the time of decay, τ_3 is the duration of the event

Fig. 4.20 MN spectra of the working aerodynamic pipe (the analyzed band of 1 kHz): *1* is normal service conditions with a model and a holder, *2* is without a model and a holder, *3* is the operation of ejectors only; resonance peaks of curve *1* at frequencies of 140 and 200 kHz are probably related to the conditions of a model and a holder streamline

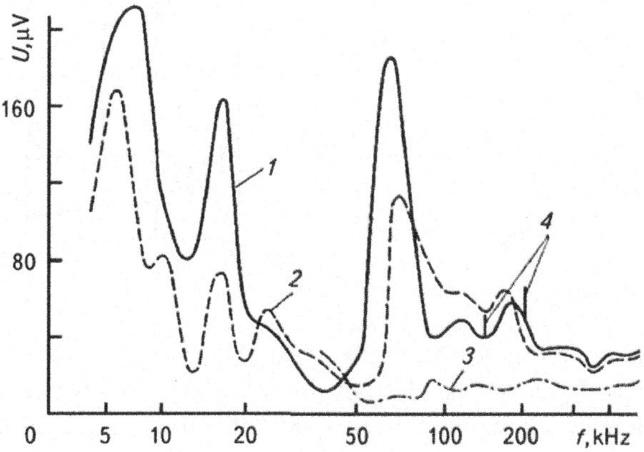

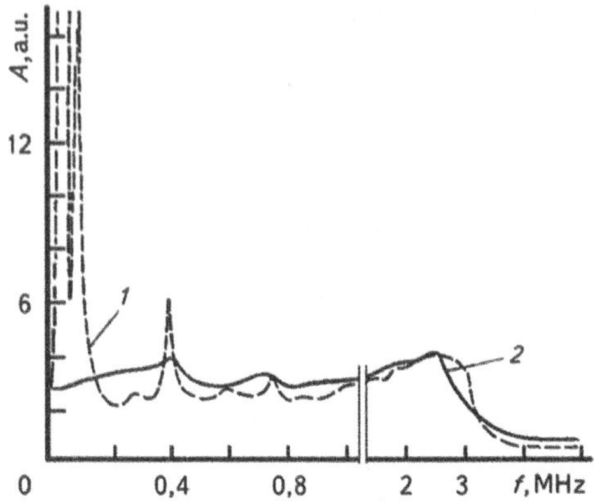

Fig. 4.21 Spectrum of noises of nuclear power reactor "Dresden-1": *1* is diameter of a circulating water collector of 560 mm; a flow speed is 7.6 m/s, high-pass filter (HPF) with a cut-off frequency of 20 kHz, amplification of 80 dB; *2* are noises of a testing system, with the AET placed on an external surface; amplification of 80 dB; HPF with a cut-off frequency of 20 kHz

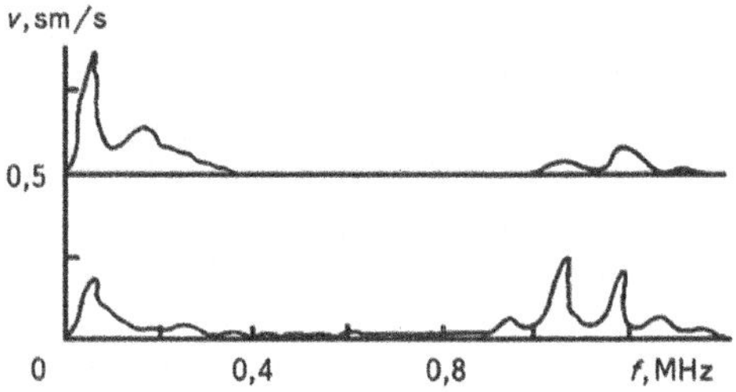

Fig. 4.22 Spectrogram of a roll bearing with the normal (*upper*) and insufficient lubrication (*v* is the speed of rotation)

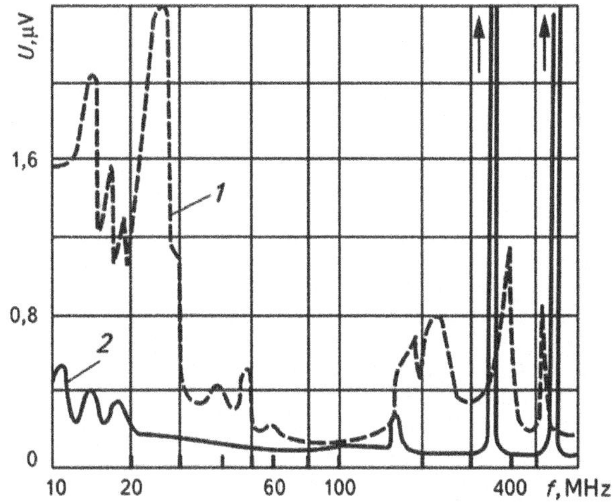

Fig. 4.23 Spectra of noises of testing machines (the analyzed band of 0.4 kHz): *1* is an electromechanical drive, and *2* is a hydraulic drive (frequency peaks at 340 and 590 kHz are electrical interferences)

hydraulic machines—20…400 kHz, and the vibration devices generate noises in the frequency band from dozens of Hz to several dozens of kHz. Electric interferences have the widest spectrum: from dozens of kHz to the units of MHz.

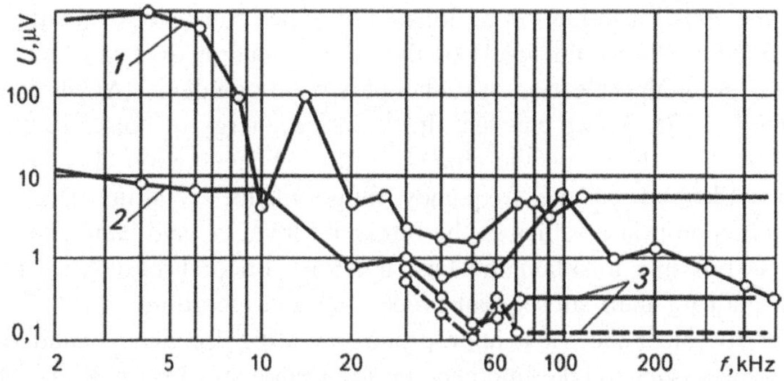

Fig. 4.24 Spectra of noises of cyclic loading testing machines (the analyzed band of 20 kHz): *1* is URM-2000, *2* is MUP-20, *3* is UMU-02 (mechanical drive)

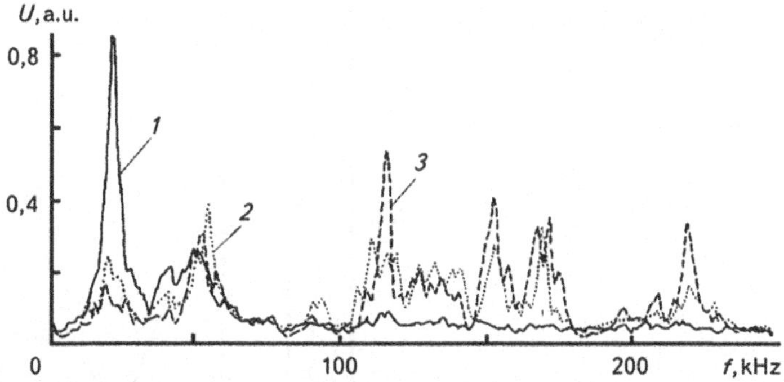

Fig. 4.25 The effect of backpressure P_0 on the spectrum of leakage noises in a hydraulic two-positional sleeve-valve at a flow pressure of 20 MPa: *1* $P_0 = 0$, efflux velocity $v_0 = 20$ ml/min; *2* $P_0 = 0.35$ MPa; $v_0 = 21$ ml/min; *3* $P_0 = 0.45$ MPa; $v_0 = 19.8$ ml/min

Fig. 4.26 Typical AE charts during leakage in a bullet valve of a diameter of 101.6 mm with the roughly treated layer and a new saddle: *1* is pressure drop $\Delta P - 0.072$ MPa, efflux velocity $v_0 = 2000$ ml/min; *2* $\Delta P = 0.045$ MPa; $v_0 = 2000$ ml/min; *3* $\Delta P = 0$; $v_0 = 0$

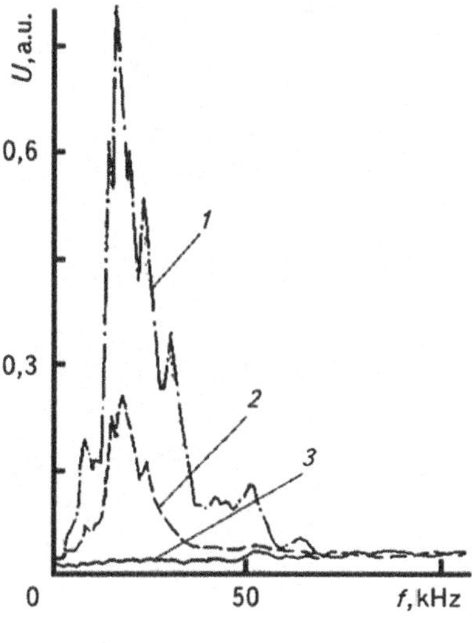

Partly, methods for limitation of noise of all types consist of evaluating the level of discrimination (threshold level) of the AES selection over the level of noises (Fig. 4.19c) and in the selection of such a pass band of a device Δf where noises are the least (Fig. 4.27). As we can see, the radiation energy of noises in the range of high frequencies is low, and the change in the AE generation character does not substantially depend on the frequency range of its recording (Fig. 4.28). By selecting the appropriate values of the threshold level U_g and band pass of the AE device, it is possible to attain the best transmission of useful AES through the measuring path for their subsequent processing and recording.

In order to select the most appropriate operating frequency band of the AE device, it is necessary to take into account the frequency spectra of the AES which can radiate during the expected defect development. For instance, the method of

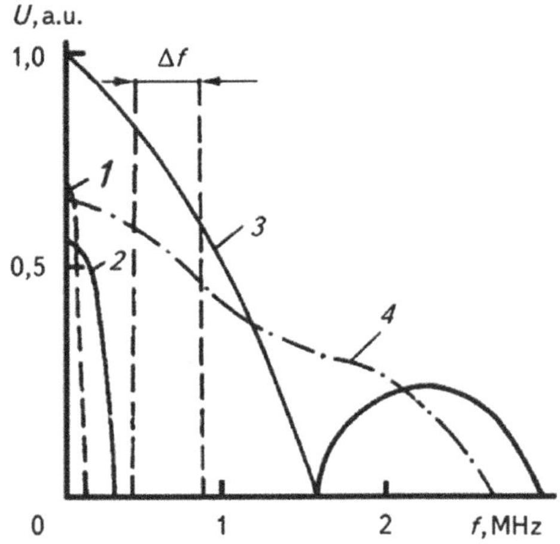

Fig. 4.27 Dependence of the noise spectrum amplitude on frequency: *1* are vibration machines, *2* are mechanical machines, *3* is leakage, and *4* are the AE useful signals

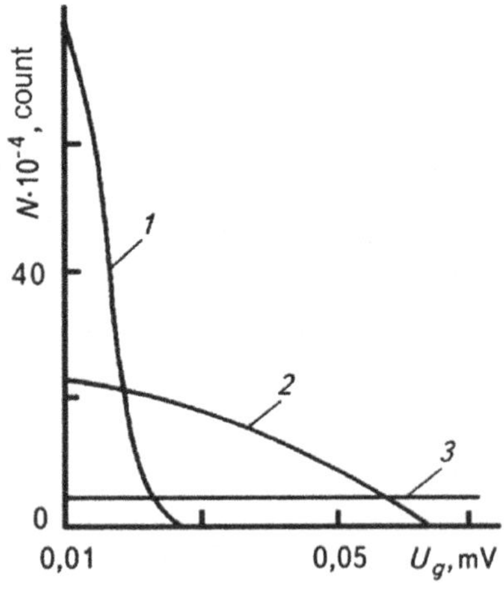

Fig. 4.28 Dependence of the mean value of the cumulative count of AES, *N*, on the level of discrimination and low cut-off frequency of the HPF: *1* for 360 kHz, *2* for 110, *3* for 11

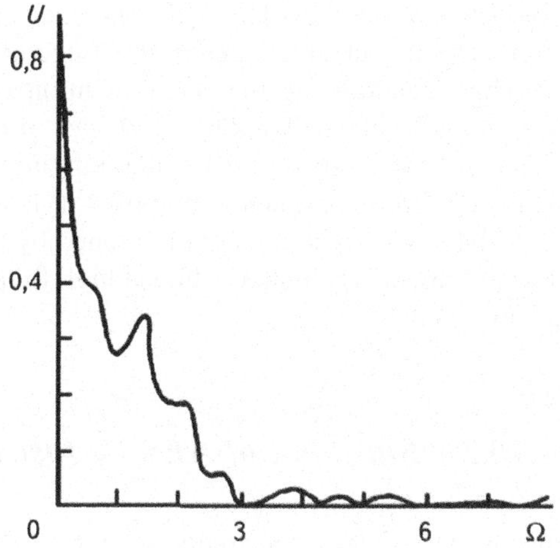

Fig. 4.29 Amplitude of the longitudinal wave frequency spectrum in the far-zone of radiation

preliminary theoretical estimation of the AES frequency spectrum, which can be generated during sub-critical jump-like crack propagation, is presented in paper [32]. It was modelled according to the formation of a penny-shaped flaw at the macro-crack front [5]. The solution to the corresponding dynamic problem of the crack theory made it possible to obtain a distribution of amplitudes shown in Fig. 4.29 in the frequency spectrum of the normalized displacement vector module.

Here, $\Omega = 2\pi f r_0 / c_2$, where r_0 is the radius of a penny-shaped crack. As shown in Fig. 4.29, under the condition $U_0 \approx 0.7U$, the width of the main maximum $\Delta\Omega \approx 0.15$. If $c_2 \approx 3 \times 10^3$ m/s (for steel) and $r_0 = 2 \times 10^{-5}$ m, we get the frequency band $\Delta f = 0 \ldots 3.6$ MHz.

Some researchers, when investigating the initiation and development of materials failure, have experimentally determined the amplitude-frequency characteristics of AES. For example, it is shown in [88] that the AE radiation spectrum under mechanical testing of the welded specimen with a faulty fusion has specific features. The authors claim that AES can be identified by the amplitude and spectrum depending on a radiating source. Powerful AES during plastic deformation are recorded in the range of 0.8…3 MHz with maximal amplitude of about 0.06 V, when the micro-crack grows—0.35…1 MHz with amplitude to 0.15 V, and when the MC propagates—0.01…0.6 MHz with an amplitude up to 0.4 V.

The change of the frequency spectrum was also established experimentally in [56], where brass and steel specimens were tested. It was found that for these materials, the frequency spectrum is in the range of 0.1…1.0 ÌHz, and the most optimum operation frequency range is 0.15…0.3 MHz. We should note that the frequency spectra of AES are related to the mechanism of deformation, and the obtained investigation results do not depend on the material type, AET, specimen geometry, and the method of spectral analysis, although some areas of a frequency spectrum can be explained by the AES resonances that do not affect the main frequency harmonics. The AES frequency spectrum in a brittle fracture of electronic chip ceramic substrates lies in the range of up to 10 MHz [89].

Thus, generalizing the above-mentioned, in order to optimize the width of the operating frequency band of AE device, it is necessary to take into account that the AES frequency spectrum is wide-ranging: from units of Hz to dozens of MHz. However, for experimental purposes, it is used as a rule within the range of 0.1… 2.0 MHz. Such a choice is conditioned by the fact that at lower frequencies, a high level of noise is revealed, while at high frequencies a strong decay of the AE waves is observed.

4.10.2 Filtration of AES by Instrumental Facilities

In the AE devices, the frequency range of input circuits is formed by setting the cut-off frequencies of filters, taking into account the selected operating frequency band and amplitude-frequency response of the AET. The width of the pass band is chosen by a low-cut filter (LCF) and HPF; the first one sets the upper limit, and the second sets the lower limit. Therefore, if the values—for instance 500 and 150 kHz—are set on them, this means that a pass band of measuring path frequencies is 150…500 kHz. In such devices, as a rule, the operating frequency band of the preamplifier is higher than the similar one in the amplification unit. Therefore, it is necessary to coordinate the pass-bands of its filters only. Hence, the filters should satisfy the following requirements [90]: (1) a high slope of the pass band cut-off; (2) a discrete adjustment by the choice of the upper and lower values of frequencies; and (3) avoid getting in resonance at cut-off frequencies.

It follows from the above that the frequency filtration of AES is an important link in the chain in the AE investigations and the key moments of its realization consist of the following sequence of measures:

- To preliminarily estimate the frequency spectrum of background noise of the working equipment that effects the IO;
- To preliminarily evaluate the amplitude-frequency characteristics of elastic wave radiation from the expected AE sources; and
- To choose an optimum pass band of filters, taking into account the frequency properties of the AET, noises, and useful AES.

4.10.3 Application of the "Dead Time" Mode

According to [29, 33], the AE signals can be classified as waves of a continuous shape, similar to "white noise," and as waves of a pulsating shape, i.e., pulses. The AES of the first type have a low energy level and many low-frequency components. They arise, as a rule, during plastic deformations. In the AES of a pulse, i.e., in a discrete type of wave, the level of energy is high and the spectrum is characterized by high-frequency components. Such AES arise, for example, at a brittle fracture of materials. A proposed classification is conditional, because in the presence of a great number of discrete AE signals—for instance, at high rates of loading, when the time of their appearance is less than the duration of the AE event —it is very difficult to separate them. However, at slow jump-like sub-critical macro-crack growth [6], the signals originating from its jump-like propagation are well distinguished and are recorded as discrete AE.

In Fig. 4.30, the AE signal obtained at the AET output and its processing is represented schematically. It is clear that the selection of the AES threshold level affects the cumulative count and, consequently, the AE count rate. Therefore, during the noise cut-off, a certain error is introduced into the results of the AE research that is to be taken into account. Some methods of reducing the effect of such errors are known. For instance, the electronic units of the AES processing are

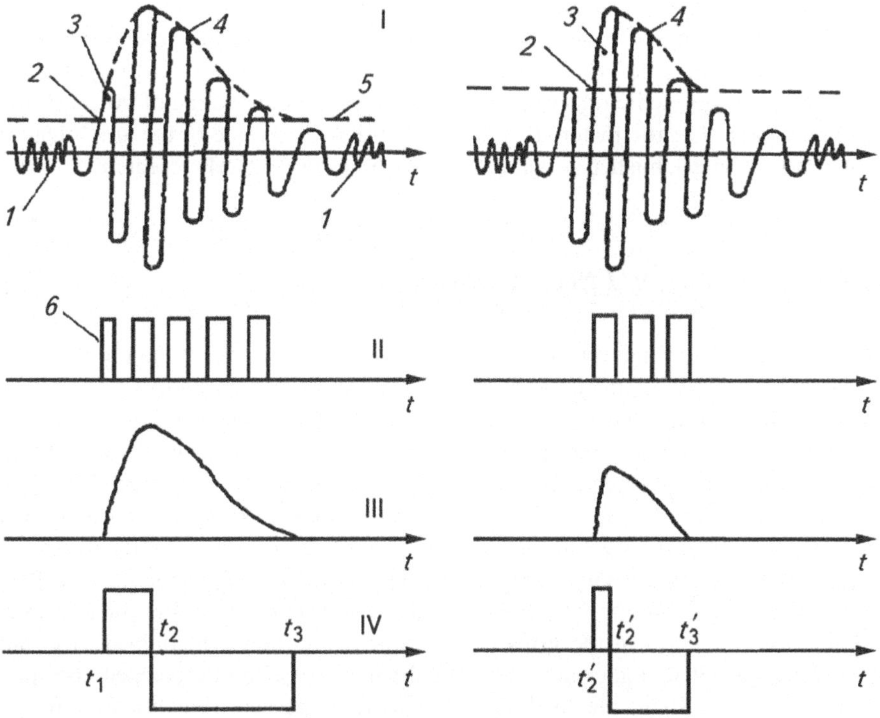

Fig. 4.30 AES processing by devices: *I* is the AE signal against the background noises; *II* is the count of pulses that exceed the threshold level; *III* is envelope selection; *IV* is formation of the "dead time" mode; *1* is a noise level, *2* is the moment that the count starts, *3* is the first pulse, *4* is the envelope level, *5* is the threshold level, *6* is the formation of the width of pulses by their duration; $t_1 - t_2$ $(t_1^1 - t_2^1)$ is a period of time when an AE channel is open; $t_2 - t_3$ $(t_2^1 - t_3^1)$ is when it is closed

adjusted to count the number of events that are estimated by the envelope curve formed by a peak detector (Fig. 4.30, Variant III) or the count is done by the first AE pulse obtained that exceeds the threshold level, and then the condition of “dead time” is applied (Fig. 4.30, variant 4). This, evidently, does not solve the problem of assessing the real number of counts (cumulative count of AE). However, the application of turn-off of the AE measuring path device for some time (“dead-time”) enables us to distinguish spurious oscillations arising for these reasons. At first, the elastic AE waves that assist fracture due to the small specimen sizes are repeatedly reflected from its lateral surfaces, causing new types of elastic AE waves. All elastic waves, which are additionally excited in the specimen, are superposed and recorded by the AET. While applying the resonance AET, additional spurious vibrations caused by the resonance phenomena appear in the AET. In order to exclude their count from useful AES, an interval of “dead time” is used after the pulse reaches the threshold value of t_1 or t_1'.

When a certain threshold level for the selection of AES is specified, it is necessary to correctly choose the amplification factor of the AE path, which also affects the AES transmissions and their processing results. The amplification factor of the AE path is the sum of amplification factors in dB of the preamplifier and the measuring amplifier, i.e., the power amplifier. Our experience and the published data show that for ductile materials, the amplification factor can be of 90…120 dB, and in the case of brittle materials, it is up to 80 dB. The value of the amplification factor should be established separately for each case of the AE research.

4.10.4 The Kaiser Effect Application

In almost all cases of the AE investigations, the KE is used for reducing the noise effect and for specifying the threshold level. The essence of the approach consists of the following: During the repeated loadings at the level that does not exceed the previously reached value, the material is “noiseless,” i.e., the AE recorded at the first loading is not observed (Fig. 4.31); thus, the material “memorizes” the load previously attained. For high-strength materials, such a “memory” is kept for a relatively long time. In low-carbon steels, especially at the elevated temperatures, it is possible to reduce the damage by annealing, and thus to regenerate the AE [91]. Hence, when adjusting the AE facility for noise protection, first the trial specimen without a crack is put in the grips or on the supports of the loading device, and stresses by 20…30% higher than the critical values for a similar cracked specimen are created in it. Then, the loading is gradually decreased—but not to zero—in order to avoid motions in grips or on supports, and after that the loading is increased again to the previously attained level. If no AES are recorded in this case, it means that there are no noise sources (i.e., friction at grips or at supports, noise from loading devices, etc.). In the case of the KE violation, it is necessary to use special clamps, sound insulation, and damping to eliminate this effect.

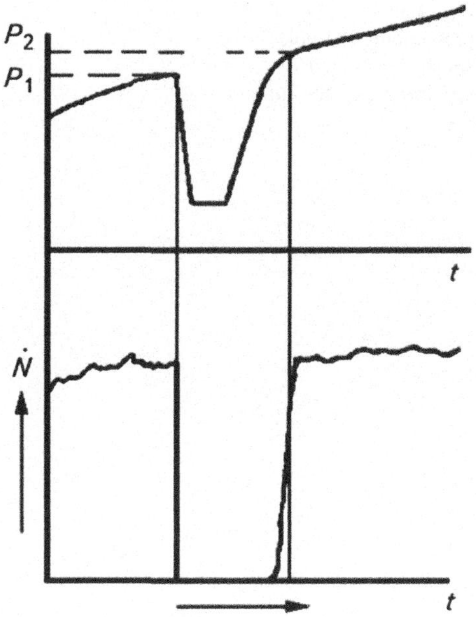

Fig. 4.31 Schematic of the KE implementation

At high levels of loading, the KE can be violated due to time-dependent processes, such as creeping. This is especially typical of the composite structures, where the KE disappears under the repeated loadings, which makes approximately 50% of a damaging failure load (the Felicity effect [4]). The KE does not hold during phase transformations, either.

Using the KE, the authors of [27] propose improving the reliability of the AE information obtained by the mechanical shunting of specimens before their testing. Shunts are mounted on the specimen's working part, and the specimen is loaded by a force that exceeds the damaging force. Thus, using the KE, the noises that would probably arise during loading in the elements of coupling, grips, and supports, are eliminated during their grinding to the specimen. Some technical approaches are known which, based on this effect, provide pressurization before testing the holes for the installation of the fingers of a loading device grip. For instance, a special device (Fig. 4.32) is used for this purpose in paper [92]. Similarly, it is possible to push the balls of a greater diameter and hardness through the specimen holes.

4.10.5 A Method of Spatial Selection of AES

To improve the reliability of the AE information obtained, the method of spatial selection is also used [93] (Fig. 4.33). Its essence is as follows: If AES 1 is first recorded by the auxiliary AET 3, and then by basic AET 4, this proves that the AE source is outside the inspected Zone I. A logical gating circuit of signal coincidence

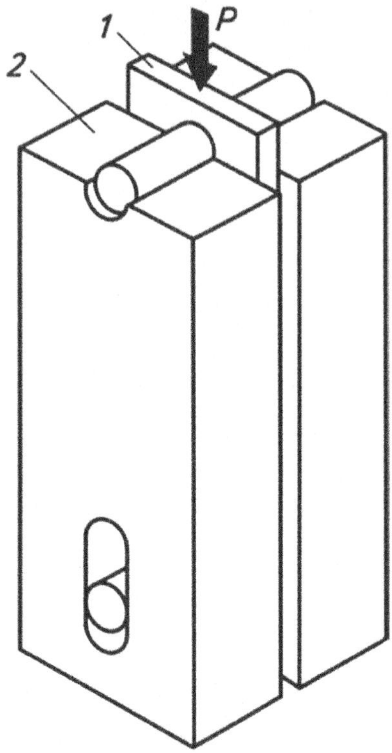

Fig. 4.32 Method of pressurizing the holes in a specimen by load *P*: *1* is the specimen, *2* is the support

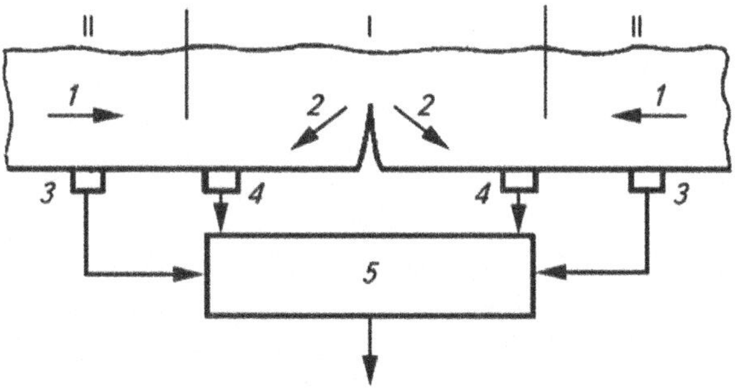

Fig. 4.33 A chart of the spatial selection method

5 in this case closes the input circuits of the AE devices and AES are not recorded. On the contrary, if at first AES 2 are recorded by PAE 4, the AE paths are open and AES are subsequently processed.

In paper [94], this method was modified: only AES reflected from the lateral surfaces of the specimen are recorded, which permits applying two arbitrary AET mounted on the specimen. The setups of spatial selection require the application of a computer technique, but at the same time they improve the reliability of AE information while investigating the material fracture. To increase the resistance of large-scale objects to noise, a method of triangulation is used; this permits determining the AE source coordinates and thus improves the reliability of information.

4.10.6 Other Methodical Approaches

There are other methods for improving the noise resistance of the AE facilities and the reliability of the AE information. For instance, the AETs, which are sensitive only to a certain type of elastic waves or detect only a longitudinal spatial wave in IO, are used.

Under a cyclic loading, a part of MN within every cycle of loading is generated, where either useful signals are absent (friction of crack faces during the compression part of a cycle), or their amount is negligibly small. In such a case, an ordinary closing of input sections of the AE facility (time gating) in certain regions of a loading cycle eliminates the noise passing to the equipment.

In some AE facilities [95], in order to eliminate the noise effect on the AE results, setting a program-controlled selection of only those AES is foreseen; its time of growth of the forefront of the envelope curve is in the range assigned by an operator. Moreover, it is possible to determine the time of the envelope decrease, the time of the pulse after-effect, or to combine these values. Usually, in order to set the ranges of variation of these parameters, it is necessary to conduct the preliminary experiments on the IO material specimens.

It is well known [96] that even under uncertain selection of the AE equipment frequency range, the nature of the AES change during the tests is mainly the same. The authors validated this idea by testing standard 4340 10 mm thick steel compact specimens; these were subjected to tension with the parallel record of AES obtained from a low-frequency AET with the pass band of 2.0...20.0 kHz, and from a high-frequency AET with the pass band of 0.3...1.0 MHz. The AE information obtained is almost identical in both cases. The authors emphasize that the AES spectra do not depend on the specimen size, which concurs with the results of [56].

Summarizing what has been presented above, it is necessary to take into account the investigation results described in [33]. Using the results of the theoretical and experimental studies known at that time, the probabilistic scheme of defect detection in steels under loading up to the stress level of 500 MPa (Fig. 4.34) was synthesized therein.

Based on a review of a number of published sources, the authors proved that the AE method, compared with other known methods of non-destructive testing, is most sensitive to the detection of initiation and propagation of defects in a solid, and they emphasized that the threshold of detection in Fig. 4.34 in the laboratory testing is shifted substantially to the left, since in these conditions the least displacement at the IO surface, which is reliably recorded by the AET, is 10^{-14} m. The modern AE facilities make it possible to record the motion of single dislocations, fracture of carbides smaller than 1 μm in steels, lamination, or fracture of such

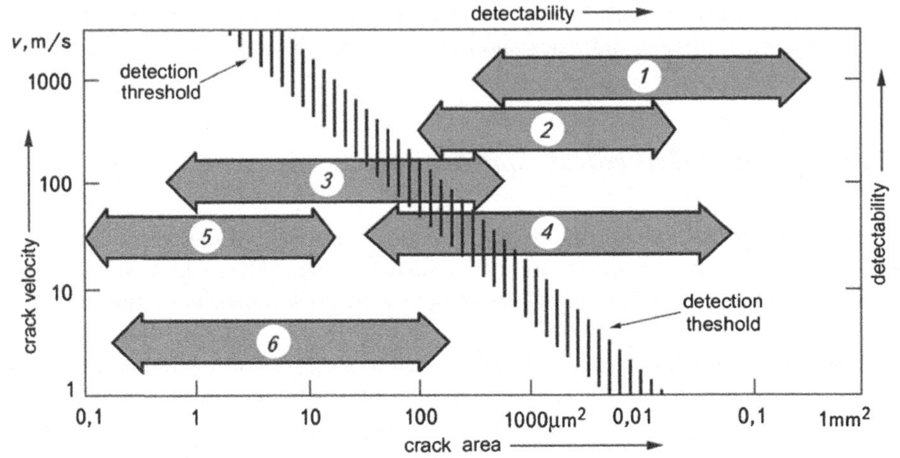

Fig. 4.34 Schematic distribution of the possibility of fracture detection in steels at different crack propagation rates *v* (the detection threshold corresponds to the AET sensitivity to the surface displacement of 10^{-13} m caused by a longitudinal wave; the distance between AET and the AE source is 0.1 m) *1* is the intergranular failure, *2* is the cleavage, *3* is the fracture of inclusions, *4* is the alternating shear, *5* is the fracture of carbides, *6* is the microvoid coalescence

inclusions as sulfides and silicides, initiation and coalescence of micropores, micro-cracks, etc.

Thus, there is a variety of methods for improving the noise resistance of the AE facilities. Their number grows steadily, and it is necessary to consider this and, therefore, to combine them in the AE studies. This will also enable correctly identifying the mechanisms of AES generation.

References

1. (1985) GOST 25.506–85. Raschety i ispytaniya na prochnost'. Metody mechanicheskich ispytaniy metallov. Opredelenie charakteristik treschinostoykosti (vyazkosti razrusheniya) pri staticheskom nagruzhenii. Vved. v deystvie 27.03.1985 g. (State Standard 25.506–85. Calculations and tests for strength. Methods of mechanical testing of materials. Determination of crack growth (fracture toughness) characteristics under static loading. Approved 27.03.1985). Izdatel'stvo standartov, Moskva
2. Ivanov VI, Belov IM (1981) Akustiko-emissionnyy kontrol' svarki i svarnych soedineniy (Acoustic emission monitoring of welding and welded joints). Mashinostroenie, Moskva
3. (1976) Standard recommended practice for acoustic emission monitoring of structures during controlled stimulation. ASTM E, pp 569–576
4. (1981) The EWAGE AE Code for acoustic emission examination of sources of discrete acoustic events. NDT Int 14(8):181–183
5. Andreykiv AY et al (1987) Teoreticheskie konzepzii metoda akusticheskoy emissii v issledovanii prozessov razrusheniya (Theoretical concepts of acoustic emission method in the

fracture processes investigations). Preprint, NAN Ukrayini, Fizyko-mechanichnyi instytut, 137(1987), L'viv
6. Andreykiv AY, Lysak MV (1989) Metod akusticheskoy emissii v issledovanii prozessov razrusheniya (A method of acoustic emission in investigation of fracture processes). Naukova Dumka, Kiev
7. Andreykiv OY, Lysak MV, Skalsky VR (1996) A method of accelerated evaluation of K_{ISCC} under stress corrosion cracking. Eng Fract Mech 54(3):387–394
8. Andreykiv AY et al (1992) Metodika opredeleniya K_{Iscc} stali v srede vodoroda s pomosch'yu metoda akusticheskoy emissii (A method of determining K_{Iscc} values of steel in hydrogen using acoustic emission). Tekhnicheskaia diagnostika i nerazrushyuschii kontrol (Tech Diagn Nondestr Test) 1:18–26
9. Andreykiv OY, Serhiyenko OM, Skalskyi VR (2001) Aktual'ni pytannya vidboru i peredachi signaliv akustichnoyi emisiyi pid chas rostu vnutrishnich trischyn (Actual problems of selection and transmissions of signals of acoustic emission during growth of internal cracks). Vidbir i obrobka informaziyi (Sel Process Inf) 15(91):51–59
10. Brown W, Srawley J (1972) Ispytaniya vysokoprochnych metallicheskich materialov na vyazkost' razrusheniya pri ploskoy deformazii (Testing of high-strength metallic materials for plain strain fracture toughness). Mir, Moskva
11. Panasyuk VV, Andreykiv AY, Kovchyk SY (1977) Metody ozenki treschinostoykosti konstrukzionnych materialov (Methods of assessment of crack growth resistance of structural materials). Naukova Dumka, Kiev
12. (1990) Metody mechanicheskich ispytaniy metallov i svarnych soedineniy. Opredelenie charakteristik treschinostoykosti (vyazkosti razrusheniya) pri staticheskom nagruzhenii (Methods of mechanical testing of metals and welded joints. Determination of crack growth resistance (fracture toughness) characteristics under static loading). SEV, Moskva
13. Shvets TM et al (1986) Akustoprozrachnye klei dlya krepleniya datchikov AE (Acoustic transparent glues for mounting AE gauges). Diagnostika i prognozirovanije razrushenija svarnykh konstruktsii 1:93–97
14. Kalemanov VI, Yaremchenko MA (1986) Osobennosti krepleniya datchikov akusticheskoy emissii (Peculiarities of AE gauges fixing). Ibid 3:86–89
15. (1990) GOST 1497-84. Metally. Metody ispytaniy na rastyazhenie, vved. 01.11. 1990 g. (Standard 1497-84. Metals. Tensile test methods. Introduced 01.11. 1990)
16. Lysak VM, Skalskiy VR, Serhiyenko OM (1994) Doslidzhennya vplyvu chvylevodu na zminu parametriv signaliv akustichnoyi emisiyi (Investigation of the influence of wave guide on the change of AE signal parameters). Fizyko-chimichna mechanika materialiv (Physicochem Mech Mater) 2:64–70
17. Skalskiy VR (2001) Okremi metodologichni zasady rozroblennya prystroyiv dlya peredavannya akustichnoyi emisiyi (Some methodological bases of development for the design of devices for acoustic emission transmission). Mashynoznavstvo 7:49–52
18. Skalskiy VR et al (1998) Prystroyi i ustanovky dlya vyznachennya trischinostiykosti konstrukziynych materialiv metodom akustichnoyi emisiyi (Devices and equipment for determination of crack growth resistance characteristics of structural materials by the method of acoustic emission). Preprint, NAN Ukrayiny, Fizyko-mechanichnyi institut, 1(1998), L'viv
19. Ermison AL, Muravin HB, Shyn VV (1986) Akustiko-emissionnye pribory i signaly (Acoustic emission devices and signals). Defektoskopia 5:3–11
20. Skalskiy VR, Demchyna BG, Karpukhin II (2000) Ruynuvannya betoniv i akustychna emisiya (Oglyad). Povidomlennya 2. Koroziya zalizobetonu. Aparaturni zasoby. AE–kontrol' ta diagnostyka budivel'nych sporud (Concrete fracture and acoustic emission (A review). Report 2. Corrosion of the reinforced concrete. Equipment. AE examination and diagnostics of building constructions). Tekhnicheskaia Diagnostika i Nerazrushyuschii kontrol (Tech Diagn Nondestr Test) 2:9–27
21. Lysak MV, Skalskiy VR (1997) Metodychnyi pidchid dlya eksperimental'noi akustyko-emisiynoi ozinky trischynotryvkosti konstrukziynych materialiv (Methodical approach for experimental acoustic-emission assessment of crack growth resistance of

structural materials). Fizyko-chimichna mechanika materialiv (Physicochem Mech Mater) 5:17–30
22. Ivanov VI (1980) Metody i apparatura kontrolya s ispol'zovaniem akusticheskoy emissii (Methods and inspection devices which use acoustic emission). Mashynostroyenie, Moskva
23. Milius PB, Hrabets I (1984) Absolyutnaya kalibrovka p'ezopreobrazovateley, ispol'zuemych dlya registrazii signalov akusticheskoy emissii (Absolute calibration of piezo-transducers, used for acoustic emission signals recording). Defektoskopia 5:27–32
24. Stryzhalo VA et al (1990) Prochnost' I akusticheskaja emissija materialov i elementov konstruktsij (Strength and acoustic emission of materials and structural elements). Naukova Dumka, Kiev
25. Vakar KB (ed) (1980) Akusticheskaya emissiya i ee primenenie dlya nerazrushayuschego kontrolya v atomnoy energetike (Acoustic emission and its application for non-destructive testing in nuclear power engineering). Atomizdat, Moskva
26. Andreykiv AE, Skalskyi VR, Lysak NV (1992) Sposob kontrolia rosta treschin v obraztsakh materialov (A test method of crack growth in material specimens). USSR Inventor's Certificate 1758545 G 01 N29/04, Bulletin No. 32, 30 Aug 1992
27. Mykytyshyn SI, Hrytsyshyn PM (1981) Sposob mekhanicheskikh ispytanij na prochnost' (A method of mechanical testing for strength). USSR Inventor's Certificate 879373, G 01 N3/00, Bulletin. No. 41, 07 Nov 1981
28. Jones MH, Brown WF (1964) Acoustic detection of crack initiation in sharply notched specimen. Mater Res and Stand 4(3):120–129
29. Greshnikov VA, Drobot YuB (1976) Akusticheskaya emissiya (Acoustic emission). Izdatel'stvo standartov, Moskva
30. Smirnov VI (1979) Ob ozenke razmerov defektov metodom akusticheskoy emissiy s poziziy mechaniki razrusheniya (On the assessment of the defect size using the method of acoustic emission from the viewpoint of fracture mechanics). Defektoskopia 2:45–50
31. Filinenko SF, Gorodiskii NI, Biriukov VS (1985) Osobennosti signalov akusticheskoy emissii pri plasticheskom deformirovanii i chrupkom razrushenii materialov (Peculiarities of acoustic emission signals under plastic deformation and brittle fracture of materials). Fiziko-chimicheskaya mechanika materialov (Physicochem Mech Mater) 6:105–106
32. Andreykiv OY et al (1990) Metodicheskie aspekty primeneniya metoda akusticheskoy emissii pri opredelenii staticheskoy treschinostoykosti materialov (Methodical aspects of application of the acoustic emission method in evaluation of static crack growth resistance of materials). Preprint, NAN Ukrayini, Fizyko-mechanichnyi instytut, 165(1990), L'viv
33. Eitzen DG, Wadley HNG (1984) Acoustic emission: establishing the fundamentals. J Res Nat Bur Stand USA 89(1):75–100
34. Breckenridge FR, Tschiegg CE, Greenspan M (1975) Acoustic emission: some application of Lamb's problem. J Acoust Soc Am 57:626–631
35. Breckenridge FR (1982) Acoustic emission transducer calibration by means of the seismic surface pulse. J Acous Emiss 1:87–94
36. Hsu N (1976) Acoustic emission simulator. USA Patent 4018084 assigned to Lockheed Aircraft Corporation, Buibouk, CA, May 1976
37. Fowler KA (1971) Acoustic emission simulation test set. Mater Res Stand 11(3):35–38
38. (1971) Now hear this. A newsletter on acoustic emission, vol 1(3). Dunegan/Endevco
39. Jones M, Green R, Hsu N (1983) Comparison of simulated acoustic emission sources. Proc Ultrasonic Int
40. Williams JH, Kahn EB, Lee SS (1983) Effect of specimen resonances on acoustic-ultrasonic testing. Mater Eval 13:1502–1510
41. Katsuyama K, Sato K (1985) Influence of acoustic emission source location on acoustic emission waveform. Saiko to khoan 31(1):2–10
42. Andreykiv OY et al (1993) The contribution of Rayleigh waves to the acoustic field arising from internal defect growth. Mater Sci 29(2):115-120

43. Skalskyi VR, Zazuliak VA, Kovchyk SY, Rybytskyi IB (1994) Sposib kalibrovky akustyko-emisiynoho vymiruval'noho traktu (A calibration method of acoustic emission measuring channel). Patent of Ukraine No. 2895, G01 N 29/14, Bulletin. No.5-1 26 Dec 1994
44. Viktorov NA (1966) Fizicheskie osnovy primeneniya ul'trazvukovych voln Releya i Lemba v technike (Physical bases of ultrasonic Raleigh and Lamb waves application in engineering). Nauka, Moskva
45. Wadley HNG, Scrubi CB, Speake JU (1980) Acoustic emission for physical examination of metals. Int Met Rev 25(2):41–64
46. Skalskyi VR, Andreikiv AY (2003) Sposib imituvannia akustychnoji emisii na objekti kontrolu (A method of imitation of acoustic emission in the inspected object). Patent of Ukraine No. 54963A, G01N29/14, G01N3/00, Bulletin No. 3, 17 Mar 2003
47. (1981) Mechanika razrusheniya. Bystroe razrushenie, ostanovka treschin (Fracture Mechanics. Rapid fracture, crack arrest). In: Goldstain (ed) Mechanika. Novoe v zarubezhnoy nauke (Mechanics. Novelty in foreign science), is 25. Mir, Moskva, pp 199-221
48. Obuhivskyi OI Rasklinivanie kvadratnogo obrazza s bokovoy treschinoy (Wedging of a square specimen with an edge crack). Fiziko-chimicheskaya mechanika materialov (Physicochem Mech Mater) 3:111-112
49. Likhtman VI, Rebinder PA, Karpenko GV (1954) Vliyanie poverchnostno-aktivnoy sredy na prozessy deformazii metallov (Influence of surface-active medium on deformation processes of metals). Izd-wo AN SSSR, Moskva
50. Bukhalo OP (1997) Obrazzovyy impul'snyy stochasticheskiy signal–primenenie, vosproizvedenie, attestaziya (A standard pulse stochastic signal-application, reproduction, certification). Izmeritelnaya tekhnika (Meas Equip) 6:24–33
51. Andreykiv OY et al (1993) Spektral'nyy analiz akusticheskoy emissii rastuschey treschiny (Spectral analysis of acoustic emission of a growing crack). Tekhnicheskaja diagnostika i nerazrushajuschii kontrol (Tech Diagn Nondestr Test) 1:75–84
52. Greshnikov VA, Braginskyi AP (1980) Ob analize signalov akusticheskoy emissii (On the analysis of acoustic emission signals). Defektoskopia 5:101–106
53. Batuyev VT et al (1973) Spektral'nyy analiz EVH na ploschadke tekuchesti (Spectral analysis of stress wave emission at the yield area). In: Tezisy dokladov nauch.-techn. konf. "Herazrushayuschiy kontrol' kachestva", Hovosibirsk, 1973 (Theses reports conf. "Nondestructive testing of quality", Novosibirsk, 1973)
54. Muravin G, Lezvinskaya LM (1982) Issledovanie spektral'noy plotnosti signala AE (Investigation of spectral density of AE signal). Defektoskopia 7:10–15
55. Acquaviva CJ (1980) Interlaboratory comparison of crack emission spectra. NDT Int 5:230–234
56. Kline R, Hartman WF (1976) Frequency analysis of acoustic emission signals. In: International conference on mechanical behavior of materials, pp 1631–1635
57. Vainberg VE, Kats MS, Purych EI (1981) Vliyanie razmera obrazza na chastotnyy spektr akusticheskoy emissii (Influence of specimen size on the frequency spectrum of acoustic emission). Defektoskopia 4:110–112
58. Gorbunov AI, Lykov YuI (1986) Vliyanie amplitudno-chastotnych charakteristik ob'ekta na spektral'nye charakteristiki signalov AE (Influence of amplitude frequency characteristics of an object on spectrum characteristics of AE signals). Ibid 9:39–45
59. Muravin G, YaV Simkin, Merman AI (1989) Identifikaziya mechanizma razrusheniya materialov metodami spektral'nogo analiza signalov akusticheskoy emissii (Identification of the fracture mechanism by the method of spectral analysis of acoustic emission signals). Ibid 4:8–15
60. Fleischman P (1975) A spectrum analysis of acoustic emission. Non-Destr Test 8(5):241–244
61. Greham JZ, Alera G (1974) Spectrum analysis of acoustic emission. Mater Eval 2:31–37
62. Stephens RWB, Pollock AA (1971) Waveform and frequency spectra of acoustic emission. J Acoust Soc Amer 50(3):904–910
63. Lysak MV (1994) Acoustic emission during jumps in subcritical growth of crack in three-dimensional bodies. Eng Fract Mech 47(6):873–879

64. Kucherov IY et al (1985) Issledovanie chuvstvitel'nosti i diagramm napravlennosti priemnikov akusticheskoy emissii (Investigation of sensitivity and directional diagrams of the acoustic emission receivers). Diagnostika i prognozirovanije razrushenija svarnykh konstruktsii 1:53–58
65. Kishi T, Okita T (1984) Ozenka tipa razrusheniya po diagramme napravlennosti izlucheniya v signalach akusticheskoy emissii (Evaluation of the fracture type by the directional diagram of irradiation in the acoustic emission signals). Nihon kindziuku gakkaisi 48(9):911–917
66. Scruby CB, Wadley HNG, Rusbridge KL (1983) Origin of acoustic emission in Al–Zn–Mg alloys. Mater Sci Eng 59(2):169–183
67. Andreykiv AY, Lysak NV, Serhiyenko ON (1990, 1991) Modelirovanie prozessov lokal'nogo razrusheniya, soprovozhdayuschegosya akusticheskoy emissiey v materialach i izdeliyach (Modelling of the process of local fracture, assisted by acoustic emission in materials and products). Tekhnicheskaja diagnostika i nerazrushajuschii control (Tech Diagn Nondestr Test) 1:9–20, 1:59–65
68. Kishi T (1985) Acoustic emission source characterization and its application to microcracking. Z Metallk 76(7):512–515
69. Lysak NV, Skalskiy VR (1993) O napravlennosti akustiko-emissionnogo izlucheniya pri razrushenii materialov i ee prakticheskoe primenenie (On direction of acoustic emission irradiation during material fracture and its practical application). Tekhnicheskaja diagnostika I nerazrushajuschii control (Tech Diagn Nondestr Test) 3:22–32
70. Achenbach JD, Harris JC (1979) Acoustic emission from a brief crack propagation event. Trans ASME, J Appl Mech, Ser E 46(1):107–112, 187–221
71. Wadley HNG, Scruby CB (1983) Elastic wave radiation from cleavage crack extension. Int J Fract 23:111–128
72. YaU Saatov, Bykovtsev AS, Khamidov LA (1985) Seysmologicheskie zadachi mechaniki razrusheniya (Seismic problems of fracture mechanics). Mekhnat, Tashkent
73. Drobot YuB, Lazariev AM (1987) Herazrushayuschiy kontrol' ustalostnych treschin akustiko-emissionnym metodom (Nondestructive testing of fatigue cracks by an AE method). Izdatel'stvo standartov, Moskva
74. Andreykiv OYe et al (2000) Deyaki metodichni zasnovki ozinki poshkodzhenosti virobiv za signalami akustichnoyi emisiyi (Some methodical pre-conditions of estimation of damages of products by the signals of acoustic emission). Fiziko-chimichna mechanika materialiv (Physicochem Mech Mater) 2:83–92
75. Nesmashniy EV et al (1984) O svyazi amplitudy signala AE s prirascheniem ploschadi treschiny (On the relation of the amplitude of AE signals with increment of the crack area). In: Sbornik tezisov i dokladov I Vsesoyuznoy konferenzii “Akusticheskaya emissiya materialov i konstrukziy” (Theses of reports 1st all-union conference. “Acoustic emission of materials and structures”, 11–13 Sept 1984, Rostov upon-Don, 1984), vol 1. Rostov-na-Donu
76. Aleksieyev IG, Kudrya AV, Shtremel MA (1994) Parametry akusticheskoy emissii, nesuschie informaziyu ob edinichnoy chrupkoy treschine (Acoustic emission parameters, containing information about a single brittle crack). Defektoskopia 12:29–34
77. Harris JG, Pott J (1984) Surface motion excited by acoustic emission from a buried crack. Trans ASME J Appl Mech 51(1):77–83
78. Kolesnikov AE (1982) Ul'trazvukovye izmereniya (Supersonic measuring). Izdatel'stvo standartov, Moskva
79. Melnychenko ZM, Shvets TM, Podychevska SA, Kharchenko LF (1983) Kley (Glue). USSR Inventor's Certificate, G1014875, Bul. No. 16
80. Appasov AM (1998) Ustroystvo dlya priema signalov akusticheskoy emissii (Equipment for receiving acoustic emission signals). Zavodskaya Laboratoria 10:25–29
81. (1972) Acoustic emission. ASTM Special Technical Publication 505. Baltimor
82. Andreikiv OY, Skalskyi VR, Lysak MV (1994) Sposib kontrolu rostu trischyn u zrazkakh materialiv (A method of checking the growth of cracks in the material specimens). Patent of Ukraine N2914, MPK: G01N29/14, Bul. 5-1, 26 Dec 1994

83. Sharp R (ed) Metody nerazrushayuschich ispytaniy. Fizicheskie osnovy, prakticheskie primeneniya, perspektivy razvitiya (Methods of non-destructive testing. Physical bases, practical application, prospects of development). Mir, Moskva
84. Kiriakin AV, Zheleznaja IL (1984) Akusticheskaya diagnostika uzlov i blokov REA (Acoustic diagnostics of units and blocks of electronics). Radio i sviaz, Moskva
85. Kollacott R (1989) Diagnostika povrezhdeniy (Damage diagnostics) (trans Babajevskyi PG). Mir, Moskva
86. Kisi T (1976) Acoustic emission. Kindzoku Dzaire 165:107–113
87. Birchon D (1979) Cries of stress. Spectrum 165:5–8
88. Shyn VV, Dementiev AN (1987) Metodicheskie osnovy akustiko-emissionnogo kontrolya svarnych soedineniy gazoprovodov (Methodical bases of acoustic-emission control of welded joints in gas pipelines). Diagnostika i prognozirovanije razrushenija svarnykh konstruktsii 5:46–52
89. Vehaviolos SJ (1975) SWE: a tool for non-destructive testing. Tooling Prod 41(3):52–53
90. Stern R (1972) Experimental techniques in acoustic emission detection systems. In: Ultrasonics symposium, Boston, 1972, Catalogue 372 CHO 708-8SU. IEEE
91. Arrington M (1981) An introduction to the technology and applications of acoustic emission. Tin cries-but most materials talk. Phys Technol 12:16–23
92. Dunegan HL, Tatro CA (1971) Acoustic emission effect during mechanical deformation. In: Bunshan RF (ed) Techniques of metals research, vol 5, part 2 (Chap. 12). Interscience Publications, New York, pp 273–312
93. Nakamura I (1971) Acoustic emission monitoring system for detection of cracks in a complex structure. Mater Eval 29(1):8–12
94. Ruby D (1980) Emission acoustique et conrole nondesructif. CET HEDEC Revue 17 (2):13-19, 205–231
95. (1995) MISTRAS 2001. AEDSP—32/16. User's manual. Rev. 1. PAC Part Number 6300-1000
96. Takahashi H et al (1981) Acoustic-emission crack monitoring in fracture-toughness tests for AISI 4340 and SA533B steels. Exp Mech 21(3):89–99

Chapter 5
Evaluation of Mechanical Characteristics and Static Crack Growth Resistance of Materials with the Use of Aes

To diagnose the products, structures or state of their separate elements, it is important to estimate the strength characteristics of materials and the parameters of their static crack growth resistance. Among the latter, as mentioned above, there is a force criterion, SIF, that characterizes both the start of the fracture and the sub-critical stages of its development.

The results of the experimental research and the analysis of the published data prove that by using the AE methods, it is possible to determine the parameters of static crack growth resistance of structural materials more effectively. Having such preliminary experimental parameters of material fracture toughness, it is possible to use the investigation results for the creation of the methods of AE testing and diagnostics of products and structural elements.

Since the most widespread and dangerous defects—from the viewpoint of fracture mechanics—are the crack-like defects, the problem has been set forth to elaborate the methods for AES evaluation of the initial stages of their development. This leads to the task of clearly determining the moment of a macro-crack start and distinguishing the AE signals generated during the development of plastic deformations in the vicinity of the existing macro-crack. These methods are discussed in this chapter.

5.1 Identification of the AES Generated During Plastic Zone Growth

It is known [1, 2] that PD and MC growth differs in the amplitude distribution of AES. In these papers we see that PD of the material, although occurring in large areas by relatively small single deformation increments, is accompanied by long AES of insignificant amplitude, while during a jump-like propagation of a crack with approximately the same consumption as for PD energy, the AE is generated in

Z. Nazarchuk et al., *Acoustic Emission*, Foundations of Engineering Mechanics,
DOI 10.1007/978-3-319-49350-3_5

the form of discrete burst pulses of large amplitudes. Such AES identification is effective in the case when there are only two mechanisms of the AE generation: PD and macro-crack growth. However, in real structural materials, additional AE sources appear, e.g., fracture of brittle inclusions or formation of micro-cracks corresponding to the linear section of the *P*-*v* diagram long before the point of its deviation from linearity [3, 4]. These processes are accompanied by the AES, whose amplitudes are higher than for those caused by PD, but smaller than during macro-crack propagation (sometimes commensurable with them). In such cases, it is quite complicated to interpret the results for evaluation of the moment of the crack start by the AES. However, in each case, in order to solve the problem of the AES identification, it is necessary to find quantitative relationships that permit distinguishing the processes of macro-crack growth among the other AE sources. This factor is taken into consideration in the method developed.

Specific features of the AES generation during the PD formation in a material were evaluated using the following method: At first, several specimens of each material with the preliminarily induced fatigue cracks were tested, up to a complete failure, recording the "*P*-*v*" and AE diagrams simultaneously. Having the value of σ_{02} of the material and determining the value of K_C (maximum value of SIF) from the experiment in a first approximation, the plastic zone (PZ) r_p^1 radius under condition $K_I = K_C$ for each material was analytically determined by Formula (5.1) [5]:

$$r_p = K_I^2/4\pi\sqrt{2}\sigma_{0.2}^2, \tag{5.1}$$

where $\sigma_{0,2}$ is conventional yield strength of the material. Thus, the maximum value of the PZ radius r_p^1 at the macro-crack tip was obtained. Now, in order to identify the AES as maximally as possible from the processes of PZ formation (initiation and propagation of pores, micro-cracks, fracture of inclusions, etc.), we have to form the PZ approximately equal to this maximum value, but only provided that there is no macro-crack sub-critical growth. For this purpose, the following calculations were made:

The approximate solution for the 3-D elastic problem for a body with a thin inclusion was found in study [6]. According to the results obtained, the distribution of stresses σ_{yy} at the tip of a notch of radius ρ can be written as

$$\sigma_{yy} = 2K_I(x+\rho)[\sqrt{\pi}(2x+\rho)^{3/2}]^{-1}, \tag{5.2}$$

where x is the distance from the notch tip.

Taking into account that yielding begins when $\sigma_{yy} = \sqrt{3}\sigma_{0.2}$ [7], we get

$$\sqrt{3}\sigma_{0,2} = 2K_I(x^* + \rho)[\sqrt{\pi}(2x^* + \rho)^{3/2}]^{-1}, \tag{5.3}$$

where x^* is the point of a body where yielding occurred, and which is located at a maximal distance from the concentrator. When the value of ρ is known, putting $x^* = r_p^1$, we evaluate the factor K_I on which the specimen with a notch of a known radius, approximately the same PZ is formed as for the specimen with macro-crack of similar thickness. Then from Eq. 5.3, we get

$$K_{\mathrm{I}} = \sqrt{3}\pi\sigma_{0.2}(2x^* + \rho)^{3/2}[2(x^* + \rho)]^{-1}. \tag{5.4}$$

Hence, using well-known formulas [8] to evaluate the static crack growth resistance of materials, we can write (for instance, in the case of prismatic specimen three-point bend testing) the dependence for evaluating the load P applied to the specimen with a notch of radius ρ in order to form the PZ of the calculated dimensions at its tip, assuming in calculations a notch to be a crack

$$P = K_{\mathrm{I}}B\sqrt{b^3}L^{-1}Y_4^{-1}, \tag{5.5}$$

where L is the distance between the supports (three-point bending), B is the thickness and b is the specimen height, Y_4 is the correction factor $Y_4 = f(l/b)$, and l is the notch depth. For a compact tensile specimen, formula (5.3) takes the form

$$P = K_{\mathrm{I}}B\sqrt{b}Y_3^{-1}, \tag{5.6}$$

where Y_3 is a similar correction factor for a compact specimen.

Having such previous data, the AES was evaluated during the PZ formation on a notched specimen. The experiments were conducted on prismatic and compact specimens containing an induced notch with a radius tip of $\rho = 0.8$ mm. Specimens were made of 38XN3MFA and 45 steels, and of aluminum D16-T and 1201-T alloys (mechanical characteristics are given in Table 5.1).

The test setup is described in [2] (Fig. 5.1). The AE was recorded by a resonance AET with a frequency of f_0= 260 kHz, mounted on the polished lateral surface of the specimen. Signals from the AET passed to a preliminary wide-band amplifier with an amplification factor of 34 dB, and then through an AE device of AVN-3 type, which contained a block of filters with the pass band of 160…350 kHz and transmission coefficient equal to 1, the unit of basic measuring amplifier (with adjusting amplification factor with a step of 1 or 10 dB) and a unit of the AES processing. Then the signals passed to a fast-acting recorder, where the loading force P was additionally recorded in real time. The amplification factor of the AE channel was 84 dB, and the level of discrimination, i.e., the threshold level of the AVN-3 device, was 65 dB. The graph plotter simultaneously recorded the fracture

Table 5.1 Mechanical characteristics of alloys

Parameter		Material		
	D16-T	1201-T	38XN3MFA	Steel 45
$\sigma_{0.2}$, MPa	138.4	330.7	1110.0	599.7
σ_b, MPa	208.96	443.6	1570.2	739.9
δ, %	9.37	4.75	12.2	1825
Ψ, %	30.56	7.54	56	65.3

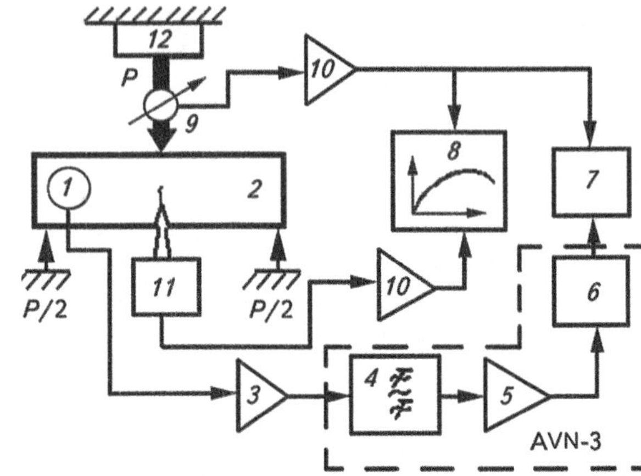

Fig. 5.1 Experimental setup: *1* is AET; *2* is a specimen; *3* is a preliminary amplifier; *4* is a block of filters; *5* is an amplifier; *6* are processing units; *7* is a recorder; *8* is a graph plotter; *9* is a strain-gauge dynamometer; *10* is a direct current amplifier; *11* is a strain gauge of displacement; and *12* is a loading device

diagram "*P*-*v*", and the record scale was 400:1. Loading force *P* was transmitted for recoding from a strain gauge dynamometer through a direct current amplifier of the U7-1 type, while the value of the macro-crack opening was measured by the strain gauge and processed similarly. The loading was applied using the SVR-5 equipment [9].

Several specimens were chosen for the recording of a fracture AE diagram from the start of loading to complete the failure of the specimen in order to evaluate the K_C. Then, using the known values of $\sigma_{0.2}$ and K_C for this material, the maximum size of the plastic zone at the crack tip, when $K_I = K_C$, was estimated. Afterwards, following the methods proposed above, the value of the force required for the formation of approximately the same PZ at the notch tip and the crack in the specimens specially prepared for this purpose was determined. The experimental results are presented in Table 5.2 and in Fig. 5.2.

It is necessary to point out that for maximal elimination of the effect of noise and casual AES, which can be generated by the supports and an indenter during testing, specimens with the notch of radius ρ = 0.8 mm were shunted first. Later on, they were loaded to the value exceeding this load level established for the test specimen or calculated preliminarily by 15…20% [10, 11]. Thereafter, taking off a shunt, the specimens were tested with a loading rate in accord with the requirements [8], simultaneously recording the "*P*-*v*" diagram and the AES (cumulative count *N*, amplitude *A*) parameters up to the necessary value, and at the moment of attaining

Table 5.2 Investigation results

Material	*B*, mm	*l*/*b*	K_I, MPa$\sqrt{m}$	*x**, mm	*P*, kN	Specimen type
D16-T	10	0.49	43.36	5.494	5.57	Prismatic
1201	20	0.49	38.59	0.856	14.95	Prismatic
38XN3MFA	10	0.52	137.98	1.969	19.1	Prismatic
Steel45	20	0.48	44.5	0.3	17.8	Compact

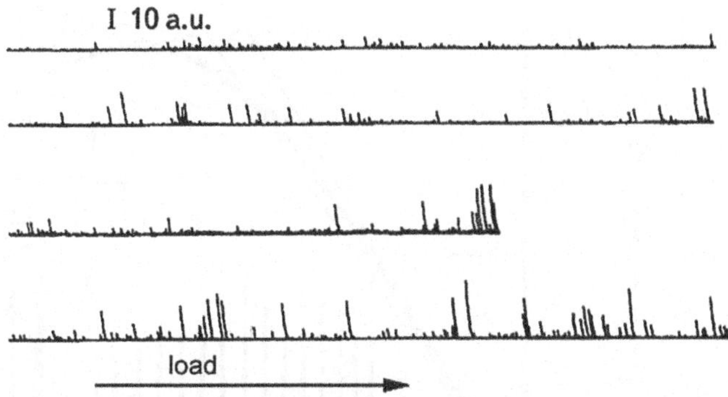

Fig. 5.2 Typical character of AES amplitude change under specimen loading up to the calculated value of *P* (*top-down*): D16-T; 1201-T; steel45; 38XN3MFA

the value, the specimen was unloaded. All the tests were conducted by our method, using the parallel AE channel [12].

As the experiments show, each of the alloys possesses its own features of elastic AE wave radiation during initiation and propagation of micro-cracks and macro-defects in the PDV. From this point of view, 38XN3MFA steel is the most active, where the AES of large amplitudes are generated almost from the start of the loading. During slight (i.e., fractions of a percent) deformations, single AES appear with amplitudes exceeding 6 relative units (a.u.). With the deformation increase, the amplitudes of discrete AES also tend to increase. Figure 5.2 shows the amplitudes of the AE event envelope curve that were detected in the time interval of 1 ms.

Low-amplitude AES are inherent to the D16-T alloys and steel 45. When the applied load approaches the calculated value of *P*, all the alloys tend to increase the AES amplitudes. This proves (as will be shown in this chapter) the jump-like character of the macro-crack initiation and propagation of the material tested.

Thus, we have preliminarily set the reference level of the AES amplitudes that accompanied the formation of the PDV and the features of their qualitative distribution. For 38XN3MFA and 45 steels, the maximal amplitudes of AES were 30 and 25 a.u., respectively, and for the aluminum 1201-T and D16-T alloys, they were 18 and 6 a.u., respectively.

5.2 A Method for Evaluating a Macro-Crack Start

It was stated in [13] that the AE radiation originates earlier than macro-crack propagation and is not directly related to this process. This means that the AE excitation depends on PD at the macro-crack tip and is proportional to the volume of PZ under the plane strain state, where the tri-axial stress state is formed and AE is generated just at the moment of the micro-crack initiation therein, when there is a

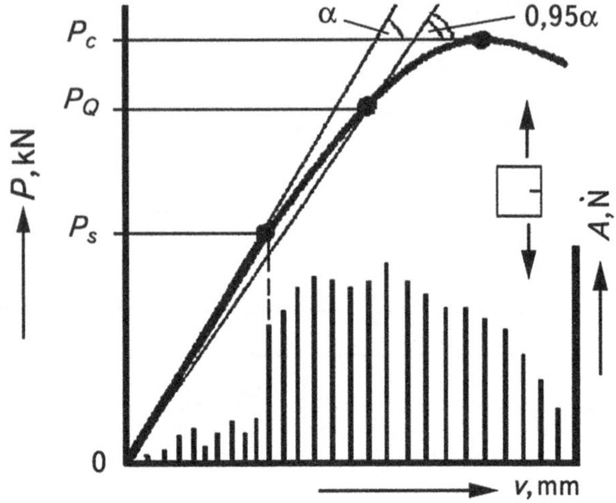

Fig. 5.3 A method for schematically evaluating a macro-crack start

cracking or delamination of the second-phase particles from the matrix surface within this PZ. These statements are based on the previously published test results [14, 15] where it is proposed to assume that the beginning of a macro-crack growth corresponds to the moment of a qualitative change in the AES count rate in the AE diagram (AG), which is recorded together with the "applied force *P*—crack opening *v*" diagram (Fig. 5.3). This value of the stress intensity factor K_{IS}, which we have proposed to take as a macro-crack start evaluated by P_S, instead of P_Q, as standard [7] requires, has lower values compared with K_{IC}. Moreover, the proposed value of K_{IS} is invariant with respect to the specimen thickness, loading mode, and testing temperature [16–18]. Thus, evaluation of the moment of macro-crack start by the AE signals, namely by the K_{IS} value, is more reliable than using the K_{IC} value.

Thus, the idea proposed for the first time in [19], which says that by using the qualitative parameters of AES it is possible to determine the macro-crack start, was further developed in [14, 15, 17, 18]. However, interpreting this idea in these studies, the authors used only the qualitative AES distribution on the AG based on a drastic increase of the AES amplitudes at that moment. Therefore, it is necessary to investigate the validity of this statement, i.e., to establish experimentally whether this fact really corresponds to the moment of the macro-crack start. For this purpose, it is necessary to have a preliminarily found maximum level of the AES and their time distribution during the formation of the plastic deformed volume at the stress concentrator. Having established these values, the tests were performed in the following way.

Prismatic, compact, and cylindrical specimens were used in the experiments (Fig. 5.4), which were loaded by three-point bending and tension, respectively, keeping to the requirements of [8]. Thus, a specimen with an induced fatigue crack was smoothly loaded, and at the same time a fracture diagram "*P*-*v*" and AG were recorded. At the moment when the AES with amplitudes higher than the reference level appeared, the loading stopped (the value of the corresponding applied load is

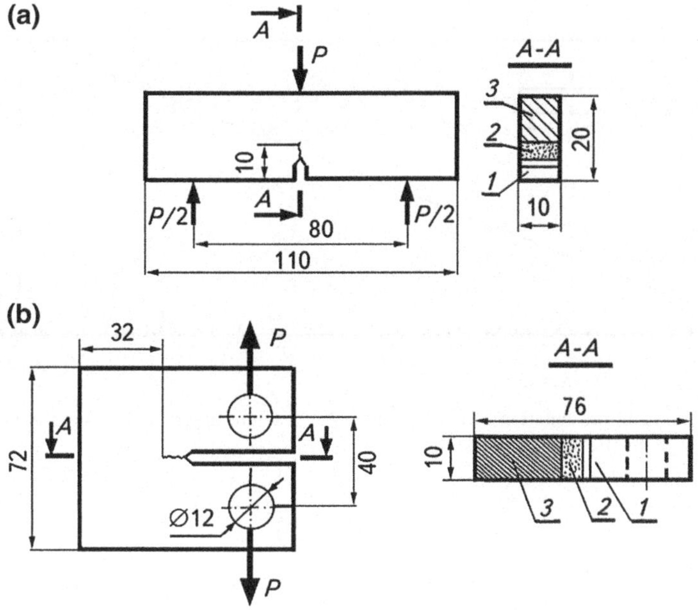

Fig. 5.4 Prismatic (**a**) and compact (**b**) testing specimens: *1* is notch, *2* is fatigue crack, *3* is specimen body

denoted by P_{AE}) and the specimen was unloaded (Fig. 5.5). In another case, the applied load slightly exceeded the value of P_{AE}. The specimen was removed from the testing machine and new surfaces were thermally tinted; then the specimen was placed in liquid nitrogen and ruptured by impact loading.

The experimental setup (Fig. 5.1) and modes of selection, processing, and AES recording were the same as for the testing of specimens with a concentrator.

Fracture surfaces, especially the juvenile tinted surfaces, were investigated, and MIM-12 or laboratory MBS-9 microscopes were used for this purpose. The length of the fatigue crack was measured using the methods described in [8]. Then, the value of K_{IS}, i.e., the SIF, was calculated for the P_{AE} point. The results of the research are shown in Fig. 5.6.

As our experiments have shown, for an applied load of P_{AE}, the thermal-paned new surfaces can be revealed by a special method that consists of the following: It is necessary to carefully examine the fracture surface with a microscope of a great magnification, illuminating the subject table both directly and at various angles. In some cases (for instance, for aluminum alloys) thermal painting must be performed in a certain environment, choosing the optimum temperature of a thermal chamber. Having combined these factors for different alloys, we succeeded in measuring the macro-crack increment for the applied loads higher than P_{AE} and proved its absence or insignificance at the point of P_{AE}. We were most successful with structural steels, especially 38XN3MFA steel.

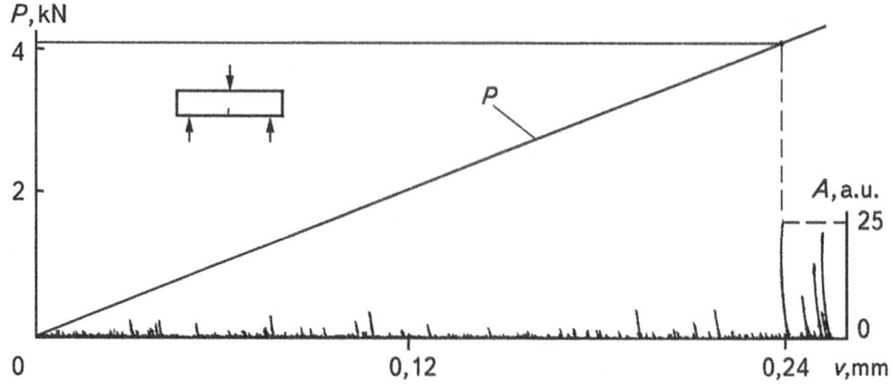

Fig. 5.5 Schematic presentation of characteristic *P*-*v* fracture diagrams and synchronous AG for 1201-T alloy prismatic specimen testing

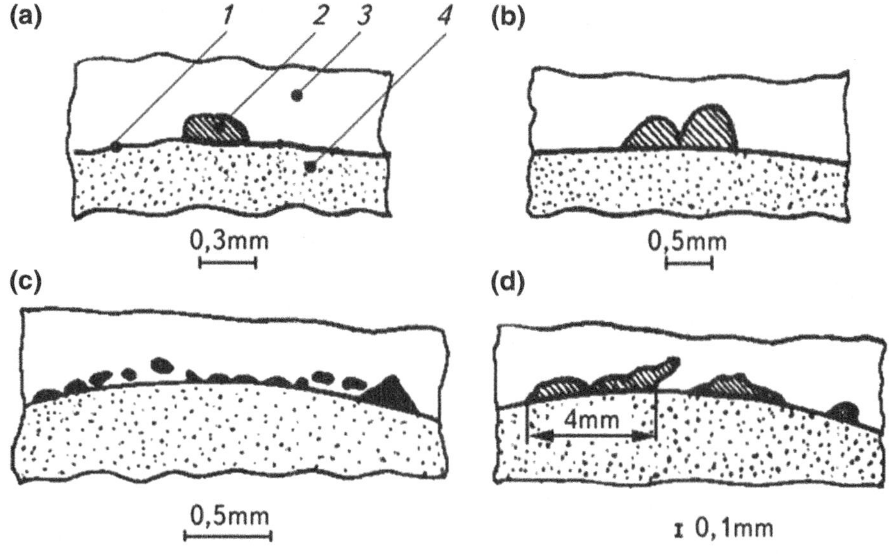

Fig. 5.6 New surfaces on the fracture surface of prismatic 9 HF (**a**, **b**) and 38XN3MFA (**c**, **d**) steel specimens for the K_I values: **a** is 1.01…1.03 K_{IS}; **b** is 1.04…1.08 K_{IS}; **c** is 1.1…1.13 K_{IS}

Thus, the analysis of the fracture surfaces of specimens confirmed a hypothesis that the method of a macro-crack start evaluation by the qualitative change in AG, namely by the moment of a drastic growth of the AES amplitudes, makes it possible to determine the actual moment of the macro-crack start. In this way, the criterion, postulated in [14, 15, 17–19], was verified.

Applying linear fracture mechanics to the estimation of the defects' sizes or the dynamics of their initiation and propagation in the material of structures is impossible without establishing the relationship between the K_{IS} parameter and the cracked specimen shape and sizes. If the K_{IS} invariance with respect to these factors is held, then we have a necessary and sufficient condition for its applicability in diagnosing the products and structures by the AE method.

In the experiments, we used structural low- and medium-alloyed steels, and aluminum alloys. The research was done this way because this class of materials had been widely studied according to the literature, and it was, therefore, important to compare the results that we obtained with data from the literature. Thus, K_{IS} was evaluated by our method, as described above. Prismatic, compact, and cylindrical specimens made of 45 and 38XN3MFA, aluminum alloys of grades D16-T and 1201-T of various thicknesses with preliminarily induced fatigue cracks in compliance with the standard requirements [8] were used. In Fig. 5.7 and in Table 5.3, the dependences of K_{IS} on the specimen thickness B for these materials are shown and test results are presented.

The data from the experiment show the invariance of the fracture toughness parameter K_{IS} with respect to the specimen thickness, its form and loading mode for the same material. It correlates well with the research results [17], where a similar invariance is shown for the 12XN4D2MFL and 20XN2D2MFL cast shell alloys (compact tensile specimens from 80 to 150 mm thick and prismatic three-point bending specimens from 25 to 60 mm thick for the first alloy, and 25 mm thick for the second), medium alloyed 15XN5MF (prismatic three-point bending specimens from 40 to 100 mm thick), and 38XN3MFA (cylindrical tensile specimen with a 20–95 mm diameter) structural steel specimens.

There are numerous reports in the literature indicating that the AES can still be observed before the beginning of the macroscopic fracture; however, there are no unified data on the applied load level and material type when AES arise. Therefore, during fracture toughness testing of specimens it was necessary to find the correlation P_{AE}/P_f (P_f is final force during macro-crack inducing) and thus to establish the moment of the AES initiation for our materials and to analyze the effect of the KE on this factor [20].

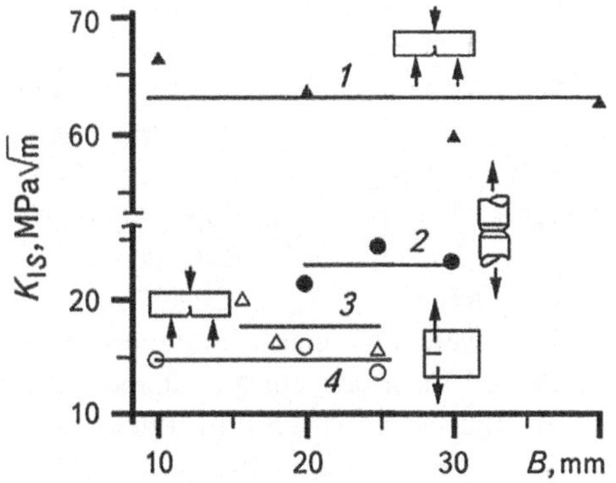

Fig. 5.7 Dependence of K_{IS} on the specimen thickness B for different materials and loading modes: *1* is 38XN3MFA steel; *2* is steel 45; *3* is 1201-T; *4* is D16-T

Table 5.3 Experimental data on the estimation of the crack growth resistance characteristics

Material	*B*, mm	*l/b*	K_{IS}, MPa$\sqrt{m}$	P_{AE}/P_f
	Ø20	–	28.49	1.28
Steel45 cylindrical	Ø25	–	30.55	1.17
	Ø30	–	24.7	1.14
Steel 45 prismatic	10	0.51	30.95	1.25
	20	0.50	31.35	1.18
	10	0.51	27.3	1.16
Steel 45 compact	15	0.46	26.4	1.12
	20	0.48	26.8	0.98
	10	0.52	69.1	0.98
38XN3MFA	20	0.51	63.4	0.88
Prismatic	30	0.53	72.36	0.92
	40	0.48	61.88	0.90
	8	0.46	19.94	1.19
1201-T prismatic	16	0.48	22.29	1.14
	25	0.45	19.2	1.01
	8	0.51	20.3	1.04
1201-T compact	12	0.50	22.4	0.93
	20	0.52	21.2	0.87
	10	0.52	19.32	1.44
D16-T prismatic	20	0.48	19.75	1.42
	25	0.48	18.1	1.21
	10	0.46	17.1	1.18
D16-T	20	0.45	17.9	1.34
	25	0.48	18.4	1.36

It follows from Table 5.3 that for the specified materials, the P_{AE}/P_f value is in the 0.88–0.98 range for 38XN3MFA steel; from 0.98 to 1.28 for 45 steel; from 1.18 to 1.44 for D16-T; and from 0.7 to 1.19 for 1201-T. That is, the generalized range of value scattering is from 0.87 to 1.44, which is given by the dispersion of the P_{AE} value in the vicinity of point P_f with an average deviation of ±28%, i.e., within the experimental error data scattering range. It is clear from Fig. 5.8 that with the growth of strength characteristics of the material, the P_{AE}/P_f has a tendency to decrease. This proves that the AE elastic waves initiate propagation prior to the moment of macroscopic macro-crack, i.e., AE is related directly to PD growth at the macro-crack tip.

The results obtained from the experiment concur with the results of [13], where, as already mentioned in this chapter, within the range of $\sigma_{0.2}$ values from 275 to 1550 MPa for the 5–25 mm. thick specimens, cut from a rolled metal at various angles to the rolling direction, the P_{AE}/P_f scattering values were within the limits of 0.8–2.0. It is possible to say that we succeeded in combining the conditions of

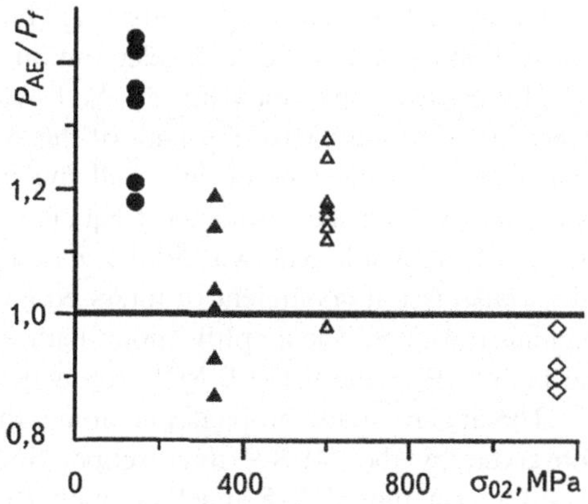

Fig. 5.8 Scattering of the P_{AE}/P_f values for materials with different yield strength. *Filled triangle* is D16-T; *filled circle* is 1201-T; *empty triangle* is steel45; *empty diamond* is 38XN3MFA

inducing the fatigue macro-crack and in performing the AE tests of fracture by the methods discussed above.

Thus, during the macro-crack propagation, a PZ develops at its tip accompanied by the AE waves reflecting the dynamics of these processes. Therefore, it is possible to experimentally estimate the moment of sub-critical macro-crack start and growth.

5.3 AE Estimation of the Stages of Sub-Critical Crack Propagation

5.3.1 Types of Specimens and Modes of AE Signals Selection

38XN3MFA structural steel. This steel is widely used in engineering and, in particular, in manufacturing the parts of the equipment for heat and nuclear power plants. Prismatic specimens sized 10 × 20 × 100 mm with an edge fatigue crack were tested. Chemical composition of the steel (in %) is as follows: 0.33…0.4 C; 0.025 S; 0.1…0.18 V; 0.17…0.37 Si; 0.25…0.5 Cu; 0.35…0.45 Mo 3.0…3.5 Ti; 1.2…1.5 Cr. Standard mechanical characteristics of the steel after quenching at 860 °C (oil) are shown in Table 5.4.

Two types of specimens were tested. They differed in the temperature of annealing—650 and 620 °C, respectively. Specimens with a higher annealing

Table 5.4 Some mechanical characteristics of 38XN3MFA steel

Annealing, °C	$\sigma_{0.2}$, MPa	σ_y, MPa	δ, %	ψ, %
670	931	1049	17	60
620	1118	1196	14	52
580	1120	1223	14	56

temperature were subjected to nitriding [21]. As a result, a brittle, near-surface layer formed on them. For the AES selection on the specimen lateral surface at a distance of 20 mm from the crack tip, an AET was mounted whose electric signals were transmitted to the measuring path of the AVN-3 device (Fig. 5.1). This transducer was located in the same place for all specimens. The characteristics of the AE path were as follows: the resonance frequency band was 0.2…0.5 MHz, amplification factor of the whole path was 50 dB, pass band of filters was 0.16…0.35 MHz with the transmission coefficient of filters equal to 1, and the threshold level of the AE signals of 0.2 V. The amplification factor of the preliminary amplifier was recorded to be 34 dB in the 0.1–1.0 MHz frequency range.

The macro-crack propagation along the lateral surface of the specimen was observed by the MBS-9 microscope. Specimens were loaded under three-point bending on the SVR-5 machine with the synchronous record of the "load P–bending f" diagram and AG.

The roll-foundry 9XF steel. Prismatic specimens of the cross-section of 25 × 50 mm with an edge crack were manufactured. They were loaded by the three-point bending. The value of K_{Ic} for this steel can be calculated using formula [22] with an accuracy of 10%

$$K_{\mathrm{I}c} = 60 - 21(r/R), \tag{5.7}$$

where r is the distance from the roll center, and R is the radius of the rolling mill roll.

The AVN-3 device with the following measuring modes was used in AE research: resonance frequency band of the TAE was 0.2…0.5 MHz, the total gain of a measuring path was 74 dB, and the working frequency band was 160…350 kHz. The results of the research are shown in Fig. 5.10a.

Gray, malleable and high-strength cast irons were used after special melting (Table 5.5) [23]. Malleable cast iron was obtained by a double process: melting in a cupola, and then in the arc furnace. Gray and high-strength (modified by ligature) cast irons were melted in a cupola furnace only.

Prismatic specimens of the cross-section of 12 × 18 mm (7…9 specimens from each cast iron grade) containing an edge fatigue crack were subjected to three-point bending. The specimens were loaded in the UME-10TM testing machine with the grip speed motion of 0.5 mm/min. At the same time, the "loading P—specimen bending f" and the "loading P—crack opening δ" diagrams, as well as AG, were recorded. The experimental setup is shown in Fig. 5.1.

The AE parameters were simultaneously recorded by AVN-3 and ARUP-3 devices by two methods. According to the first method, two AET were mounted on the specimen symmetrically with respect to the fatigue crack. Their outputs were connected to the AVN-3 device (the amplification factors of the measuring path with the preamplifier were 74 and 84 dB, the working frequency band was 120…350 kHz, the transducer resonance frequency was 250 kHz) and the ARUP-3 device (the amplification factor of the path was 60 dB). The ARUP-3 device was alternately connected to the transducers with resonance frequencies of 150, 230,

Table 5.5 Heat treatment conditions and some mechanical characteristics of cast irons

Cast-iron	Type of heat treatment	Structure of matrix	$\sigma_{0.2}$	σ_b	Hardness, HB
Gray	Cast	50% ferrite, 50% pearlite	–	182	187
	Annealing 900 °C, cooling rate 20°/h	Ferrite	–	151	131
Malleable	Annealing 900 °C	Ferrite	204	352	121
	Annealing 900 °C, normalization from 920 °C	Lamellar pearlite	294	604	–
	Annealing 900 °C, normalization from 920 °C, annealing 680 °C, 3 h	Granular pearlite	297	510	179
	Annealing 900 °C, cooling rate 100°/h	Pearlite, 20% ferrite	246	452	183
High-strength	Annealing 900 °C, 3 h cooling in a furnace	Ferrite	290	456	179
	Annealing 950 °C, 3-h normalization	Lamellar pearlite	625	804	269
	Annealing 950 °C, 3-h normalization, annealing 680 °C, 6 h	Granular pearlite	502	681	241
	Annealing 950 °C, 3-h cooling rate 100^0/h	Pearlite, 20% ferrite	482	604	229

450, and 800 kHz. For every transducer, the proper working band of the ARUP-3 device was set: 100…300 kHz, 0.3…1.0 MHz, and 1.0….2.0 MHz. According to the second method of recording, the preamplifier of the AVN-3 device (the amplification factor—84 dB, working frequency band—120…350 kHz) and ARUP-3 device (the amplification factor—60 dB, working frequency band—100…300 kHz) were connected to one AE transducer with a resonance frequency of 250 kHz. The recording by these methods was done alternately on the specimens made of the same material. The parameters of AE signals from the AVN-3 device (the amplitude, the power of signals, the cumulative count, and the AE count rate) and ARUP-3 device (the AE count rate) were transmitted to the fast-acting plotter of the N-338/6P type. An oscilloscope was used to observe the level of background noise and emission signal passing. The results of the experiment are summarized in Fig. 5.10b.

Titanium Ti3Al alloy. Plate specimens with the cross-section of 10 × 60 mm containing one- or two- edge cracks were manufactured [24]. Specimens were loaded by the URS-50/50 testing machine. Simultaneously with the "*P*–*δ*" diagram, the AE diagrams were also recorded. The operating parameters of the AVN-3 device were as follows: total gain was 64 dB, frequency was 120…360 kHz, threshold level—0.4 V. The results of the experiment are shown in Fig. 5.10c.

Table 5.6 Composition of basalt ash and slag concrete B-15

Cement, kg	Sand, kg	Macadam, kg	Ash and slag, kg	Basalt fibre, kg	Water, kg	SBG, %	SK, cm
250	350	1050	300	150	175	0.29	1…4

SK shrinkage of cone; *SBG* sulphide-barm grain

Basalt, ash, and slag concrete. Prismatic specimens of a cross-section of 100 × 100 mm were used (Fig. 5.10d and Table 5.6 [25]).

Specimens contained 21 mm deep notches with the angle of the notch tip at 60° and a radius that did not exceed 0.1 mm. They were subjected to three-point bending in the test machine FP-100/1 (Fig. 5.1). The AET with a resonance frequency of 250 kHz received the AES. The amplification factor of the AE path was 84 dB in the frequency band of 120…360 kHz, and the device threshold level was 0.4 V.

5.3.2 Interpretation of Investigation Results

The AG of 38XN3MFA steel can be divided by the AE activity into three regions (Fig. 5.9): The first AE signals of large amplitude appeared in an elastic area at early stages of loading the specimens. This was assumed to be the beginning of the first region. A further growth of cumulative count during specimen-bending occurred more or less linearly. The appearance of the second region in the AE diagram is characterized by a jump in the N–f dependence. The increase in N is similar to that in the previous region. An abrupt increase in N, growing up to the maximum load, occurs in the third region. Note that at lower amplification factors of AE devices, the qualitative change of the count rate was the same [26].

To compare the AE signals with the processes of local failure of 38XN3MFA steel, the following research was conducted: Specimens were being loaded until Regions I, II, and III appeared in AG. Then they were unloaded and fractured by impact loading. After each loading-unloading cycle, a trace was formed on the specimen fracture surface that corresponded to the crack front location during unloading. These experiments showed that there is no crack growth in Region I. The AES in this region are obviously caused by the fracture of carbides [23]. The beginning of Region II coincides with the crack start, and its propagation is slow in this area.

For 9XF steel, Region I is less distinctive. An abrupt growth of the amplitudes of the AE signals and their cumulative count takes place at the moment that the crack starts (Fig. 5.10). The unloading of specimens, their thermal tinting and final failure, as described above, confirmed the start of the crack after an abrupt growth of these parameters.

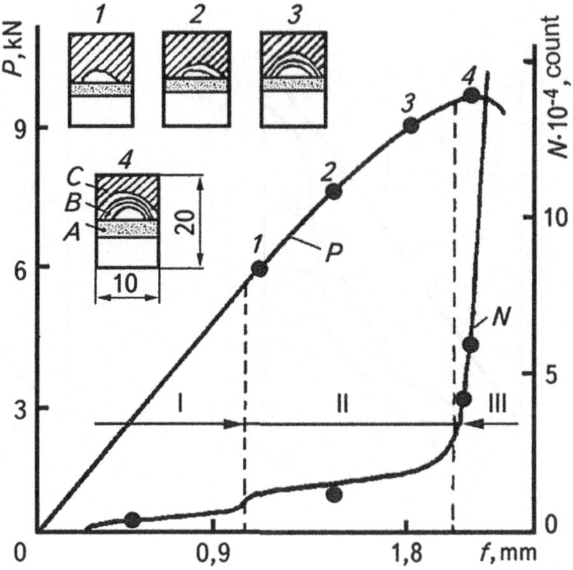

Fig. 5.9 The "load *P*—bending *f*" diagram and a change in the cumulative count of AES *N* under static loading of 38XN3MFA steel specimens: *A* is a pre-induced fatigue crack area; *B* is a sub-critical crack growth area; *C* is a fracture area

For every type of cast iron, depending on the matrix composition and graphite shape, the parameters of AE signals were different. However, for all of them in various conditions of AE measurements, three regions of the AE activity were also distinguished, although the nature of their origin is somewhat different from that of steels. The first of them appeared immediately after load application, which means that AE was not caused by the development of defects in the material but by grinding the indenter and supports to the specimen. Therefore, the signals in this region were not taken into account. The second region with the parameters of signals similar to the first one appeared soon after the start of loading (Fig. 5.10b). Fractography analysis showed [27] that the corresponding AES were generated due to the delamination of graphite from a metallic matrix, graphite inclusions failure, and, possibly, due to the friction of crack faces during opening. The third region differed mainly by the parameters of AE signals. To detect the sources of signals corresponding to it, the sub-critical crack front on the surface of the specimen fracture was thermally tinted. Before the appearance of the third region—and immediately after it—the specimens were unloaded and the macro-crack was cyclically extended. As a result, it was found that the signals of the third region were conditioned by a sub-critical crack growth.

Note that estimation of the crack start and the corresponding SIF for cast irons is an extremely important task from the point of view of fracture mechanics. This is explained by the fact that micro-defects intensively initiate in graphite inclusions and at the graphite-metallic matrix interfaces during the loading of cast irons. As a result, the behavior of cast irons becomes nonlinear-elastic. During the loading of ferrite-based cast irons, the "*P*-δ" diagram is dome-shaped, making the application of a standard method of K_{Ic} estimation impossible [8]. In this case, the specimens

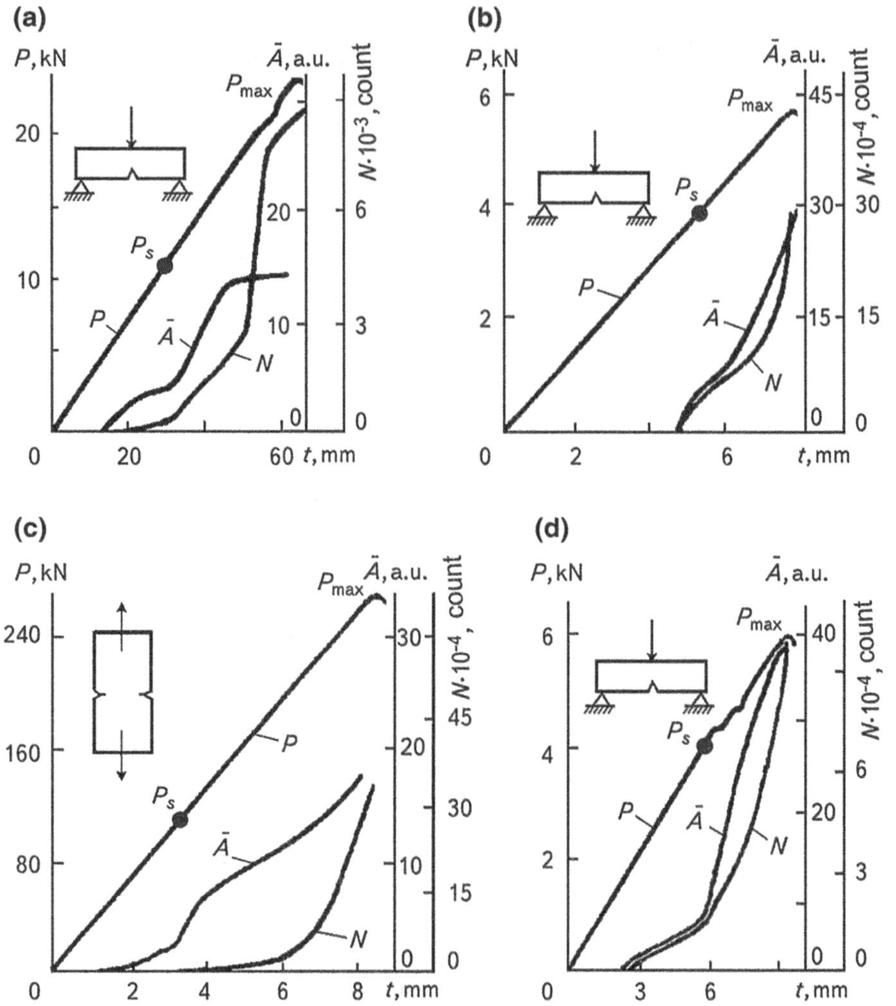

Fig. 5.10 Time dependence of loading P, medium amplitude $\bar{A}$ and cumulative count N of AE signals for: 9XF steel (**a**); gray cast-iron (**b**); Ti3Al alloy (**c**); ash and slag concrete (**d**)

did not fail spontaneously, rather following a time of slow, sub-critical macro-crack growth. Fracture surfaces of such specimens were smooth and without stretches.

AE diagrams of Ti3Al titanium alloy were similar to those of 38XN3MFA steel. The second region was more distinct, which is characterized by an abrupt growth of the AES amplitudes and their cumulative count (Fig. 5.10c).

The first AE signals appeared due to insignificant loading, possibly the grinding of the indenter and supports to the specimen. They were characterized by low amplitudes and small cumulative count.

Structural changes in a concrete, such as a fracture of slag inclusions or basalt fibres, the delamination of macadam from the matrix, etc., caused the appearance of the second group of signals in the AE diagram. In this case, the medium amplitude increases 1.4…1.7 times, and the AE count rate increases 2…3 times. The dependence of the cumulative count N on SIF at this stage of loading can be expressed by the empiric dependence $N = \beta(K_I - K_{I\mathrm{AE}}) + N_0$ where $K_{I\mathrm{AE}}$ is the value of K_I during the appearance of the AE signals of the second group, N_0 is the N value at $K_I = K_{I\mathrm{AE}}$, β is the proportionality factor. The mean values of these parameters are $\beta = 0.55 \cdot 10^4/\mathrm{MPa}\sqrt{\mathrm{m}}$, $N_0 = 0.4 \times 10^4$. The third region of the AE activity, which begins somewhat lower than the point of deviation from linearity in the curve "P-δ", is characterized by a further 2.5…3 times growth of A and N. It is caused by the beginning of sub-critical crack growth that can be visually observed on the lateral surface soon after the appearance of the AE activity.

Thus, the AE signals caused by the crack and plastic deformations growth, by local fracture of structural elements, and by grinding the indenter and supports to the specimen, are selected from a whole AE information by analyzing the qualitative changes in their parameters. This made it possible to evaluate the load P at the moment of the crack start that corresponds to it (Fig. 5.3), and to calculate the stress intensity factor K_{Is} by the known crack length. Conventional characteristics of crack growth resistance K_{Ic} and $K_{I\max}$ (as well as K_{Is}) were evaluated according to [8], using the values of P_c and $P_{\max}$ known from the diagram "P-δ" and from the corresponding crack lengths. The values of these parameters are given in Table 5.7.

Table 5.7 Parameters of crack growth resistance of materials

Material	K_{Is}, $\mathrm{MPa}\sqrt{\mathrm{m}}$	K_{Ic}, $\mathrm{MPa}\sqrt{\mathrm{m}}$	K_{Imax}, $\mathrm{MPa}\sqrt{\mathrm{m}}$	Comment
38XN3MFA	45	–	110	–
	57	–	180	Nitrided
9XF	19.7	31	40	–
Gray cast iron	25.4	26.3	39.2	50% ferrite, 50% pearlite
	27.5	28.1	40.1	Ferrite
Malleable cast iron	37	42	64	Ferrite
	36	45	87	Lamellar pearlite
	32	40	84.8	Granular pearlite
	35	36	78.8	Pearlite, 20% ferrite
High-strength cast iron	47	57.5	94	Ferrite
	29.5	34	34	Lamellar pearlite
	54	61	82	Granular pearlite
	53	56	76.1	Pearlite, 20% ferrite
Titanium alloy	50	63.5	11.7	1st edge crack
	46.5	64.8	85.6	2nd edge crack
Basalt ash and slag concrete	0.09	–	0.16	Grade 200

For the steels investigated, the value of K_{Is} K_{Is} is two times less than $K_{I\max}$. For cast irons, depending on the matrix composition and graphite form, the ratio of $K_{I\max}$ to K_{Is} is in the range of 1.15–2.42. Thus, the method developed permits determining the important parameter K_{Is} using the "P-δ" diagram and AE data.

5.4 Estimation of a Macro-Crack Length Increment and SIF Increase Under Static Loading

To use the AE method for the crack growth examination, the following must be done:

- Estimate the information parameters that indicate the crack start and growth;
- Find the dependencies between the value of crack length increment and the AE signals; and
- Develop the corresponding experimental methods.

The problem of selecting the informative AE parameters for detecting the crack growth was studied above, so we will not discuss it here. Instead, we will consider the second and third problems.

5.4.1 Some Theoretical Bases for AE Estimation of Macro-Crack Propagation Parameters

The increment of a through-crack length under its sub-critical propagation is related to the stress intensity coefficient K_{I} in the following way [28]:

$$\Delta l = -\alpha g(K_{\mathrm{I}}), \tag{5.8}$$

where $\alpha = \alpha_1 E K_{\max}^2 / 2\sigma_{0.2}^3$; α_1 is the dimensionless factor that depends on Young's modulus E, the yield strength and Poisson's ratio ν; the function $g(\ldots)$ is defined by the relationship

$$g(K_{\mathrm{I}}) = \left(K_{\mathrm{I}}^2 - K_{\mathrm{IS}}^2\right)/K_{\max}^2 + \ln\left[\left(K_{\max}^2 - K_{\mathrm{I}}^2\right)/\left(K_{\max}^2 - K_{\mathrm{IS}}^2\right)\right]., \tag{5.9}$$

$K_{\mathrm{I}s}$ and $K_{\max}$ in (5.9) are the initial and maximal values of K_{I}.

Based on the relationship (5.10)

$$N = -\beta_{ls} f(K_{\mathrm{I}0},\ K_{\mathrm{I}},\ K_c), \tag{5.10}$$

where

$$f(K_{\mathrm{I0}}, K_{\mathrm{I}}, K_c) = 2g(K_{\mathrm{I0}}, K_{\mathrm{I}}, K_c)/a\left(K_{\mathrm{I}}^2 + K_{\mathrm{I0}}^2\right), \beta_{Is} = \gamma\alpha_{Is}, \tag{5.11}$$

if N is equal to the number of crack jumps $\gamma = 1$ the expression for K_I estimation by N for significant AE signals before the crack start can be written by formulas:

$$N = -\beta_l t_l(K_{\mathrm{I}}) + N_{0l}, t_l(K_{\mathrm{I}}) = 2g(K_{\mathrm{I}})/a\left(K_{\mathrm{I}}^2 + K_{\mathrm{I0}}^2\right), \tag{5.12}$$

where a is the material constant, i.e., the proportionality factor that relates the length of a crack jump and the square of SIF, N_{0I} is the value of N at the crack start.

For non-through-cracks, the increase of the area Δs is related to the averaged value of the stress intensity factor $K_{I\mathrm{eq}}$ along the defect contour by the relationship:

$$\Delta s_{II} = s - s_0 = -\alpha_{ss} F_{ss}\left(K_{\mathrm{I0}}, K_{\mathrm{I}eq}, K_c\right), \tag{5.13}$$

where

$$F_{ss}\left(K_{I0}, K_{I\mathrm{eq}}, K_c\right) = \left(K_{I\mathrm{eq}}^4 - K_{I0}^4\right)/2K_c^4 + g\left(K_{I0}, K_{I\mathrm{eq}}, K_c\right),$$

$$g(K_{I0}, K_I, K_c) = \left(K_I^2 - K_{I0}^2\right)/K_c^2 + \ln\left[\left(K_c^2 - K_I^2\right)/\left(K_c^2 - K_{I0}^2\right)\right],$$

$\alpha_{SS} = \alpha_2 E K_c^4/2\sigma_T^5$, K_{I0} and K_C are the values of K_{Ieq} that correspond to the crack start and its critical value, respectively; S_0 is the initial area of a macro-crack. Then, we get

$$\Delta s = -\alpha_s f\left(K_{\mathrm{Ieq}}\right), f\left(K_{\mathrm{Ieq}}\right) = \left(K_{\mathrm{Ieq}}^4 - K_{\mathrm{I}s}^4\right)/2K_{\max}^4 + g\left(K_{\mathrm{Ieq}}\right). \tag{5.14}$$

Here, the function $f(K_{\mathrm{Ieq}})$ is given in (5.13); $\alpha_s = \alpha_2 E K_{\max}^4/2\sigma_{0,2}^5$; α_2 is the dimensionless factor depending on E, ν; $K_{\mathrm{I}s}$ and $K_{\max}$ are the values corresponding to the crack start and the maximum of load.

The cumulative count of AE signals N is expressed by K_{Ieq} as follows:

$$N = -\beta_s t_s\left(K_{\mathrm{Ieq}}\right) + N_{0s}, t_s\left(K_{\mathrm{Ieq}}\right) = 4f\left(K_{\mathrm{Ieq}}^4\right)/\pi a_s\left(K_{\mathrm{Ieq}}^4 + K_{\mathrm{IS}}^4\right), \tag{5.15}$$

where $\beta_s = \alpha_s \gamma$ is the proportionality factor between the cumulative count of the AES and the number of crack jumps; and α_s is the material constant that represents a relationship between the area of a single crack jump and K_{Ieq}^4.

5.4.2 *Test Results*

The tests were carried out on 8XN3MFA steel specimens: gray, malleable, and high-strength cast irons; titanium Ti3Al alloy, and ash and slag concrete under static three-point bending with simultaneous recording of the AES (Sect. 5.3).

As described above, two types of specimens made of 38XN3MFA steel were tested—nitrated and non-nitrated. Sub-surface layers in nitrated specimens were brittle. As a result, a crack propagated in them almost linearly over the whole cross-section with some advance at the lateral surfaces of the specimen. It is possible to observe the defect propagation with a microscope estimating the size of its increment visually. For non-nitrated specimens, a crack propagation occurs inside the specimen without reaching its lateral surface. In this case, the value of the crack length increment (change in the crack area) is estimated using the AE method.

In the AG for 38XN3MFA steel the appearance of three regions of the AE activity was noticed (Figs. 5.9 and 5.11). Unloading and failure of the specimens after the appearance of Regions II and III showed that they were caused by sub-critical crack growth. To evaluate the length or the area of the crack increment, after unloading, several specimens were reloaded again to the value that exceeded the preliminarily attained level. After every cycle of "loading-unloading." a trace was formed on the fracture surfaces that corresponded to the crack front location at the moment the unloading started, which made it possible to evaluate the defect increment on the fracture surface.

It has been established that the beginning of Region III coincides with the crack start, when a slow growth of the defect takes place. Theoretical and experimental data prove this, as is shown in Fig. 5.12. Theoretical curves were calculated according to Eqs. (5.8) and (5.14). In this case, it was set: $\alpha = 8.44$ mm, $K_{Is} = 57$ MPa$\sqrt{m}$, $K_{max} = 180$ MPa$\sqrt{m}$, $\alpha_s = 5.29$ mm, $K_{Is} = 45$ MPa$\sqrt{m}$ K_{max}. Region III is related to an abrupt crack length increment in the nitrided specimens and to its

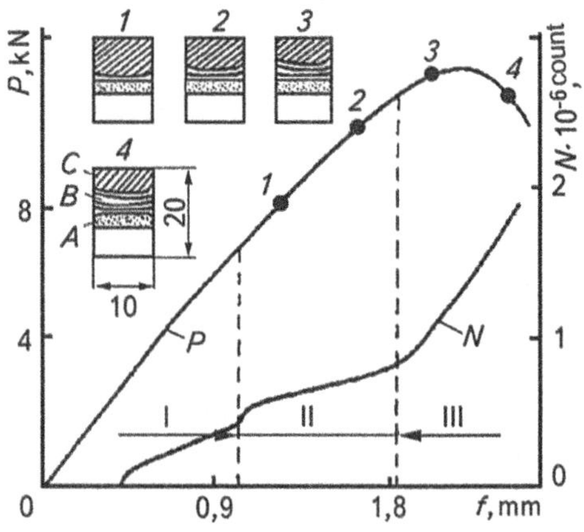

Fig. 5.11 Diagram "*P*—*f*" and AE diagram while loading the nitrated specimens of 38XN3MFA steel: *A* is the fatigue crack area; *B* is the sub-critical crack growth area; *C* is the fracture zone

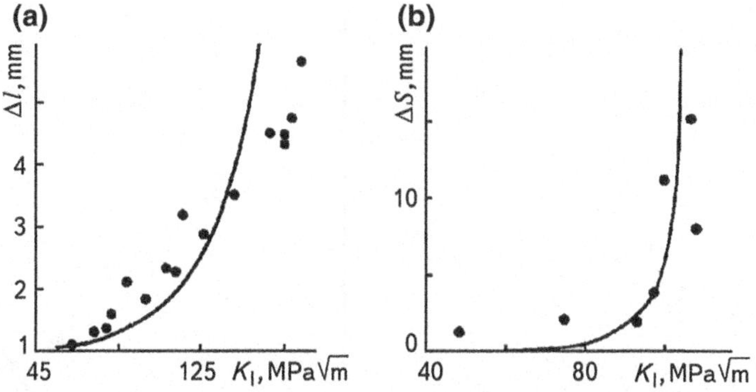

Fig. 5.12 Crack length Δl (**a**) and area Δs (**b**) increments against the SIF K_I increase for nitrated and non-nitrated 38XN3MFA steel specimens, respectively

area increment in non-nitrided steel; this region, for the first type of specimens, is not distinguished as clearly as for the second one.

Figures 5.13 and 5.14 illustrate experimental (dots) and theoretical (solid lines) data obtained by Eqs. (5.12) and (5.15) for steel specimens. Here: $\beta_l = 0.21 \times 10^{11}$, $K_{Is} = 57$ MPa$\sqrt{\text{m}}$, $K_{\max} = 180$ MPa$\sqrt{\text{m}}$, $N_{0l} = 57 \times 10^4$; $\beta_s = 0.88 \times 10^{13}$, $K_{Is} = 45$ MPa$\sqrt{\text{m}}$, $K_{\max} = 110$ MPa$\sqrt{\text{m}}$, $N_{0s} = 10^4$; $d_s = 262$.

As can be seen, the largest increase of the crack area, and cumulative count of AE, occurs when the SIF attains the maximum value. This regularity can be used for diagnosing a 38XN3MFA steel fracture by the AE method.

Three groups of the AE signals were also observed in the cast iron specimens. However, only one-third of them was caused by sub-critical crack growth. In Fig. 5.15, the N change versus K is shown for various cast irons from the very start of sub-critical crack propagation up to a complete spontaneous failure of specimens. The curves were built according to Formula (5.12) (solid line) and using

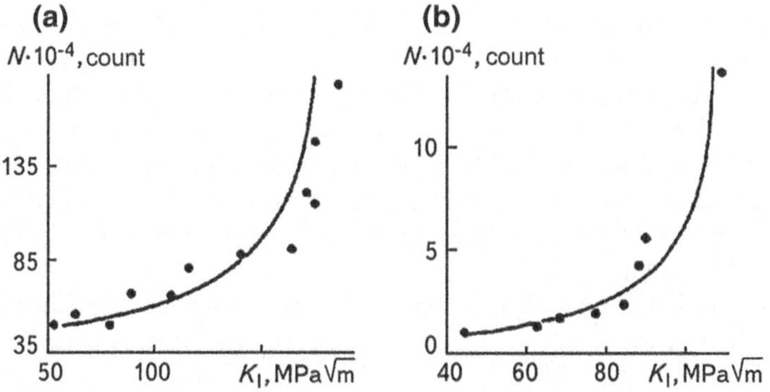

Fig. 5.13 Theoretical (*solid lines*) and experimental (*dots*) dependencies of cumulative count of the AE signals on SIF K_I for the specimens of Type 1 (**a**) and 2 (**b**)

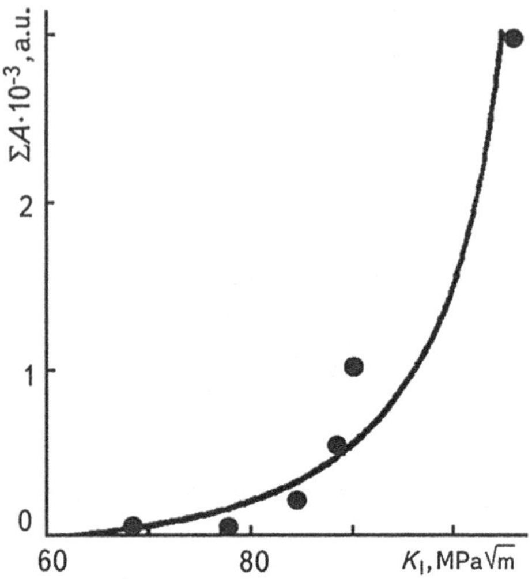

Fig. 5.14 Dependence of the sum of the amplitudes of AE signals on SIF for 38XN3MFA steel

experimental data (dots). In the experiment, an AVN-3 device was used, based on the first (*a*) and the second (*b*)—(*d*) methods for recording the parameters of AE signals.

The amplification factors η and proportionality factors β_l were: (a) high-strength cast iron: $\eta = 74$ dB, $\beta_l = 20.7$ (lamellar pearlite, Curve *1*); $\eta = 74$ dB, $\beta_l = 12.9$ (granular pearlite, Curve *2*); $\eta = 50$ dB, $\beta_l = 12.2$ (ferrite, Curve *3*); (*b*) high-strength cast iron: $\eta = 84$ dB, $\beta_l = 4.0$ (granular pearlite, Curve *1*); $\eta = 84$ dB, $\beta_l = 2.7$ (20% ferrite, 80% pearlite, Curve 2), (*c*)—malleable cast iron: $\eta = 84$ dB, $\beta_l = 9.7$ (granular pearlite, Curve *1*); $\eta = 74$ dB, $\beta_l = 5.3$ (lamellar pearlite, Curve 2); (*d*) gray cast iron: $\eta = 84$ dB, $\beta_l = 11.1$ (50% ferrite, 50% pearlite, Curve *1*); $\eta = 84$ dB, $\beta_l = 5.2$ (ferrite, Curve 2).

Thus, the analysis of AG obtained during cast iron fracture showed varying AE activities of specimens depending on their fracture mechanism (it is determined by the graphite shape, composition of matrix, size of inclusions, and a distance between them). With the growth of AE activity, the fracture mechanisms of cast irons can be classified as follows: ductile, mixed (brittle-ductile), and brittle.

While investigating the specimens made of the same type of material having the same graphite shape, depending on the matrix composition, the AE activity increases in the following progression: ferrite, granular pearlite, ferrite-pearlite, and lamellar pearlite. In cast irons of similar types of matrices, the growth of emission is determined by the shape of graphite in the following progression: spherical, vermicular, and lamellar.

It has been found that during sub-critical crack growth, the dependence of *N* on K_I is described by the same formula for all types of cast irons. An abrupt growth of AE cumulative count *N*, when K_I attains K_{max}, takes place on the Ti3Al titanium alloy specimens (Fig. 5.16). Therefore, if during the loading of this material there is

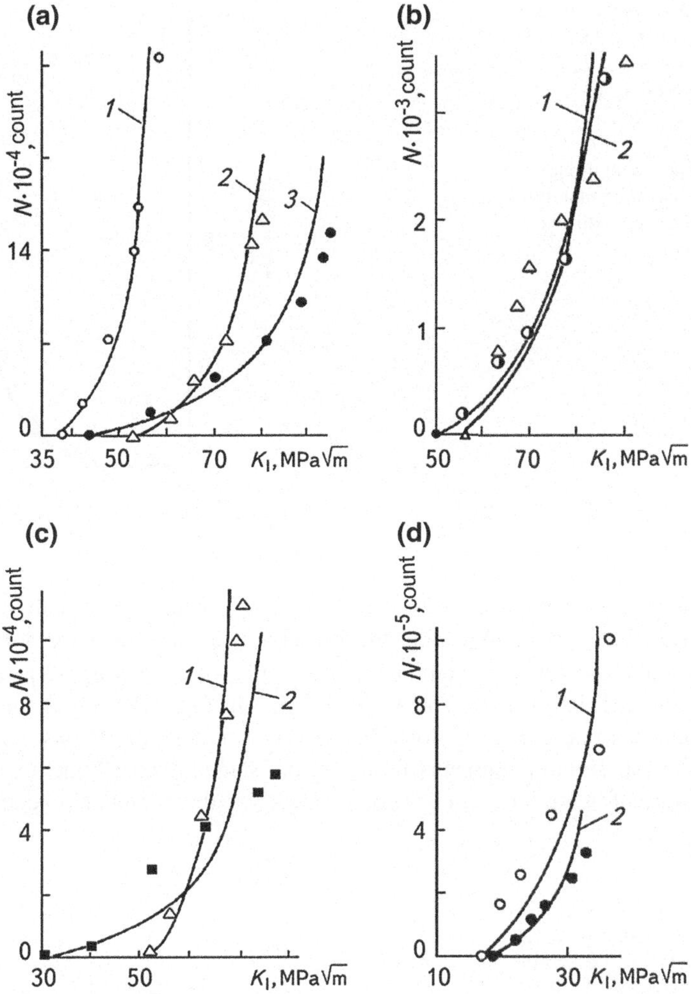

Fig. 5.15 The dependence of N on K_{I} for some structures of cast iron matrix and AE measurement modes

a drastic increase of N, it is an indication that a crack is in the region close to the sub-critical propagation stage.

The verification of Formula (5.12) application for concretes is similar to that for cast irons. For this purpose, it is necessary to show that at the sub-critical crack growth stage, the value of each macro-crack jump is proportional to K_{I}^2. The macro-crack growth in ash and slag concrete occurs in two ways: According to the first, there is a breaking of a linkage between the crack (formed due to delamination of the matrix from the filler) and the notch. In this case, the linkage size is proportional to K_{I}^2 [29]. In the second case, the macro-crack initiation in front of the

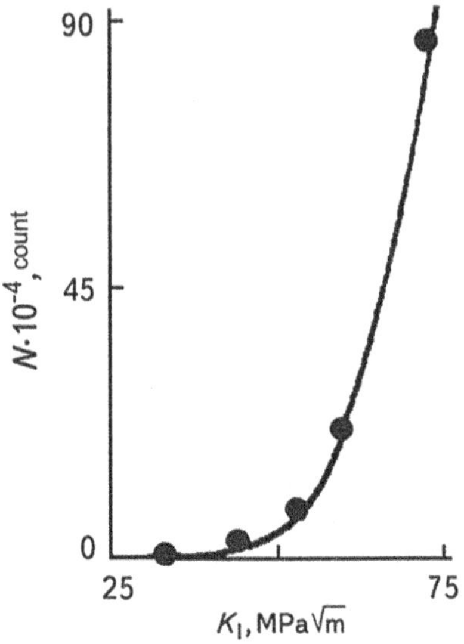

Fig. 5.16 Dependence of AE cumulative count N on the stress intensity factor K_I, evaluated according to Formula (5.12) for $\beta_l = 40 \cdot 103$, $N_{0l} = 0$ (*solid line*) and experimental data obtained on lamellar titanium (Ti3Al) specimens with two lateral cracks (*dots*)

notch in the region of maximal stresses, and its further coalescence with the notch is possible. Since the maximum tensile stresses act at a distance from the notch tip that is proportional to K_I^2 [30], the value of a single jump of a crack is $\sim K_I^2$. Therefore, the dependence (5.12) is also valid for concretes. In Fig. 5.17, the theoretical and experimental data are compared for ash and slag concrete specimens.

Thus, the AE signal parameters represent the stages of local fracture under the static loading of materials. An abrupt increase of the cumulative count of the

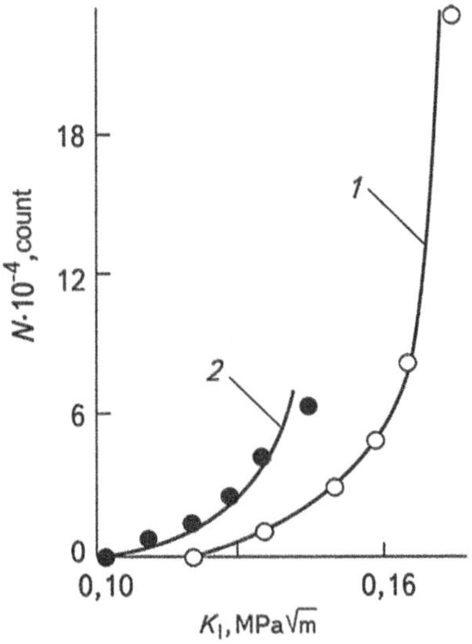

Fig. 5.17 The N vs K_I dependence obtained theoretically (*solid line*) and experimentally (*dots*) on two ash and slag concrete specimens for $\beta_l = 9.6 \cdot 10^4$ (curve *1*) and $\beta_l = 5.7 \cdot 10^4$ (curve *2*), $N_{0l} = 0$

AES reflects the change in the mechanism of local fracture or its acceleration. It was established that at the early stages, sub-critical crack grows slowly. When the value of K_I attains a maximum, an abrupt increase of the crack area occurs.

The established theoretical and experimental dependencies make it possible to evaluate the crack start, its size, and its stress intensity factor in terms of the cumulative count of the AES. In this case, they are universal for a number of structural materials.

Thus, new methods of experimental AE evaluation of the parameters of sub-critical crack growth in structural materials under their static loading in air have been developed. They are based on the established analytical dependencies.

5.5 AE Estimation of Strength Characteristics of Structural Materials

5.5.1 Investigation of Concrete Hardening by AE Signals [31]

A number of important state documents, which are aimed at providing the safe exploitation of buildings, constructions, and engineering structures, have been recently adopted in Ukraine [32, 33]. Concrete, which in the modern home construction industry is one of the most widespread materials used, turned out to be very helpful in implementing the above-mentioned measures. Most of the elements of bridge constructions are also made of concrete [34]. Therefore, quality maintenance of concrete products is an especially urgent task. It can be solved by using new progressive non-destructive testing and TD methods for in-process measurements of products made of concrete, in particular, for the selection of optimum conditions of hardening and its inspection during the manufacturing of industrial constructions. In fact, during the hardening of concrete, a certain crystalline structure is formed and micro-cracks can appear.

Among non-destructive testing methods, which are usually used for the quality rating of concrete products, the most widespread method is ultrasound [35]. However, it is frequently substituted by the AE method, which is particularly effective in checking the hardening of composite materials [36–38]. By its sensitivity, it exceeds ultrasonic methods by two orders of magnitude [36]. This method permits detecting the structural defects of concrete at the stage of their initiation and propagation, and determines the dynamics of their propagation in the concrete. Thus, the AE information permits adjusting the parameters of heat and wet treatment of concrete while manufacturing the elements of industrial concrete structures.

The state of the problem. Paper [36] investigated the hardening of alabaster with and without a filler. The AVN-1 M device, supplemented with a S8-9A memory oscilloscope, recorder N-117, and frequency counter Ch3-38, was used.

The research was carried out in the frequency band from 0.06 to 1.2 MHz with an amplification factor of 80 dB.

In the case of alabaster and gravel filler composition with a particle size of 1…2 and 4 cm, the ratio of the mass of water and alabaster was 0.54:1. The specimens were in the form of cubes of dimensions 70 × 70 × 70 cm. The AET was placed directly into the center of the form, and in the other case it was located on the AE waveguide. Identical results were obtained in both cases.

According to the experimental results obtained at the hardening of alabaster without a filler, the AE signals are generated sporadically and are characterized by low amplitude. The exothermal reaction showed the relaxation of excess energy during phase transformations at very low frequencies. In the authors' opinion, an AE of the explosive type is typical of the creation of fine defects in a structure being formed.

At the hardening of a two-composite material, the AE was generated with considerably higher parameters in comparison with the hardening of alabaster only. Amplitudes of the AE signals, the number of pulses, the AE count rate, etc., increased, and it was shown that the energy characteristics of the AE greatly depend on the granule sizes. Thus, when fine particles were used, the AE amplitude increased 1.5…2 times, and 3 times for coarse macadam. Therefore, authors in [36] concluded that the "matrix-filler" contact interface is the dominant factor of the AE generation during the hardening of a two-composite material. Micro-cracks begin to appear due to the change in a matrix volume during the hardening, when there are phase transformations that cause mechanical stresses. Every break of bonds between the alabaster and macadam leads to the AE. It is noted that exactly during alabaster hardening, its bonds with a filler are disrupted, and the end of hardening can be verified by the moment of AE energy stabilization. At the beginning of hardening, the adhesive processes are more intensive and are accompanied by the generation of high-frequency harmonics of the AE signals, which have prevailing amplitudes.

In [37], similar research was conducted on 20 × 20 × 20 cm cubes and on 500 × 150 × 20 cm plates made of grade 400 concrete that hardened in natural conditions and under hydrothermal treatment. The conditions of the AE selection and processing were approximately the same as in [36]. It is shown that AE in a concrete mixture is recorded already 10…20 min after its preparation; then the clinker minerals interact at the surface of the cement grains, and crystalline hydrides of varying composition are formed. The end of this process is clearly determined by the disappearance of the high-frequency component of the AE signals after 3…6 h. After that, the AE activity that accompanies the period of the formation of crystallization becomes centered, and a single cement stone decreases approximately by one order; this stage lasts 25…30 h. The AE generation is caused by the mutual impact of the growing crystallization grains, their fracture under high concentration, and the breakage of adhesive bonds between cement stones and concrete filler.

As a result of the research [37], the authors claim that at the stage of formation of the concrete of a thixotropy structure, the AE is caused by the fracture of the ettringite shell around a cement grain and by the fracture of separate grains of crystals. During crystal nucleus formation, the dominant factor of the AE is the breakage of adhesion bonds between the matrix and the inclusion, and during structure hardening, i.e., the fracture of the nuclei of hydrate inclusions that accompanies the crushing of the crystalline grains of concrete. The spectral composition of AE signals enables identifying the structural state of a concrete, and the total AE data permit a simple estimation of the moment of a rapid increase in the strength and formation of a stable crystalline structure of a concrete, as well as to obtain information concerning the effect of various factors on the concrete hardening and on the formation of strength properties.

In [39], using modern facilities of AE testing, the hardening of concrete at low temperatures was investigated. The authors related the parameters of the AE signals generated from the concrete that hardens at low temperatures, with similar signals that accompany the failure of concrete cubes during their compression, and obtained quantitative relationships of the AE activity for the frozen and naturally hardened concretes.

Thus, it is possible to state that there is a relationship between the formation of a concrete structure, the rate of strength increase, and the AE signals. This enables developing a non-destructive testing method for the quality inspection of concrete products—in particular, to select the technological conditions for a concrete hardening with optimum mechanical characteristics. However, such methods of non-destructive testing should be worked out based on the experimental research of hardening of a certain type of concrete and the products made of it. As can be seen, there is not enough data of this kind in the literature; therefore it is necessary to develop the methods of the AE estimation of micro-cracks initiation and propagation in the concrete bridge structures during their hardening.

Experimental results and their interpretation. A review of the published results show that an important characteristic in AE investigation is the range of working frequencies of AE devices. It depends on the material type, its porosity, the extent of damage, the degree of elastic waves damping, etc. The amplitude-frequency response of AET is an important factor that effects the selection of the AE working frequency band. For concrete, the frequency characteristics of the AE signals lie in the range of 1–500 kHz. Therefore, taking into account the effect of external noises and amplitude-frequency properties of the AET, we selected the working frequency band of 0.1…0.5 MHz.

The experimental setup was as follows: The mold (100 × 100 × 100 mm) was filled with the modified concrete mixture with the composition given in [40], and the waveguide of the AE signals was located in the mixture. The sizes of the waveguide were selected and optimized according to the method described in [41]; then the eigenfrequencies of the waveguide with the optimum sizes were calculated, using a finite element method.

Calculation of the eigenfrequencies of waveguides. A problem with the propagation of elastic waves, estimation of eigenfrequencies, and modes of vibrations in waveguides, which are of the shape of a body of rotation elongated along the rotation axis, in particular of a cylinder, was considered in many studies [42–45]. It is difficult to find an exact solution to this problem, even for a cylindrical waveguide of a finite length. Therefore, eigenfrequencies and vibration modes can be found, considering the approximate formulation of the problem and the methods of its solution. In particular, in papers [46, 47] it was suggested to use the method of limiting interpolation [48] in order to find the approximate eigenfrequencies of a cylindrical waveguide of finite length. The limiting cases of interpolation are exactly taken into account by this method: a short rod (disk) and a long rod whose eigenfrequencies are known [49]. The formula obtained in [46] enabled finding the approximate eigenfrequencies of vibrations of a cylindrical waveguide of finite length.

However, in the case of a more complicated form of a waveguide, it is impossible to calculate the eigenfrequencies using this approximation formula. In this case, numerical methods of calculation seem to be very promising—in particular, the finite element method.

The finite element method [50, 51] is based on the assumption that a body can be represented as a set of elements connected with each other only at their nodes. A relation of the node forces and node displacements is determined by the stiffness matrix [**k**] of an element. By combining the stiffness matrixes of separate elements in a global stiffness matrix, the problem of eigenfrequency estimation is reduced to the solution of the system of algebraic equations:

$$[\mathbf{K}]\{\phi_i\} = \omega_i^2[\mathbf{M}]\{\phi_i\}, \tag{5.16}$$

where [**K**] is the global stiffness matrix, [**M**] is the global matrix of the masses, ω_i is the angular frequency, and $\{\phi_i\}$ is the waveform of the i-th mode. In general, the number of eigenvalues, i.e., eigenfrequencies ω_i that satisfy Eq. (5.16), is equal to the order of square matrices included in it. A partial solution, i.e., eigenfunction $\{\phi_i\}$, which determines the individual distribution, a mode, of displacements in the region occupied by a body of a given form, corresponds to each respective eigenvalue. Taking into account that Eq. (5.16) are homogeneous, eigenfunctions are not determined ambiguously, although for each eigenfunction it is possible to find a ratio of its elements. Therefore it is reasonable to suppose that the largest eigenvalue is equal to one and thus determine the other elements with respect to it.

In experimental investigations, the possibility of using the waveguides of two types in the form of a finite cylinder and a rotation body elongated with respect to the rotation axis was considered. The waveguides are shown in Fig. 5.18a, b, respectively.

The calculations of eigenfrequencies and the corresponding vibration modes using the method of finite elements were first verified for a cylindrical waveguide. In this case, the equation for evaluation of eigenfrequencies can be written as follows [49]:

$$(x-1)^2\varphi(h'a) + (1-ex)[x-\varphi(k'a)] = 0, \tag{5.17}$$

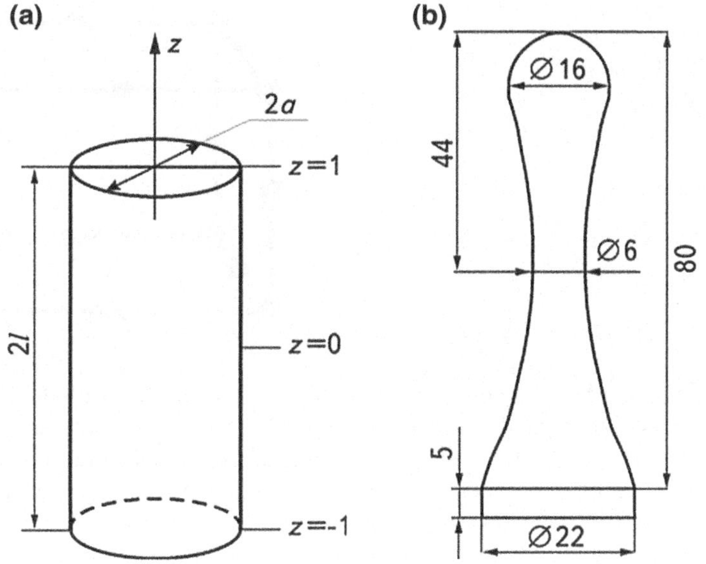

Fig. 5.18 Cylindrical waveguide (**a**) and a waveguide in the form of an elongated rotation body (**b**) of typical sizes (mm)

where$\varphi(y) = \frac{yJ_0(y)}{J_1(y)}$, $e = \frac{1-2v}{1-v}$, v is the Poisson's ratio, $h' = h^2 - \gamma^2$, $k' = k^2 - \gamma^2$, $h = \omega c_1$, $k = \omega c_2$, $J_0(\cdot)$ and $J_1(\cdot)$ are the Bessel functions of the first kind of the zero and the first order, respectively, ω is the angular frequency, a is the waveguide radius, and $2l$ is the waveguide length. Taking into account the fact that the cylinder is bound by two plane cuts with a distance between them of $2l$, the allowed values of γ are $\gamma_q = \frac{\pi q}{2l}$, q is the positive odd number.

Using Eq. (5.16), eigenfrequencies can be found from the following relationship:

$$f_{pq} = \frac{\pi q c_2}{2l}\sqrt{x_p/2}, \tag{5.18}$$

where x_p are the solutions of the Eq. (5.17).

The value of the eigenfrequency f_{pq} of the first mode of vibrations ($p = q = 1$) obtained from the dependencies (5.18) for $v = 0.28$, Young's modulus $E = 2.07 \times 10^5$ MPa, $\rho = 7800$ kg/m^3, $a = 0.005$ m and $l = 0.05$ m was 12,870 kHz.

The finite element calculations of the basic frequency of a cylindrical waveguide natural vibrations performed by Eq. (5.16), for the same values of Poisson's ratio, Young's modulus, density, radius, and length of a cylinder made it possible to find that $f = 12{,}950$ kHz. This means that the relative error of calculation in comparison with those, obtained from the dependence (5.18), did not exceed 0.62%.

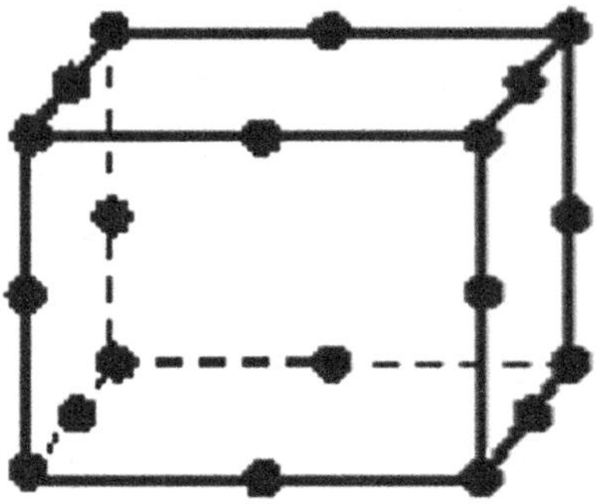

Fig. 5.19 The view of a three-dimensional finite element with 20 nodes

The calculation by the finite element method was carried out with the use of 3-D finite elements with 20 nodes. A view of this finite element appears in Fig. 5.19.

A cylindrical waveguide—after its meshing by these 3-D finite elements—is shown in Fig. 5.20. As it is seen in this figure, the cylinder along the generating line was divided into 15 segments, along the radius of the basis into 5 segments, and along the basis contour, into 8 segments. A double and triple increase in the number of divisions yielded relative errors of the basic frequency change of 0.19%. Therefore, in order to reduce the calculation time, the mesh shown in Fig. 5.20 was used.

The analysis of the results was done for those modes of a cylindrical waveguide whose vibrations occur along its longitudinal axis. It showed that there were 96 eigenfrequencies in the frequency range of 200…450 kHz, corresponding to the operating frequency range of the AET used in experiments for recording the AE signals. The distribution of normal displacements at the end surface of a waveguide for $f = 4.38 \times 10^5$ Hz is shown in Fig. 5.21.

Similar calculations by means of a finite element method were also done for the waveguide shown in Fig. 5.18b. Its view, after meshing with the 3-D finite elements presented in Fig. 5.19, is shown in Fig. 5.22.

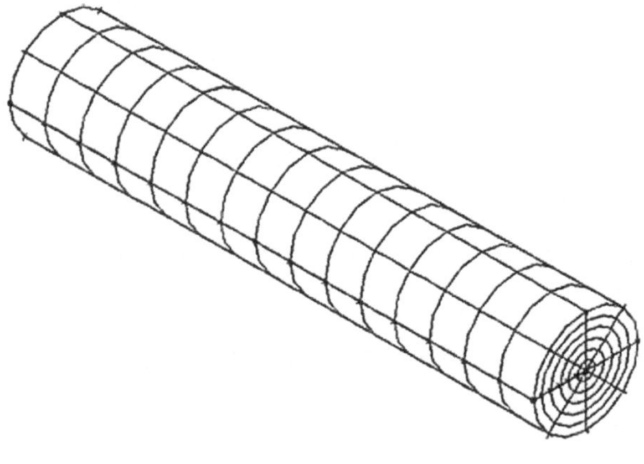

Fig. 5.20 Meshing of a cylindrical waveguide for calculations using a finite element method

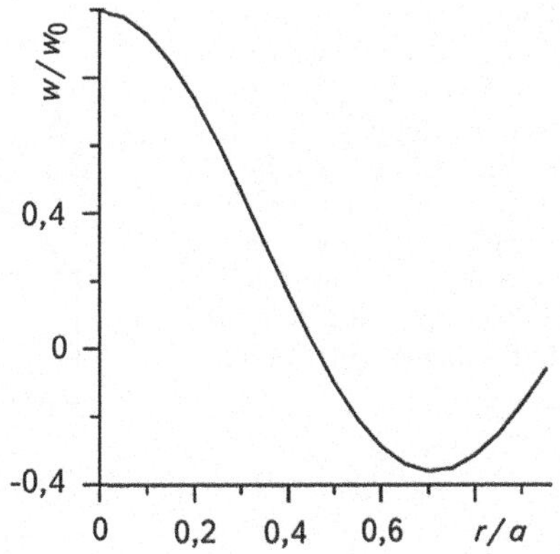

Fig. 5.21 Distribution of normal displacements at the end surface of a waveguide for $f = 4.38 \times 10^5$ Hz

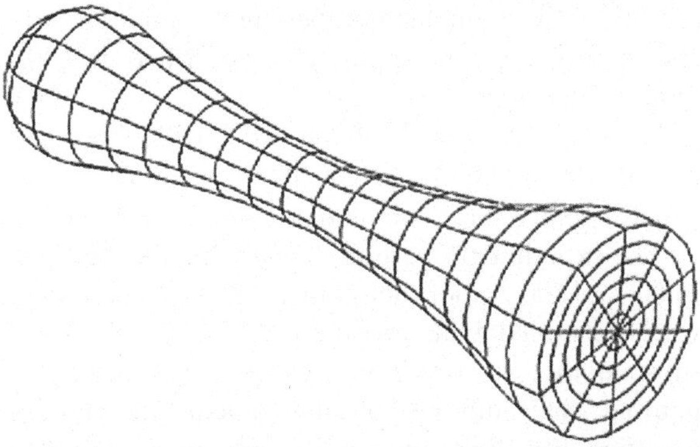

Fig. 5.22 3-D finite elements meshing of a waveguide in the form of a rotation body

The two-time increase in the number of divisions gave the relative error of the found eigenfrequencies that did not exceed 0.91% within the frequency range of 200…450 kHz. Calculations were done for the same values of the Poisson ratio, Young's modulus, and density. In total, 119 eigenvalues were found in this frequency range. Calculations showed that the longitudinal modes of vibrations of a waveguide are in the 100…360 kHz frequency range, and agree with the resonance frequencies of a quasi-resonance AET [40], which was used for experimental research. Taking into account that the compression stresses appear during the hardening of concrete, it is necessary to choose the form of a waveguide that does not cause the formation of additional stress concentrators; therefore we selected the waveguide shown in Fig. 5.18b. Its general view, which takes into account the structural features, is shown in Fig. 5.23.

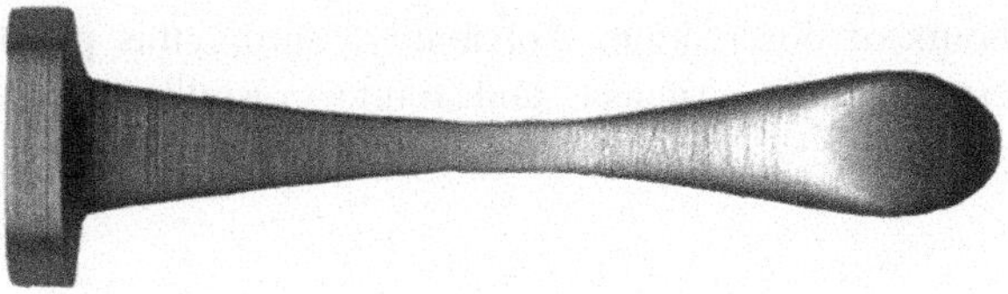

Fig. 5.23 The AE signals waveguide

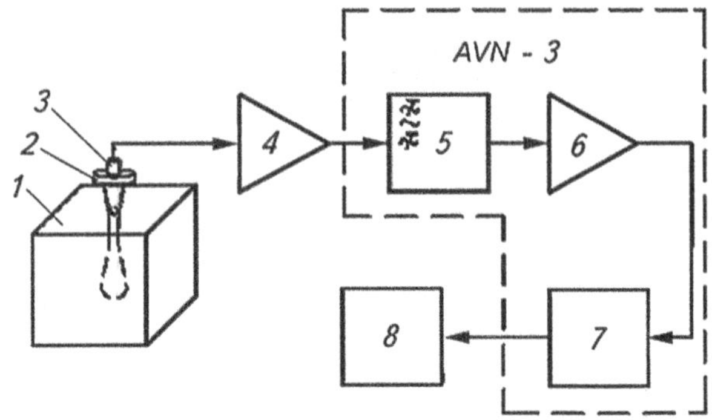

Fig. 5.24 Experimental setup: *1* is a specimen; *2* is an AE waveguide; *3* is a TAE; *4* is a preamplifier; *5* is a block of filters of high and low frequencies; *6* measures the amplifier of power; *7* are the units for processing the AE signals, and *8* is a fast-acting recorder of the N-338/4 type

Tests and discussion of the results. The concrete cubes hardened in natural conditions at an ambient temperature of about 300 K. The AE signals were selected and recorded by the equipment described in [40], having similar amplification modes (Fig. 5.24).

An acoustic contact of the AET with the contact plane of a waveguide was provided by adding a contact epoxy resin layer. Time distribution of the amplitudes of AE signals, their cumulative count N, and the rate of count $\dot{N}$ were recorded.

The experiment's results showed that the AE appeared in the mixture of concrete in 600…1200 s after the moment of its preparation. At first, individual AE signals were generated with amplitudes of 10…30 μV and then the AE activity grew substantially. Its maximum values were reached approximately one hour after pouring water into the Portland cement. The authors in [37] showed that exactly at that time, the clinker minerals at the surfaces of cement grains interacted and there crystalline hydrides of various compositions were formed. The authors proved that at that time a mineralogical heterogeneity of cement grains, as well as the defects on their surfaces, caused a periodic peeling of primary crystallites into the pore space under the action of shear stresses. There they formed the elements of a secondary permolecular structure, blocks, from which crystalline nuclei were subsequently formed. The number of sub-microcrystals continuously increased, and the distance between them decreased. When it attained $10^{-6}…10^{-7}$ m, a primary structure was formed with coagulation and crystalline contacts due to intermolecular interaction.

The results of the experimental data processing are presented in Fig. 5.25. As it follows, the most active with the greatest ratio of $\dot{N}/\dot{N}_{max}$ AE period is observed during the first six hours of observation. Precisely during this period, the concrete mixture reaches the strength of $0.08R$, where R is the ultimate strength of the concrete, and the process of the formation of its crystalline structure is mainly completed. Furthermore, the AE activity significantly decreases in comparison with the first two hours of observation. Apparently, during this period [37], individual centers of crystallization connect, and single crystalline nucleus cement stone is formed.

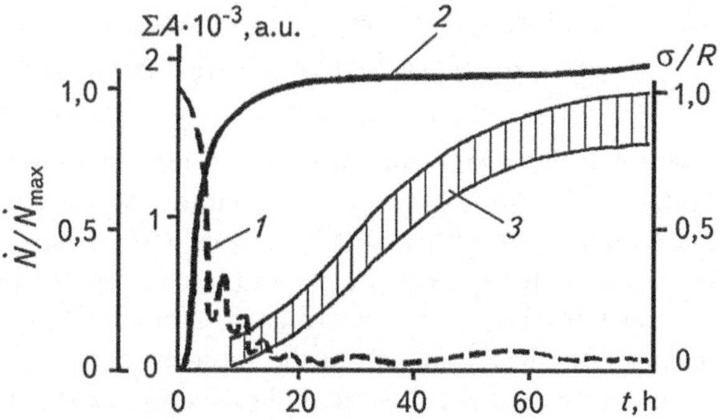

Fig. 5.25 Time variation of activity (*1*), sum of the amplitudes of AE signals (*2*) (in arbitrary units a.u.), and the values of the scattering range of the strength index (*3*) of a concrete mixture

Subsequently, the crystalline nuclei join together for 25…30 h. An increase in the strength of the concrete mixture is accompanied by a slight release of the AE energy. By analogy with the data of [37], this is conditioned by the nuclei contact and fracture, and by the breaking of adhesion between the cement stone and the filler. As a result, shrinkage cracks appear.

The end of the stage of intensive strength increase coincides with a decrease in the activity. During this period, according to the data in [52–54], pores are filled with the hydration products and the structure is packed, but the surface of cement grains is at the same time shielded, and thus the formation of a crystalline structure is considerably slowed down. The AE activity drops to the minimum value at the strength of 0.7R. Its fluctuations are caused by the micro-crack initiation, due to crystallization or osmotic pressure. Obviously, the AE growth can be also conditioned by the formation of large crystals of Portland cement in pores causing their failure.

Thus, in this study, the main objective of the investigation has been met, and it is shown that the development of the AE methods for obtaining the quantitative characteristics of the AE parameters related with the strength indexes of mixtures of concrete used in bridge construction, and civil and highway engineering, is quite promising for further investigations. Therefore, the AE waveguides are effective in the estimation of concrete hardening, when their eigenfrequencies are calculated and can be agreed with the eigenfrequency characteristics of the AET with the account of structural features of the device construction for transmission of the required AE information. The use of highly sensitive quasi-resonance AET enables quantitatively estimating the stages of crystallization and the increase of the strength of a concrete mixture and, therefore, optimizing the technological conditions of concrete mixture hardening and developing non-destructive test methods for a quality inspection of the concrete elements in bridge structures—provided that the necessary experimental data base is available.

5.5.2 AE Estimation of AES Amplitudes at a Fracture of Concrete in the Bridge Structure

There is a rather branched network of roads with more than 20,000 road-transport bridges in Ukraine. Most of the structures are made of reinforced concrete, which is quite often used with various damages and defects [55]. The most widespread are crack-like defects, which cause a brittle fracture of bridge structure elements made of concrete. However, macro-crack initiation and propagation does not occur instantly. It means that from the moment of its nucleation to the beginning of a spontaneous fracture of structural elements, a certain amount of time passes, which is determined by the period of sub-critical crack growth. Therefore, it is a very important scientific and technological undertaking to establish the moment of macro-crack initiation or the period of its jump-like sub-critical growth. Its solution will greatly contribute to predicting the reliability and durability of bridges, and protection against accidental failure with considerable material and human losses; in fact, the number of reinforced concrete bridges in Ukraine increases every year, with a simultaneous increase in transportation traffic intensity, weight of loaded transport vehicles, environment corrosivity and so on.

To solve the above-mentioned problem, various methods of non-destructive testing have been recently used [56] to obtain the necessary diagnostic information. Among them, the AE method is characterized by great sensitivity. This method, due to its substantial advantages, keeps its leading place among other technical approaches on the research and diagnostics of both separate elements of bridge structures and the bridges themselves [57, 58].

Analysis of the state of AE investigations of fracture of concrete cubes. Modern concepts on the physics of concrete fracture prove the existence of three stages of concrete deformation. In the first stage, in the process of concrete compression, a closure of the available pores occurs. In the second, large pores and micro-defects become the nuclei of fracture, and sub-micro-cracks are initiated. In the third stage, sub-micro-cracks coalesce into micro-cracks, the material loosens, its properties change irreversibly, and an avalanche-like macro-crack develops.

At the initial stages of applying the AE phenomenon to the investigation of the aforementioned deformation stages [59, 60], the AES generation during the formation of a deformation was not detected at the first stage. This means that the stage of elastic deformation of a material seems not to have been accompanied by AE. This proved not to be true, and the statement was false due to the use of equipment of low engineering standards available at that time. It should be noted, however, that in some early research, despite the absence of a required experimental base, valuable scientific results were obtained. For instance, by using a small microphone planted into the concrete specimen, it was possible to investigate the effect of various factors, such as hydrometric, thermal, chemical, and dynamic under force action, on strength characteristics in relation to the components of a concrete and its aggregate states [61] by the AE method. It was shown that the AE method permitted recording the processes of the internal initiation of micro-cracks, as well as effectively determining plastic strains and fracture. The alternating character of the AES was illustrated at different stages of the material integrity breaking, and the results were comprehensively presented with the analysis of about 230 published sources. The Aerojet Corporation, as well as a number of experiments by other researchers carried out with advanced facilities [62–64] that tested the processes of fracture under tension and compression of concrete specimens, proved the efficacy of the method.

The results of the experiments [65] were obtained somewhat later. They also confirmed the generation of AES under stresses of about $\sigma = 0.5R$, where σ is the compression stresses, and R is the ultimate strength of a concrete. AES were detected in the frequency range of 2…20 kHz. Only authors of studies [66–69] were able to describe AES at all stages of concrete deformation, including the early ones. The authors of these papers found, basing their findings on fractography and ultrasonic analysis, that at the first stage, the AE is weak and is caused by material compacting due to the coalescence of pores, the development of individual defects, and the collapse of initial micro-cracks. At the second stage, due to micro-crack initiation, the AE amplitudes increase, although the intensity of the AE is constant for a constant strain rate, while at the stage of failure, i.e., the third stage, the AES amplitudes and their intensity increase considerably. It is illustrated that the total AE energy increases simultaneously with loading, reaching its maximum value at the moment of failure. For the deformation stages indicated, this index is 10, 30, and 60% of the maximal value, respectively. The same tendencies of the AE energy distribution, as well as manifestation of deformation stages, were typical of the sand-cement solution [68].

A somewhat different approach to this problem is proposed in [70–72]. The authors of [71], based on the AES analysis, concluded that there are four stages of deformation of porous materials. Elastic deformation of concrete occurs at the first stage. At the second stage, when $0.2R < \sigma < 0.75R$, the micro-crack initiates and propagates, and at the third, at the values of $0.75R < \sigma < 0.96R$, a macro-crack initiates. The authors consider the overcritical stage of a macro-crack growth, when the loading reaches $\sigma \geq 0.96R$, to be the fourth stage. In [72], it is believed that there is an unstable fracture propagation at this stage, and the cumulative count of the AE pulses with volume deformation is related by a power dependence. If we take into account the effect of the age of the concrete on the initiation of micro-cracks, it should be noted that this factor has an effect on the energy characteristics of the AE radiation [73].

Important results were obtained in [74]. The AES were analyzed in combination with the data of strain gauge measurements and optical microscopy. The results of testing the concrete plates to be 150 $\times$ 70 $\times$ 8 mm with the open structure of the material, which was subjected to quasi-static axial compression by a specially developed method were obtained. It was shown that a jump-like crack propagation is caused by the material microstructure. A number of individual jumps characterize the micro-crack length comparable to the sizes of structure, and they can be used as the basis for classification of the initiating cracks. At this stage, the AES were recorded, from which, using special methods, it is possible to extract individual pulses corresponding to individual jumps of a crack. For the device noises of 5×10^{-6} V and for a signal: noise ratio of 10 dB in the frequency range of 10…200 kHz, cracks were detected that initiated at the micro-, meso- and macro-structural levels of a concrete, the number of pulses in the AE event being dependent on the stress level.

The authors showed that it is very difficult to visually identify the propagation of such cracks at the micro-structural level due to their great number, small size, and specific structure of the cement stone. At the meso-structural level, for a crack length less than 5 mm, cracks are characterized by the AES having 10…12 pulses, and it is possible to compare the AES with the results of other non-destructive test methods. The duration of propagation of such cracks is approximately 10…30 ms. It is stressed that AES at the stage of macro-crack propagation should be analyzed particularly carefully, even when its numbers are small. The amplitude spectra of AES are of a clearly expressed exponential character and when the loading grows, they shift towards large amplitudes, the AES form being preserved.

Data obtained in [75] are of considerable scientific interest and practical value. The tests were conducted on 7 × 7 × 28 cm prismatic specimens made of grade 400 concrete loaded by a hydraulic press of high stability. Applying a standard method [76] and strain measurements, the specimens were centered. The method of a scanning electron microscopy was used to obtain data on the changes in the form, size, and location of structural elements after deformation and fracture. Alongside the methods mentioned, ultrasonic examination of elastic wave velocity by the "BETON–8" device was also carried out. To study a dispersion-crystalline structure, the samples were prepared from the previously tested specimens. Plates were sawed out from the non-fractured prisms, and during fracture, the samples were prepared from the material pieces with a juvenile fracture surface. For the AES selection, the AET with the resonance frequency of 1.35 MHz was used. The device for the AES processing of the AVN-1 M type operated in the frequency band of 0.06…1.5 MHz with an amplification factor of 80 dB. The spectral analysis of AES was performed using a specially developed method.

At the stage of linear creep, $\sigma < 0.55R$, the process of fracture development proved to be closely related to the sizes and mutual placement of crystals, and fracture stopped when the orientation of the neighboring grains differed substantially. Most large cracks were stopped by gel, which grew into lamellar crystals of Portland cement. When the loading reached the limit of $\sigma = 0.55R$, fractography showed the appearance of many-branched cracks that propagated and coalesced, forming large-size steps.

For $\sigma > 0.55R$, a coalescence of cracks that grew from single pores, as well as their merging, was observed. A bulk micro-cracking changed into local fracture near the main ruptures. First, unstable fracture surfaces appeared, and the AE intensity decreased in comparison with the previous period due to a reduction of the material volume involved in deformation and due to an increase of its porosity. The further load growth, higher than $0.7R$, resulted in an abrupt increase in the number of the main cracks on fractograms, which merged at an angle of 30°, forming characteristic planes of a macro-fracture. This process was accompanied by a considerable growth of amplitudes and the rate of the AES count.

The analysis of the AES energy spectrum showed the prevalence of harmonics of signals with a frequency of 1.2…1.6 MHz. At a different loading, no change was observed in the type of spectral characteristics with the change of the fracture mechanism, and the energy of fracture differed as well. A non-linear creep and mass fracture of crystals was accompanied by the AE with the amplitudes 1.5…2 times larger than at the stage of elastic deformation. Therefore, it is possible to distinguish the moment of intensive crack formation by an abrupt change in the AES intensity and by the growth of their amplitudes.

In [77], concrete prisms of 150 × 150 × 600 mm made of B-25 concrete six months of age were subjected to compression. The loading was stepwise by 1.0 MN with a 10-min holding at each step. At the initial stage, specimens were centered, with additional final loading up to 2.0 MN and further unloading. A one-channel AVN-3 analyzer of AES and a four-channel SPRUT-4 analyzer were used to estimate the AE parameters. Both devices operated on the of 100… 750 kHz frequency band, with an amplification factor of 70 dB. The AES were selected by a quasi-resonance AET mounted on the CTS-19 ceramics, and a typical diagram of fracture was obtained. The analysis of deformation energy showed that it consisted of the sum of energies: elastic, which is accumulated in the specimen, and fracture energy. Since the AE radiation is related to local rearrangement of the material structure and to the development of fracture processes, it is assumed that there is a functional relationship between the AE energy and plastic deformation work. For this purpose, a pair regression analysis of the radiation energy and the energy used for fracture was conducted.

Similar investigations were carried out in later studies—for example in [78–80], where the methods described above were partly repeated. It is worth noting that the data were recorded using more advanced devices, yet they concurred with the data from the known literature.

Thus it is possible to confirm that there is a functional relationship between the deformation of concrete and, therefore, between the initiation and propagation of micro- and macro-cracks and the AES parameters. This enables elaborating on new methods of non-destructive testing and technical diagnostics of the state of bridge concrete structural elements using the analysis of AES emitted during the structure loading. However, such methods cannot be developed without a preliminary estimation of the values of the AES amplitudes that accompany the early stages of fracture development in concretes used in bridge constructions. The data published in the literature are too scant; besides, it seems incorrect to apply the results obtained by other authors, even if quantitative indices of the AES are minimal for modern types of concretes and technology of their production in Ukraine.

As it follows from the above, in order to correctly perform the diagnostics of the state of a bridge structure, it is important to select the quantitative characteristics of the AES that accompany the initiation and propagation of micro- and macro-cracks in concretes. To partially solve the problem described, concrete cubes were tested under compression with simultaneous recording of the AE signals.

Test results and their interpretation. Cubes with the dimensions of 100 × 100 × 100 mm of a modified concrete (grade 400) possessing improved

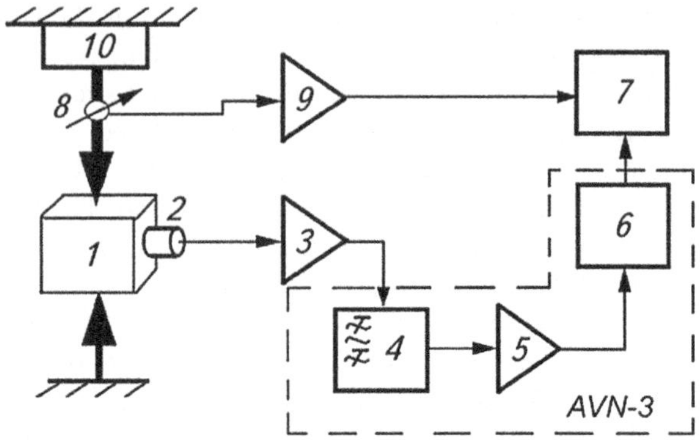

Fig. 5.26 Block diagram of the experimental tests

strength characteristics were tested. The composition of the concrete mixture (in kg/m^3) was as follows: Portland cement of 400–480 grade; sand—480; macadam of the fraction of 5…10–550; macadam of the fraction of 10…20–720; water to cement ratio—0.41; and plasticizer content—1.5%. The age of concrete cubes was 90 days.

The experimental setup is presented in Fig. 5.26. The loading was carried out by a quasi-static force uniformly distributed over the cube *1* faces. The force was exerted by the devices of hydraulic press P-250 *10* and was measured by a strain gauge dynamometer *8*. To select the AES, the AET *2* was mounted on the lateral surface of a cube through an epoxy resin layer and then fixed with a resin shock absorber with a compression force of 3…5 N. Electric AE signals from a quasi-resonance TAE *2* were transmitted to a preliminary amplifier *3*, block of filters of high and low frequencies *4*, measuring amplifier of power *5* and to the units of the AES processing *6* of device AVN-3. The AES were amplified in the frequency range of 0.1…0.3 MHz, which is in accord with the frequency characteristics of the quasi-resonance AET (Fig. 5.27).

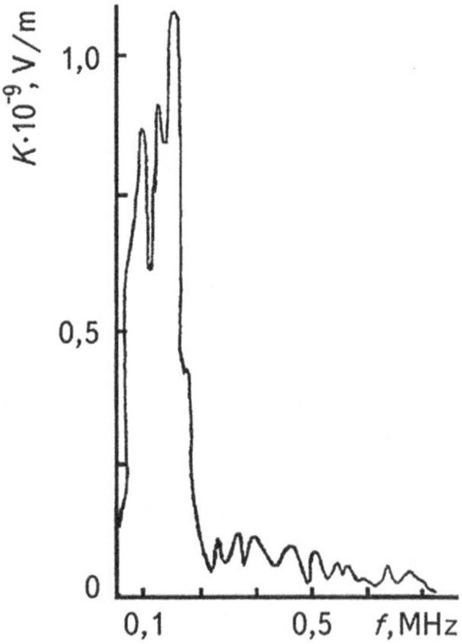

Fig. 5.27 Frequency dependence of the AET transmission factor

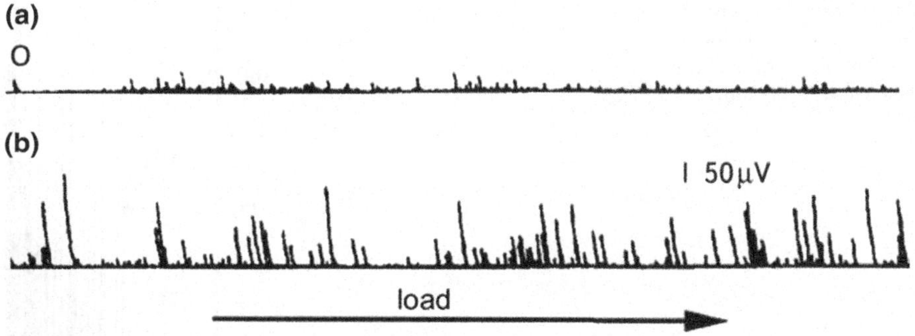

Fig. 5.28 Typical AES amplitudes recalculated in a common scale with respect to the amplitudes at the AET output for the stage of the formation of micro-cracks (**a**) and initiation and propagation of meso-cracks (**b**)

The unit of the main measuring amplifier *5* made it possible to set the amplification factor stepwisely with a step of 1 and 10 dB, while the transfer ratio of filters was about 1. The general amplification factor of the AE path was 70 dB. From the processing unit of AES *6*, electric signals were transmitted in analog form to a fast-acting recorder *7* of N-338/4 type, where the AES enveloping curve was recorded for the period of 0.1 ms, their cumulative count, etc., as well as the loading force *P*. The latter one was recorded as an analog electric signal from the strain-gauge dynamometer of force *8* and was passed to recorder *7* through the direct current amplifier *9* of the U7-1 type.

The tests showed (Figs. 5.28a, 5.29) that for this concrete, the micro-crack initiation was accompanied by AES with amplitudes not higher than 20 μV. They appeared on the linear section of the diagram "applied load *P*—time *t*" almost immediately after the application of forces (point O in Fig. 5.28a). AES with low amplitudes were typical of the applied load in the $0 \leq \sigma \leq 0.15R$ range. This corresponds to interval 0-a (the first stage) on the integral AE diagram (Fig. 5.29) of the amplitude time distribution of AES. Furthermore, the AES amplitudes increase with the growth of the compression load applied to the cube. As seen in Fig. 5.29, after point "a" there begins a period of a slight growth of the AES amplitudes within the range of 20…40 μV, which is accompanied by the AES with amplitudes typical of the previous stage (a period between points *a–b* in Figs. 5.29 and 5.30).

Figure 5.28b (the second stage) shows a typical distribution of AES for this period of loading, which is in the range of $0.15R \leq \sigma \leq 0.48R$. Then follows the moment of abrupt growth of the AES amplitudes >100 μV. It lasts up to the stage of supercritical macro-crack propagation when a spontaneous fracture of concrete cubes takes place. These signals also alternate with the AES typical of the previous two stages. Besides, starting at this moment, the number of AE events abruptly grows until the final failure of a concrete cube (point *b* in Fig. 5.30).

Analyzing the obtained results and taking into account the published data, one can hypothetically assume that the first period is characterized by the micro-cracks

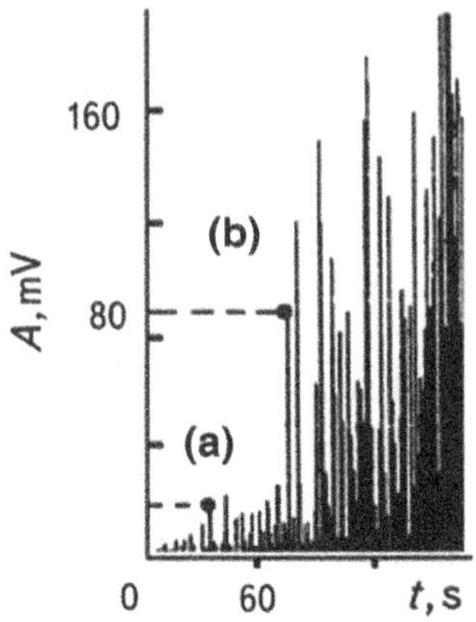

Fig. 5.29 A fragment of integral time amplitude distribution of AES during concrete cube loading

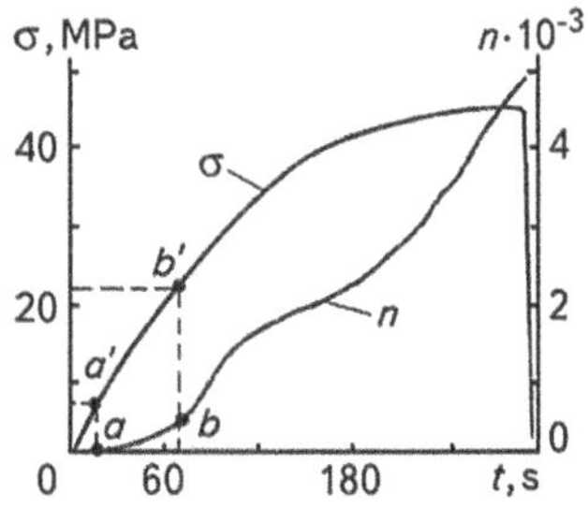

Fig. 5.30 The change in conditional stress σ and the number AE events n with time of concrete cube loading (points a and b correspond to similar ones in Fig. 5.29)

initiation and propagation, while during the second period, the similar processes occur at the meso-level. The moment of an abrupt growth of the AES amplitudes corresponds to the moment of a macro-crack propagation start, which continues until a spontaneous failure of concrete cubes.

Thus, the investigation's results correlate well with the known published data and allow us to state that the goal of the research, i.e., a quantitative estimation of the AES amplitude distribution at early stages of micro- and macro-crack initiation and propagation, was achieved. Further investigations in this direction should be conducted to establish similar quantitative amplitude characteristics of the AES for concretes with various admixtures and plasticizers at various rates and types of loading, etc.

Thus, the AE during the fracture process of concretes of bridge structures provides information on the initiation and dynamics of the fracture process development in a concrete, which is very important for diagnosing real objects operating in service conditions or during technological inspection of their state. Quasi-resonance AET increases the sensitivity of the AE method and permits revealing the early stages of concrete fracture. The results of investigations showed that micro-cracks initiation and propagation were accompanied by low-amplitude AES, which are 1.5…2 times less than similar signals caused by meso-crack formation, and almost one order less than those caused by the initial propagation of macro-cracks.

5.5.3 AE Estimation of Mechanical Characteristics of Steels

In the history of technological progress there are many cases of sudden catastrophic failures of bridges, steam boilers, cannons, water and gas turbines, oil- and gas pipelines, railways, etc. To prevent these failures, the engineers try to introduce the safety factor into the calculations. However, very often due to the requirements of either economy or service conditions, it is necessary to have a structure that possesses a required durability and reliability, is as light as possible, and material that "works" at the ultimate strength value. Thus, the use of any safety factor should be scientifically grounded. For this it is necessary to have more accurate characteristics of the strength of structural materials, which are mainly found by the tensile testings of specimens. These characteristics include:

(a) *The limit of proportionality (conditional)* σ_p, which is the stress at which the deviation from a linear dependence of load P on the elongation Δl reaches such a value that the tangent of a slope angle, formed by a tangent to the deformation curve P—Δl at the point P_p with the axis of loading, increases by 50% of its value for a linear elastic region [81];
(b) *The elasticity limit (conditional)* $\sigma_{0.05,}$ which is the stress, at which the residual elongation reaches 0.05% of the length of the specimen region, which is equal to the strain gauge base;
(c) *The yield strength (physical)* σ_y, which is the least stress at which a specimen is deformed without visible increase of a tensile stress;
(d) *The yield strength (conditional)* $\sigma_{0.2}$, which is the stress at which the residual elongation attains 0.2% of a calculation length;
(e) *The temporal resistance* σ_b, which is the stress that corresponds to the maximal load $P_{\max}$, at which the specimen failure begins;
(f) *The true failure resistance* s_k, which is the stress that is determined by the ratio of load P_k at the moment of rupture to the specimen cross-section area in the place of rupture F_k;
(g) *The relative elongation after rupture* δ, which is the ratio of the increment of the calculated specimen length after rupture to its initial size; and
(h) *The relative narrowing after rupture* Ψ, which is a relative decrease of the specimen cross-section area in the place of rupture (normalized with respect to the initial area of the specimen cross-section).

State of-the-art of the problem. In design calculations and at certification of metal products, the characteristics of resistance to small plastic deformations were used; these are determined by specimen testing under tension according to respective standards—for example [81]. The most widespread characteristics of the strength of structural materials, which are used in project-design works, are proportionality limit, elasticity limit $\sigma_{0.05}$, and yield strength $\sigma_{0.2}$.

Modern understanding of the processes of metal plastic deformation is based on the concept of intensive formation, and movement of the crystal lattice defects under its loading. Therefore, it is thought that dislocations make a basic contribution to these processes. However, the way of scientific searches regarding the establishment of a final version of a modern hypothesis was not simple, and the application of the AE method described above greatly contributed to the study of these processes, which has been proved in a number of monographs and reviews.

Slipping and twinning were considered in the literature to be two determining mechanisms of plastic deformation. They are both related to the initiation and motion of dislocations and are accompanied by the generation of elastic waves, which are actually AE [82]. Generally speaking, in the research mentioned, it is also noted that the conditions of deformation, type, and state of a material, type of crystalline lattice, and many other factors have a great effect on the AE parameters. In further studies, the authors of [83] show that the accelerated motion of dislocations and their abrupt arrest is the source of the AE. In the last case, additional friction tensions of about 5 MPa are formed—for example, for crystal Cu-10% Zn. Actually, this is almost a low limit of the yield strength of the alloy, and the mechanism of the AE suspension is caused by shear stresses higher than the frictional stress. The authors suppose that these particular mechanisms are typical of plastic deformation, and in a crystal that contains many local dislocation fields, the character of the movement of dislocations changes from quasi-static to dynamic with the growing average speed.

When carrying out such investigations, one has to take into account the very important fact that sometimes it is difficult to eliminate the sources of extraneous noise [84]. This fact somewhat slowed down the development of experimental AE methods of testing and they were considered to be just the methods supplementary to the known technologies of non-destructive testing, or simply as a method indicating the fracture process. For instance, the AE presence or absence indicated taking into account a deviation from linearity in the "load-displacement" diagram that there is either a crack propagation or a plastic deformation development at its tip [85, 86]. Still, the great potential capabilities of the AE method made it possible to effectively use it not only for diagnostics of structures in which sub-critical crack growth or stress corrosion cracking take place [86], but also in fundamental studies.

For instance, the dislocation mechanisms of the AES generation in lithium fluoride single crystals deformed by a load applied to the slip planes were studied in [87]. The displacement of 25×10^{-8} m was accompanied by AE pulses of relatively large amplitudes, while smaller displacements of 0.5×10^{-8} m caused a synchronous excitation of pulses of small amplitudes that were always observed in tests, while some durable AE pulses did not correspond to any displacements. Based on these results, the mechanism of the AES excitation was considered and the dislocation group velocities were estimated. The results concur with the dislocation velocity measurements in lithium fluoride. The mechanism enables the interaction of the accumulated dislocation groups and barriers that cause them, around which local stress energy increases. When the minimum value of the energy required for detachment and accelerated movement of a part of a dislocation group is reached at the head of a dislocation group, this local stress energy is enough to cause crystal lattice vibrations leading to the AE initiation. The process of the secondary AE formation as a result of collisions between the coherent groups of moving dislocations and obstacles in their slip planes is considered as well. Finally, the emission with an alternative type of vibrations is observed at high stresses after a vast plastic deformation when stress concentration at the head of a dislocation group causes the initiation and propagation of micro-cracks. In their turn, the authors of [88], having investigated the deformation of monocrystals of copper, magnesium, iron and polycrystals of brass, and Cu-7% Al alloy, concluded that the AE bursts correspond to the process of the formation of slip bands by avalanche dislocation motion. They estimated an average strain ε_{av} per one AE pulse. For example, for a magnesium monocrystal $\varepsilon_{av} = 8 \times 10^{-7}$, and a specimen elongation of 4×10^{-8} m that corresponds to it, while for a manganese monocrystal, these parameters are, respectively, $\varepsilon_{av} = 5 \times 10^{-7}$ and 2×10^{-8} m. Using electron microscopy, they found the length of a free path of dislocations at the stage of easy slipping, which was 10^{-4} m, and every dislocation source consisted of approximately 40 dislocations. The authors showed that for each source of this kind there was the value of deformation $\varepsilon_{av} \cong 2.5 \times 10^{-11}$, and during a slip band formation, a single AE pulse was generated with the participation of about 2×10^4 dislocation sources. At that time, the results of the AE investigations were usually too rough, and the evaluation of an average value of strain for each AE pulse by dividing a total strain by the same total number of pulses was not quite correct. In fact, it is now well known (as will be shown below) that the character of the AE varies at the stage of strain hardening. Besides, the results presented above testify to the appearance of considerable AE during a non-stationary movement of dislocations at the section of a free path, which is determined as a distance between dislocation groups. In this study, an empiric estimation of the displacement per one AE pulse due to an avalanche movement of dislocations is by a few orders of magnitude higher than a similar value for the twinning [89]. At the same time, the amplitude of the AES pulses, and hence the value of displacement in twinning is considerably higher in comparison with the dislocation slipping [90, 91]. Experimental data [90] show that the beginning of twinning during deformation of annealed polycrystalline zinc of technical cleanness causes the AE intensity increase approximately by one order of magnitude as compared with slipping, and an abrupt growth of this parameter verifies the beginning of cracking. In [91], having examined the nature of continuous and discrete AE, the authors tend to conclude that the first one originates due to dislocation accumulation and detachment, while the second one is of a varied nature.

It should be noted that in [92], the first attempt was made to estimate the time of the AE pulse growth based on the data on the dislocation movement speed at free path. According to the data, this time is about 3 μs. It is worth nothing that at the early stages of the AE investigations, the quantities presented in various works were not reliable enough, due to imperfect measuring systems and facilities. Moreover, a number of authors do not present their specifications of even such important parameters of selection and processing of the obtained results as amplification factors of the AE path, sensitivity and frequency parameters of AET, threshold level of the AES, pass band of a channel, etc., in their publications. This, in turn, does not enable comparison of the results and somewhat diminishes their scientific value.

Despite the above, researchers are intensely proceeding to apply the AE method for further studies.

Significant results, which are helpful in understanding the nature of AES sources, are published in [92–94], where the KCl and LiF alkaline-halogen crystals were investigated. By applying the etching method to dislocations in LiF monocrystals, the authors demonstrated [94] that at the initial stages of plastic deformation, the AE source is a detachment of dislocations from strong fixation points. In annealed crystals, this process stipulates the beginning of micro-yielding, and during subsequent deformation of crystals it brings about macro-yielding, when the detachment of dislocations from strong fixation points begins. According to the research results, it was found that the simultaneous movement of a huge number of dislocations caused the generation of each AE pulse. If we sum up the length of dislocations, the value will be of a few thousand centimeters. A hypothesis is proposed in the paper, according to which the AES intensity depends on the degree of deformation that is to be determined by the function of distribution of the length of dislocation segments in a crystal. From the above, it follows that at certain moments, a simultaneous detachment of many dislocation segments of identical length occurs. It was assumed that a few groups of segments, having the amount of dislocations from 10^5 to 10^6 in each of them, could detach from the fixation points due to the effect of elastic vibrations during acoustic wave propagation. Taking into account the summing up of AE pulses, one can consider the stimulation of detachment of other dislocation segments of a smaller length due to stress wave propagation in a crystal. From this point of view, one can explain, for example, such features of the plastic deformation processes as discontinuous crystal yielding, the formation of slip bands, and low temperature brittleness of crystals, and determine their kinetics by the amplitudes of the AES of the discrete (burst) type as well as the frequency of their arrival. Later on, high efficiency of the AE method was also proved by experimental results obtained while studying the processes of deformation of molybdenum monocrystals [95].

In [96, 97], one sees the possibility of determining the elasticity limit or the yield strength of steels and alloys in tensile tests by the AE method. A drawback of these studies is the lack of systematic comparison of stresses σ_{AE} that correspond to an abrupt growth of the AE activity, with characteristics of resistance to small plastic deformations obtained by standard methods. Besides, there are almost no publications that deal with the effect of technological factors, i.e., heat treatment, contamination by non-metal inclusions, etc., on the regularities of the AE change in steels, and possible variations of the relation between σ_{AE} and σ_p, $\sigma_{0.05}$, $\sigma_{0.2}$. In [98], the regularities of the AES change under tensile deformation of medium-carbon structural steel specimens are investigated after their quenching and tempering at various temperatures, giving a different level of material strength. Special attention was paid to a comparison of characteristics of resistance to small PD obtained according to the national standard and the AE method. The experiments were conducted on steels that are widely used in engineering: Steel 40 and 38XN3MFA steel, which contain a similar amount of carbon (0.4%). Specimens of Steel 40 were cut out of the hot-rolled metal ingots along the roll direction and out of 38XN3MFA steel along and across the roll direction. The specimen treatment conditions were tempering from 860 °C in oil and annealing at 200…600 °C, to provide a wide range of strength and plasticity of steels.

Experimental estimation of some strength characteristics of structural steels. The sizes of cylindrical smooth specimens in a working part were conventional ($d_0 = 5$ mm, $l_0 = 25$ mm), and the heads of specimens were of a larger diameter for the AET mounting and had circular plane sites to reduce the effect of friction at the grips at the AES level. To this end, the heads were preliminarily crimped to a maximum load, which was expected in testing.

In accordance with the national standard, a universal 1958U-10 machine with the traverse movement rate of 2 mm/min was used in testing. During the testings, the diagrams "load—elongation" (P–Δl) and "AE signals amplitude—elongation" (U_{AE}–Δl) were synchronously recorded.

The AE were recorded by an AET of a resonance type ($f_0 = 145$ kHz), having a sensitive factor of 10^9 V/m, which was mounted on the specimen edge surface and was connected through a preliminary amplifier with an AKT–1 device [99]. In order to plot the U_{AE}-Δl diagram, the scanning along the X axis of a plotter was carried out by a signal from the strain gauge of elongation, and the other plotter simultaneously recorded the "P-Δl" diagram. The scales of elongation recorded by two plotters were 400:1. An oscilloscope controlled the shape of the AE signals.

The test results were processed as follows: Using a "P-Δl" diagram, and taking into account the initial area of the specimen cross-section, the values of $\sigma_{0.05}$, $\sigma_{0.2}$, σ_y or σ_{pp} were calculated using a standard method for the case of the presence of a yielding region (marked below as σ_i). In U_{AE}-Δl and P-Δl diagrams, the loads were found that corresponded to the beginning and the maximum of the AE signals near the yield strength region, and then critical stresses σ^H_{AE} and σ^M_{AE} were calculated. The point at which the U_{AE} value twice exceeded the AE background level in the elastic region of the strain curve was taken as the beginning of the AE peak. To compare σ_i and σ_{AE} values, the deviations were calculated:

$$\Delta\sigma_i^H = \left|\frac{\sigma^H_{AE} - \sigma_i}{\sigma_i}\right| \cdot 100\%;$$

$$\Delta\sigma_i^M = \left|\frac{\sigma^H_{AE} - \sigma_i}{\sigma_i}\right| \cdot 100\%.$$

The results of the AE and mechanical tests for two temperatures are presented as diagrams in Fig. 5.31. An irregular character of AE is a common feature for steel deformation. A relatively high level of the AE signals was recorded in the elastic region of loading. In this case, in similar research on "clean" iron, the AE in this region was not actually observed. Close to the yield strength point, an abrupt growth of AE, a peak value, was observed.

The investigations showed that the values of U_{AE} in the elastic region and close to the values of the yield strength depended on the structural state that changed, due

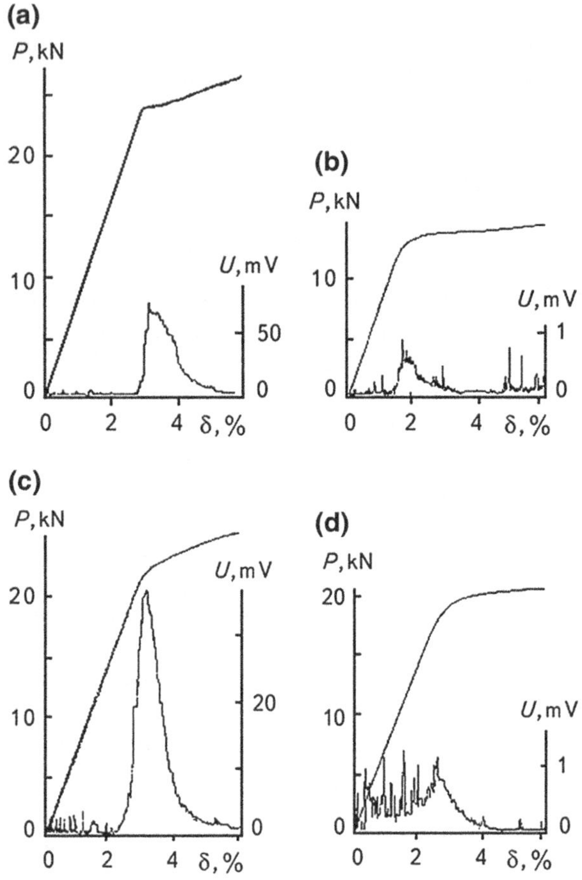

Fig. 5.31 Tensile diagram and the dependence of the AES amplitudes on the strain of Steel 40 and (**a**, **b**) and 38XN3MFA (**c**, **d**) steel; tempering temperature: **a**, **c** is 350 °C; **b**, **d** is 600 °C

to steel tempering at different temperatures. In the strengthening area, the value of the AE signals decreased to the instrumental noise level. By general features, these regularities of the AE change were identical for both tested steels [98].

The most interesting thing in determining σ_p, $\sigma_{0.05}$, $\sigma_{0.2}$ by the AE method was a regular peak of AE at the yield strength point. In the case of steel alloying, the results were obtained for both orientations of the specimen cut-off. The AE peak maximum corresponds to the elasticity limit $\sigma_{0.05}$, and its beginning corresponds to the proportionality limit σ_p. The yield strength $\sigma_{0.2}$ of 38XN3MFA steel considerably differs from the values σ_{AE}^{M} and σ_{AE}^{H} for all temperatures of tempering.

A comparison of experimental data made it possible to determine the deviations of σ_p and $\sigma_{0.05}$ for structural steels using the AE method. For instance, a measured deviation of the average values of σ_p and $\sigma_{0.05}$ was 2.7% for 38XN3MFA steel. Although the results obtained for Steel 40 showed a number of specific features, for tempering temperature of 350 °C, the values of σ_p, $\sigma_{0.05}$ and $\sigma_{0.2}$, σ_{AE}^{M} did not differ substantially.

Thus it is shown in [98] that the AE method enables detecting the early stages of the failure materials at any fracture mechanisms. Therefore, from the positions of linear fracture mechanics, and owing to its advantages, the method is more frequently used in fundamental and applied investigations of both the processes of plastic deformation and fracture initiation and propagation in various structural materials, products, and structures. Besides, the investigation of physical and mechanical characteristics of structural materials, their volume damaging, and the dynamics of fracture by the AE signals is impossible without using the correct calculation models and original devices and equipment that provide the necessary testing conditions.

5.5.4 AES Generation Under Reinforced Concrete Beam Bending

Some theoretical ideas. Reinforced concrete beams of 100 × 200 × 2100 mm were tested. They were prepared from the concrete mixture with the following composition: cement (grade M 500)—2.20 kg, sand (the fineness modulus M_k = 1.784)—0.814 kg, macadam (fractions 5 … 20 + 30% of fraction 20…40)—1.226 kg, a water/cement ratio—0.4.

A beam framework consisted of reinforcement beams with a diameter of 12 mm of Class III and a length of 2.08 m. In their edge parts, a transversal reinforcement was provided by beams of a diameter of 3 mm of Vr-1 class and the length of 0.18 m, step of 0.1 m. The total number of beams was 16. The upper reinforcement was made of two Vr-1 class beams of 3 mm in diameter and 0.73 m in length. The percentage of a cross-section reinforcement of the structure was 0.642%. The construction of the tested beam and the reinforcement chart are shown in Fig. 5.32. Basic characteristics of 35 GS steel working reinforcement were as follows: The diameter of beams was 12 mm, cross-section area A_s = 1.313 sm^2, elasticity modulus = 20.6 × 104 MPa, ultimate strength σ_{in} = 621 MPa, yield strength σ_y = 436 MPa, and relative elongation δ = 24.24%.

The theoretical calculations of a beam were done as follows: Taking into account its construction and loading mode, basic force and deformation parameters were calculated: bending moment *M*, bending moment at the beginning of crack initiation M_{crc} and deflection *f* in the middle section of the beam [99]. The strength value of the cubes of 100 × 100 × 100 mm made of concrete described in [40, 100] was preliminarily experimentally evaluated. The concrete compression strength obtained by means of the above method was R_b = 35 MPa, E_b = 34 × 10^2 MPa.

The bending moment *M* was calculated according to this expression: $M = bh_0^2 R_b \xi(1 - 0.5\xi)$ = 8.67 kN × m, where b = 100 mm and h_0 = 176 mm are the working width and height of the beam cross-section, respectively, the factor $\xi = x/h_0$, and the variable x was defined as $x = R_s A_s / R_b b$, where *Rs* is a standard tensile resistance of reinforcement.

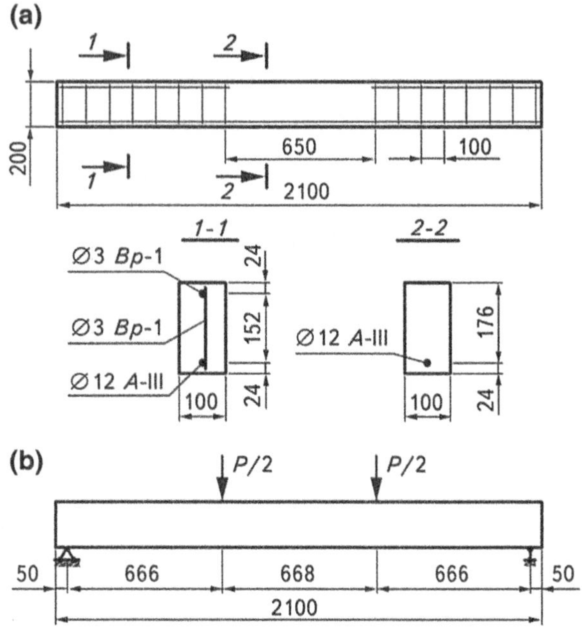

Fig. 5.32 Reinforced concrete beam (**a**) and a chart of its four-point bending (**b**)

Theoretically evaluated permissible failure force P_{max} with no account of the beam's own weight was $P_{max} = 6\ M/l_0 = 26$ kN. The bending moment M_{crc} of the beginning of the crack initiation in a concrete under condition of elastic strain in a compressed area was determined by the formula $M_{crc} = R_{btm}W_{pl} = 3.88$ kN × m, where $R_{btm} = 0.18\sqrt[3]{R_b^2}$, W_{pl} is the elastic-plastic moment of the cross-section resistance in the stretched area for reinforced concrete elements of a rectangular, non-pre-stressed cross-section that is determined from the dependence $W_{pl} = (0.293 - 1,5E_sA_s/E_bbh)bh^2$ [101].

The value of the theoretically calculated load of the crack initiation P_{crc} was determined as $P_{crc} = 6M_{crc}/l_0 = 11.5$ kN. The calculated maximum deflection f in the mid-section of the beam was established using the methods of structural mechanics [102], and in our case its value was $f = 3.9$ mm.

Results of experimental research. The concrete mixture was prepared for 3… 4 min in a concrete mixer of a forced action equipped with the mass metering devices of a volume of 0.5 m^3. The specimens were made of concrete from one batch in a rigid metallic cassette casing that was designed for eight beams, and the mixture was compacted on the vibrating bench. After preparing the specimens, they were put into a steam-curing chamber to acquire strength, where no additional pressure was created, and a stationary temperature of +15 °C was maintained. To avoid the negative action of condensate that gathered on the lower side of the steam-curing chamber lid, specimens were preliminarily covered with a polyethylene film. There was no wetting of the specimens, and the casing was removed after 8…10 days, when the specimens were formed.

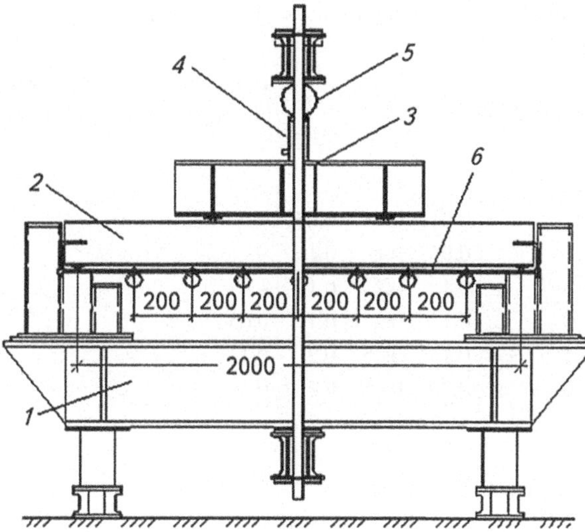

Fig. 5.33 Testing of a reinforced concrete beam on a force stand: *1* is basis; *2* is beam; *3* is distributing traverse; *4* is jack; *5* is dynamometer; *6* is frame with indicators

Reinforced concrete beams that were 460…500 days old were subjected to four-point bending. They were loaded on a power stand (Fig. 5.33) by two concentrated forces using a hydraulic jack. The force was created stepwise (holding at each step not less than 15 min before taking the recordings from devices) according to the following scheme: The first three steps of approximately 30% of M_{crc}, the following three steps—10% of M, the other steps—20% of M. The total time of loading step duration was 50 min, and its level was controlled by a calibrated circular dynamometer that was set above the distributing traverse, and by the readings of manometer indicators of a pumping facility. The readings of dynamometer indicators were taken immediately after reaching the necessary loading level and were controlled during each loading step.

In testing, a beam deflection under bending was measured by a frame with clockwise indicators with a scale factor of 0.01 mm. Deflection was measured before and after taking the readings at a loading step.

The AE signals, which occur in a beam during the initiation and development of fracture, were recorded by a programming AE device, AKEM [103]. The AET was mounted through the layer of an acoustic transparent lubricant ("Ramzai") on the lower beam face at a distance of 0.8 m from its edge and pressed to it by a rubber ribbon with a force of 20…30 N. Signals were amplified and transmitted to a computer for processing and for subsequent analysis with the use of the AKEM software. The amplification factor of AE path was 90 dB in the frequency band of 0.1…2.0 MHz.

The crack opening was optically measured using the MPB-2 microscope (×24, scale factor—0.05 mm). Measurements were conducted at each loading step from the moment of crack initiation to the value of 0.7…0.8 P_{max}. The macro-crack opening displacements, the crack propagation, orientation, and the formation of new cracks were recorded at each loading step (Table 5.8).

Table 5.8 Experimental values of crack opening displacements at the beam surfaces in the working area

Load, kN	Crack opening displacement, mm							
	Number of the detected crack							
	1	2	3	4	5	6	7	8
5	–	–	–	–	–	0.01	–	–
7.5	0.01	0.05	0.05	–	0.1	0.05	0.07	–
10	0.05	0.1	0.1	–	0.2	0.09	0.1	0.06
15	0.1	0.2	0.12	–	0.13	0.1	0.15	0.1
20	0.1	0.2	0.15	0.05	0.2	0.15	0.15	0.17
25	0.1	0.45	0.2	0.1	0.25	0.25	0.25	0.3
30	0.25	0.45	0.4	0.1	0.3	0.4	0.4	0.3

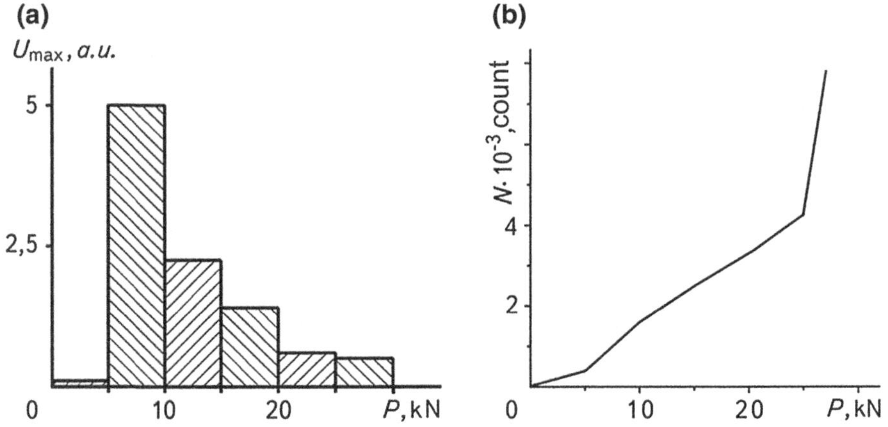

Fig. 5.34 The dependence of distribution of maximum amplitudes of the AES U_{max} on the applied load (in arbitrary units, a.u.) (**a**) and their cumulative count N (**b**)

In Fig. 5.34, a histogram of the distribution of the AE signal maximum amplitudes U_{max} (in arbitrary units, a.u.) and a cumulative count are shown for a certain range of loading of a beam.

As follows from Fig. 5.34, the low-amplitude AE is mainly generated in the interval of loads from 0 to 5 kN. Later on, the single bursts of discrete AE occur with relatively high amplitude that confirms that a crack is initiated in the beam bulk. This is confirmed by a visual observation of its surfaces (Table 5.8). At the same time, if the visually discovered surface cracks did not open during this interval of the applied load, their opening and growth would be observed farther, i.e., their surface initiation and possible propagation into a beam body are also obvious. The subsequent loading leads to a decrease of the amplitude values of signals with an increase of their number. Such a tendency is observed at the load step of 10…15 kN and in the ranges presented in Fig. 5.34 and in Table 5.8.

Figure 5.34b presents the clearly distinguishable three intervals. At the first one, in the range of the applied load of 0...5 kN (the first stage of fracture), an insignificant growth of the cumulative count *N* of the AES is noticed. Then, in the range of 5...25.5 kN, this value grows more rapidly (the second stage of fracture) and then again abruptly increases in the range of 25.5...27 kN (the third stage of fracture). According to the published data [38], a jump-like crack initiation is observed at the first stage. It increases under further loading with simultaneous propagation and coalescence of the cracks. At the third stage, the fracture is avalanche-like (Table 5.8). It should be noted that integrally, this process makes its own contribution to the delamination of reinforcement from a concrete matrix. In our opinion, it should be investigated by special experiments, because in the elastic area of loading of the inspected objects, this factor can substantially affect the interpretation of the results.

Thus, the stages of fracture initiation and development are well detected using the AE method. Combining it together with other methods of non-destructive testing provides a satisfactory correlation of the results obtained.

References

1. Skalskiy VR (2003) Development of methods and facilities for estimation of volume damaging and fracture of materials and products by the parameters of acoustic emission. Theses of Doctor's Degree in Engineering, Karpenko Physico-Mechanical Institute of the NAS of Ukraine, Lviv
2. Skalskiy VR, Andreikiv OY (2000) Do problemi avtomatizatsiyi diagnostuvannya konstrukziy metodom akustichnoyi emisiyi (To the problem of automation of diagnostics of constructions by the acoustic emission method). Tekhnicheskaja diagnostika i nerazrushajuschii control (Technical diagnostics and nondestructive testing) 4:3–9
3. Pollock AA (1979) Quantitative evaluation of acoustic emission from plastic zone growth. Int Advances in NDT 6:239–262
4. Kisi T, Ono T, Kuribajasi T (1981) Generation of acoustic emission and mechanical conditions during fracture toughness testing in elasto-pastic region. Khihakai Kensa 30 (11):896–902
5. Irwin GR (1960) Plastic zone near a crack and fracture toughness. In: Proceedings of 7th sagamore conference
6. Stadnyk MM (1988) Ob odnom metode priblizhennogo resheniya trechmernoy uprugoy zadachi dlya tela s tonkim vklyucheniem (On one method of approximation solution of 3D elastic problem for a body with a fine inclusion). Fiziko-chimicheskaya mechanika materialov (Physicochemical Mech Mater) 1:53–65
7. Cherepanov GP (1974) Mechanika chrupkogo razrusheniya (Brittle Fract Mech). Nauka, Moskva
8. (1985) GOST 25.506–85. Raschety i ispytaniya na prochnost'. Metody mecha-nicheskich ispytaniy metallov. Opredelenie charakteristik treschino-stoykosti (vyazkosti razrusheniya) pri staticheskom nagruzhenii. Vved. v deystvie 27.03.1985 g. (State Standard 25.506–85. Calculation and testing for strength. Methods of materials mechanical testing. Determination of crack growth resistance characteristics (fracture toughness). Implemented 27.03.1985). Izdatel'stvo standartov, Moskva

9. Skalskiy VR et al (1998) Pristroyi i ustanovki dlya viznachennya trischinostiykosti konstrukziynich materialiv metodom akustichnoyi emisiyi (Devices and equipment for assessment of crack growth resistance of structural materials by the method of acoustic emission). Preprint, NAN Ukrayiny, Fizyko-mechanichnyi institut, 1(1998), L'viv
10. Andreykiv AY et al (1990) Metodicheskie aspekty primeneniya metoda akusticheskoy emissii pri opredelenii staticheskoy treschinostoykosti materialov (Methodical aspects of application of the acoustic emission method for determining static crack growth resistance of materials). Preprint, NAN Ukrayini, Fizyko-mechanichnyi institut, 165(1990), L'viv
11. Lysak MV, Skalskyi VR (1997) Metodichnyi pidchid dlya eksperimental'noyi akustiko–emisiynoyi ozinky trischinotryvkosti konstrukziynych materialiv (Methodical approach to experimental acoustic-emission assessment of crack growth resistance of structural materials). Fizyko-chimichna mechanika materialiv (Physicochemical Mech Mater) 5:17–30
12. Andreikiv OY, Skalskyi VR, Lysak MV (1994) Sposib kontrolu rostu trischyn u zrazkakh materialiv (A method of checking the growth of cracks in the material specimens). Patent of Ukraine N2914, MPK: G01N29/14, Bul. 5–1, 26 Dec 1994
13. Kishi T (1985) Acoustic emission source characterization and its application to micro-cracking. Z Metall 76(7):512–518
14. Novikov NV, Likhatskii SI, Maistrenko AL (1973) Opredelenie momenta stragivaniya treschiny akusticheskim metodom pri ispytanii obrazzov s nadrezom na vnezentrennoe rastyazhenie (Determination of the crack start moment by the acoustic method during eccentric tensile testing of specimens with notches). Problemy prochnosti (Prob Strength) 9:21–25
15. Vainberg VYe, Sosedpov VN, Kushnir AM (1975) Issledovanie rosta treschin metodom akusticheskoy emissii (Investigation of the crack growth by the acoustic emission method). Defektoskopia 3:127–129
16. Andreykiv OY et al (1998) Determination of threshold values of stress intensity factor using acoustic emission method. In: Proceedings of 18–th symposium on experimental mechanics of solids. Jachranka near Warsaw, Poland, 14–16 Oct 1998
17. Smirnov VI (1979) Ob ozenke razmerov defektov metodom akusticheskoy emissii s pozizii lineynoy mechaniki razrusheniya (On estimation of defects sizes by the acoustic emission method from the positions of linear fracture mechanics). Defektoskopia 2:45–50
18. Korovkin ED, Skoblo AB, Dudina LP (1980) Opredelenie starta treschiny akusticheskim metodom (Determination of the crack start by the acoustic method). Zavod Lab 9:865–867
19. Jones MH, Brown WF Jr (1964) Acoustic detection of crack initiation in sharply notched specimens. Mater Res Stand 4(3):120–129
20. Kaiser J (1950) Untersuchungen uber das Auftreter Geraushen beim Zugversuch. Ph. D. Thesis, Technische Hochschule München
21. Lysak MW, Andrejkiw OJe, Skalskyj WR (1998) Schallemissionsmessungen beim Rißfortschritt in Stah unter statischer Belastung. Mater und Werkstofftechnik 29(2):90–93
22. Zazuliak VA et al (1984) Treschinostoykost' materiala krupnych opornych valkov prokatnych stanov (Crack growth resistance of large support rolls of rolling mills). Fiziko-chimicheskaya mechanika materialov (Physicochemical Mech Mater) 5:95–96
23. Tkach AN et al (1988) Metodicheskie osobennosti opredeleniya staticheskoy treschinostoykosti chugunok (Methodical peculiarities of determination of static crack growth resitance of cast irons). Ibid 1:68–73
24. Andreykiv OY et al (1990) Zastosuvannya akustychnoyi emisiyi dlya viznachennya statichnoyi trischynostiykosti lytych tytanovych splaviv (Application of acoustic emission for determination of static crack growth resistance of cast titanium alloys). Ibid 1:103–107
25. Lysak NV, Skalskiy VR, Luchko YY (1989) Akusticheskaya emissiya i razrushenie betona pri staticheskom nagruzhenii (Acoustic emission and fracture of concrete under static loading). Izvestia Vyshykh Uchebnykh Zavedenii. Stroitelstvo i Arkhitektura 12:48–51
26. Lysak NV, Skalskiy VR (1986) Issledovanie subkriticheskogo rosta treschin s pomosch'yu akusticheskoy emissii (Investigation of subcritical crack growth using acoustic emission). Fiziko-chimicheskaya mechanika materialov (Physicochemical Mech Mater) 4:113–114

27. Lysak NV, Skalskiy VR, Serhienko ON (1989) Ispol'zovanie metoda akusticheskoy emissii dlya issledovaniya razrusheniya chugunov (Application of the acoustic emission method for investigation of cast irons fracture). Technicheskaja diagnostika i nerazrushajuschij kontrol (Tech Diagn Nondestr Test) 3:37–45
28. Andreykiv AYe, Lysak NV (1989) Metod akusticheskoy emissii v issledovanii prozessov razrusheniya (A method of acoustic emission in investigation of the fracture processes). Naukova Dumka, Kiev
29. Aravas N, McMeecking RM (1985) Finite element analysis of void growth near blunting crack tip. J Mech Phys Solids 13(1):25–49
30. McMeeking RM (1977) Finite deformation analysis of crack tip opening in elastic-plastic materials and implications for fracture. J Mech Phys Solids 25(5):357–381
31. Skalskiy VR et al (2004) Doslidzhennya tverdinnya betonu za sygnalamy akustychnoi emisii (Investigations of concrete hardening by acoustic emission signals). Fizyko-chimichna mechanika materialiv (Physicochemical Mech Mater) 5:104–106
32. Postanova Kabinetu ministriv Ukrayiny "Pro zabezpechennya nadiynosti i bezpechnoyi ekspluataziyi budivel' sporud ta inzhenernych sporud" No 409 vid 05 travnya 1997 r. (Resolution of the Cabinet of Ministers of Ukraine "On providing reliability and safe exploitation of buildings and engineering constructions" No. 409, May 05, 1997)
33. Normativni dokumenty z pytan' obstezhen', pasportyzaziyi, bezpechnoyi ta nadiynoyi ekspluataziyi vyrobnychych budivel' ta sporud. Vvedeni 01 grudnya 1997 r. (Normative documents on the problems of inspections, certification, safe and reliable exploitation of production buildings and constructions. Implemented on December 01, 1997)
34. Iosylevskyi LI et al (1986) Zhelezobetonnye proletnye stroeniya mostov industrial'nogo izgotovleniya (Konstruirovanie i metody rascheta) (Iron-concrete span constructions of prefabricated bridges (Design and methods of calculation)). Transport, Moskva
35. (1995) DSTU 2865-94. Kontrol' neruynivnyi. Terminy ta vyznachennya (Standard 2865-94. Nondestructive testing. Terms and definitions. Izdatel'stvo standartov, Kiev)
36. Muravin GB, Mochulskyi VA, Pavlovskaya FS (1989) Issledovaniya mechanizma izlucheniya voln napryazheniy pri tverdenii dvuchkompozitnoy smesi (Investigations of the irradiation mechanism of elastic waves during hardening of a two-composite mixture). Transpornoye Stroitelstvo 5:28–29
37. Muravin GB, Pavlovskaya GS, Shchurov AT (1984) Issledovanie akusticheskoy emissii tverdeyuschego betona (Investigation of acoustic emission of hardening concrete). Defektoskopia 10:77–81
38. Skalskiy VR, Demchyna BG, Karpukhin II (2000) Ruynuvannya betoniv i akustychna emisiya (Oglyad). Povidomlennya 1. Statychne navantazhennya i vplyv temperaturnogo polya (Fracture of concretes and acoustic emission (a review). Report 1. Static loading and temperature field effect). Tekhnicheskaja diagnostika i nerazrushajuschii kontrol (Technical diagnostics and nondestructive testing) 1:12–23
39. Muravin GB, Sniezhytskyi YuS, Pavlovskaya GS (1989) Issledovanie prozessa tverdeniya betona pri nizkich temperaturach metodom akusticheskoy emissii (Investigation of the process of concretes hardening at low temperatures by acoustic emission). Defektoskopia 10:9–15
40. Skalskiy VR, Lototskyi YL (2004) Ozinka amplitud sygnaliv akustychnoyi emisiyi pid chas ruynuvannya betonnych kubiv (Estimation of acoustic emission signals amplitude during concrete cubes fracture). Fizychni metody ta zasoby kontrolyu seredovysch, materialiv ta vyrobiv (Physical methods and facilities for testing environments, materials and products) 9:54–61
41. Skalskyi VR (2001) Okremi metodologichni zasady rozroblennya prystroyiv dlya peredavannya akustychnoyi emisiyi (Some methodological bases for designing devices for acoustic emission transmission). Mashinoznavstvo (Mech Eng) 7:49–52
42. Grinchenko VT (1978) Ravnovesie i ustanovivshiesya kolebaniya uprugich tel konechnych razmerov (Equilibrium and steady-state vibrations of elastic bodies of finite dimensions). Naukova Dumka, Kiev

43. Chernyshov VV, Shegai VV (1977) Sobstvennye kolebaniya tverdych zilindrov konechnoy dliny (Natural vibrations of solid cylinders of infinite length). Acusticheskii zhurnal (Acoust J) 23(4):627–631
44. Komissarova GA (1975) O podchode k issledovaniyu nestazionarnych kolebaniy zilindra konechnoy dliny (On an approach to investigation of non-stationary vibrations of a cylinder of finite length). Voprosy matematicheskoi fiziki i teorii kolebanii 3:96–102
45. Imenitova Y, Chernyshov VV, Shegai VV (1976) O raschete svobodnych kolebaniy uprugich zilindrov konechnoy dliny (On calculations of free vibrations of elastic cylinders of finite length). Doklady AN SSSR. 226(2):315–317
46. Lysak MV, Skalskiy VR, Serhiyenko OM (1994) Doslidzhennya vplyvu chvylevodu na zminu parametriv sygnaliv akustychnoyi emisiyi (Investigation of the waveguide effect on the change in acoustic emission signals parameters). Fizyko-chimichna mechanika materialiv (Physicochemical Mech Mater) 2:64–70
47. Andreykiv OYe et al (2001) Analysis of acoustic emission caused by internal cracks. Eng Fract Mech 68(7):1317–1333
48. Andreykiv OYe (1982) Prostranstvennye zadachi teorii treschin (Spatial problems of the crack theory). Naukova Dumka, Kiev
49. Kikuchi E (ed) (1972) Ul'trazvukovye preobrazovateli (Ultrasonic transducers). Mir, Moskva
50. Zienkiewicz OC (1975) Metod konechnych elementov v technike (The finite element method in engineering science). Mir, Moskva
51. Steng G, Fink GJ (1977) Teoriya metoda konechnych elementov (An analysis of the finite element method). Mir, Moskva
52. Mironov SA, Malinina LA (1961) Uskoreniya tverdeniya betonu (Acceleration of concrete hardening). Academy of construction and architecture of the USSR, Moskva
53. Sheikin AYe (1974) Struktura, prochnost' i treschinostoykost' zementnogo kamnya (Structure, strength and crack growth resistance of cement stone). Strojizdat, Moskva
54. Krivenko PV (ed) (1993) Budivel'ni materialy (Building materials). Vyshcha shkola, Kiev
55. Koval PM (2003) Vykorystannya metodu akustichnoyi emisiyi pry doslidzhenni mostiv (The use of acoustic emission method for bridges investigation). Avtoshliakhovyk Ukrainy 1:34–37
56. (2003) DSTU 4227-2003. Rekomendaziji scodo akustyko-emisijnoho kontroliu objektiv pidvyschenoji nebezpeky (Standard of Ukraine 4227-2003. Recommendations for acoustic emission monitoring of high-risk facilities), Kiev, Derzhspozhyvstandart Ukrayiny
57. Koval PM, Stashuk PM, Fal YA (2003) Doslidzhennya progonovoyi budovy novogo stalezalizobetonnogo avtodorozhn'ogo mosta z vykoristannyam metodu akustychnoyi ekmisiyi (Investigations of span structure of a new steel reinforced concrete transport bridge with the use of acoustic emission method). Diagnostyka, dovgovichnist' ta rekonstrukziya mostiv i budivel'nich konstrukziy (Diagnostics, durability and reconstruction of bridges and building structures) 5:85–93
58. Rusch H (1959) Physikalishe Fragen der Betonprufung. Zement–Kalk–Gips 12(1):1–9
59. L'Hermite RL (1959) What do we know about the plastic deformation and creep of concrete. RILEM Bull RILBA 1:21–51
60. L'Hermite RG (1960) Volume changes of concrete. In: Chemistry of cement: Proceedings of the 4th Internalioral Symposium, Washington, 1960
61. Green AT (1969) Detection of incipient failure in pressure vessel materials by the stress–wave analysis technique. In: Gas–cooled reactor program semiannual progress. Report for Period Ending March 31, 1969
62. Dunegan HL, Green AT (1971) Factors affecting acoustic emission response from materials. Mater Res Stand 11(3):21–24
63. Dunegan HL, Tetelman AS (1971) Acoustic emission. Res Dev 22(5):20–24
64. Wells D (1959) An acoustic apparatus to record emission from concrete under strain. Nucl Eng Des 12(1):80–88

65. Pochtovik GYa et al (1972) Ispol'zovanie shumometricheskoy apparatury dlya ozenki energii razrusheniya zementno–peschanogo rastvora (Application of audio-noise meter equipment for evaluation of the fracture energy of cement – send solution). Energetycheskoje Stroitelstvo 4:64–66
66. Pochtovik GYa, Temnik NL, Fillipova NB (1974) K metodike ozenki energii razrusheniya betona akusticheskimi metodami (To the methods of evaluation of concrete fracture energy by acoustics methods). Nauchnye trudy Akademii komunal'nogo chozyaystva (Scientific publications of Academy of Municipal economy) 104:3–19
67. Smolenskaya NG, Pochtovik GYa, Temnik NL (1974) Issledovanie akusticheskimi metodami prozessa treschinoobrazovaniya betona pri dlitel'nom nagruzhenii (Investigation by acoustic methods of crack formation process in concrete under long-term loading). Ibid 104:20–26
68. Pochtovik GY, Tsybinoga BG, Gritsenko BS (1977) Sravnitel'noe issledovanie prozessov treschinoobrazovaniya v rastyanutom betone metodami akusticheskoy emissiii i mikroskopicheskim (Comparative investigation of crack formation process in a tensioned concrete by the acoustic emission and microscopic methods). Trudy MISI (Proceedings of MISI) 51:111–117
69. Arrington M, Evans BM (1977) Acoustic emission testing of high alumina cement concrete. NDT Int 7:81–87
70. Tomachevsky EG, Drouet A, Despreslas PJ (1975) Recherche de la dimination de resistance en fraction du beton par detection d'emission d'ondes de contrainte. J d'Etudes sur I'Emission Acoustique 17:336–360
71. Reymond MC (1980) Acoustic emission in rock and concrete under laboratory conditions. Microseismic Act Geologic Struct Mater Ser Rock Soil Mech 5:27–34
72. Sokolov GB et al (1977) Vliyanie vozrasta betona na prozessy mikrotreschinoobrazovaniya (The influence of concrete age on microcracks formation). Izvestiya VNII "Gidrotechnika" (Reports of VNII "Gidrotekhnika") 116:44–49
73. Pochtivik GYa, Pshenichkin AP, Gritsenko BS (1984) Mechanizm razrusheniya neodnorodnych tel i ego svyaz' s parametrami akusticheskoy emissii (Fracture mechanism of heterogeneous bodies and it correlation with the parameters of acoustic emission). In: Sbornik tezisov i dokladov I Vsesoyuznoy konferenzii "Akusticheskaya emissiya materialov i konstrukziy" (Proc. All-Union Conf. "Acoustic Emission Mater. and Struct.", 11–13 Sept 1984, Rostov-upon the-Don), vol 1. Rostov-na Donu, 1984
74. Muravin GB, Shchurov AF (1985) Issledovanie prirody akusticheskoy emissii pri staticheskom deformirovanii betona (Investigation of nature of acoustic emission under concrete static deformation). Mechanika kompozitnych materialov (Mekhanika kompozitnykh materialov) 3:557–560
75. Leshchinskyi MYu (1980) Ispytanie betona (Testing of concrete). Strojizdat, Moskva, Moscow
76. Bordiugov DM, Yermisson AL (1992) Energiya akusticheskoy emissii v prozesse razrusheniya betona (Energy of acoustic emission in concrete fracture process). Doklady i tezisy III Vsesoyuznoy konferenzii po akusticheskoy emissii. Teoriya i praktika (Reports and theses of the III All-union Conf. on acoustic emission. Theory and Practice), vol 1. Obninsk, 1992
77. Koval PM, Luchko YY, Stashuk PM (2001) Ozinka trischynostiykosti betoniv v mostovych konstrukziyach z vykorystannyam metodu akustychnoyi emisiyi (Estimation of crack growth resistance of concretes in bridge structures with the use of acoustic emission method). Diagnostyka, dovgovichnist' ta rekonstrukziya mostiv i budivel'nich konstrukziy (Diagnostics, durability and reconstruction of bridges and building structures) 3:91–100
78. Fal AYe (2003) Akustychna emisiya pry doslidzhenni betoniv dlya plyt proyiznoyi chastyny avtodorozhnich mostiv (Acoustic emission in investigation of concretes for the plates of roadway of the transport bridges). Perspektyvy rozvytku budivel'nych konstrukziy, budivel', sporud ta ich osnov (Prospects of development of building structures, buildings, structures and their bases) 58:406–412

79. Koval PM, Stashuk PM (2000) Doslidzhennya trischynostiykosti betonu metodom akustychnoyi emisiyi (Investigation of concrete crack growth resistance by the method of acoustic emission). Resursoekonomni materialy, konstrukziyi, budivli ta sporudy (Resource-saving materials, structures, and buildings) 5:309–315
80. (1990) GOST 1497-84. Metally. Metody ispytaniy na rastyazhenie. Vved. 01.11.90 g. (State Standard 497-84. Metals. Tensile test methods. Implemented 01.11.90)
81. Fitzgerald ER (1960) Mechanical resonance dispersion and plastic flow in crystalline solids. J Acoust Soc America 32(10):1270–1289
82. Ookawa A, Yazu K (1963) The energy radiated from a dislocation by an accelerated motion through impurity fields. J Phys Soc Japan 18:36–43
83. Brown WF(Jr), Srawley JE (1966) Current status of plane crack toughness testing. NASA TM X-52209. Cleveland, Ohio
84. Brown WF(Jr), Srawley JE (1967) Plane strain crack toughness testing of high strength metallic materials. In: American society for testing and materials. Philadelphia: Pennsylvania
85. Dunegan HL (1969) Ultrasonic acoustic emission from materials. IEEE Trans Sonics Ultrasonics 1:16–32
86. Engle RB (1966) Acoustic emission and related displacements in lithium fluoride single crystals. Ph.D. Thesis. Michigan State University
87. Fisher RM, Lally LS (1967) Microplasticity detected by an acoustic technique. Canad J Phys 45(2):1147–1159
88. Boiko VS et al (1970) Zvukovoye izluchenije dvojnikuuschikh dislokatsij (Sonic radiation of twin dislocations). Phizika Tverdogo Tela (Phys Solid) 12(6):1753–1755
89. Melekhin VP, Mints RI, Kugler LM (1971) Vliyanie mechanizmov plasticheskoy deformazii zinka na akusticheskuyu emissiyu (The influence of mechanisms of zinc plastic deformation on acoustic emission). Izvestija Vuzov. Tsvetnaia metallurgia 3:128–131
90. Schofield BH (1972) Research on the sources and characteristics of the acoustic emission. In: Acoustic emission, ASTM STP 505. Baltimor, pp 11–19
91. James DR (1971) The sours of acoustic emission in deforming single crystals. In: International conference on mechanical behavior of materials, 1971, Kyoto
92. Segwick RT (1968) Acoustic emission from single crystals of LiF and KCl. J Appl Phys 39 (3):1728–1740
93. James DR, Carpenter SH (1971) Relationship between acoustic emission and dislocation kinetics in crystalline solids. J Appl Phys 42(12):4685–4697
94. Gusiev OV (1982) Akusticheskaya emissiya pri deformirovanii monokristallov tugoplavkich metallov (Acoustic emission during deformation of monocrystals of refractory metals). Nauka, Moskva
95. Skoblo AV et al (1979) Primenenie akusticheskoy emissii dlya opredeleniya predela uprugusti konstrukzionnych staley (Usage of acoustic emission for determining the elasticity limit of structural steels). Zavod Lab 3:363–367
96. Pichkov SM et al (1980) Issledovanie deformirovaniya obrazzov s vytochkami metodom akusticheskoy emissii (Investigation of deformed specimens with notches by a method of acoustic emission). Fiziko-chimicheskaya mechanika materialov (Physicochemical Mech Mater) 3:120–122
97. Fadeiev YuI et al (1987) Opredelenie mechanicheskich charakteristik stali metodom akusticheskoy emissii (Determination of physicomechanical characteristics of steel by the method of acoustic emission). Defektoskopia 8:44–49
98. Barteniev OA, Khamitov VA, Fadieiev YuI (1982) Pribor akustiko-emissionnogo kontrolya treschinoobrazovaniya AKT-1 (Acoustic emission device for checking crack formation AKT-1). Information letter No 25-82. Udmurtskiy vezhotraslevoj terrritorial'nyj TsNTI, Izhevsk
99. Baikov VN, Sigalov EE (1991) Zhelezobetonnye konstrukzii: Obschiy kurs (Reinforced concrete structures: A review course), 5th edn. Strojizdat, Moskva

100. Koval PM, Stashuk PM (2001) Doslidzhennya zalizobetonnych konstrukziy metodom akustychnoyi emisiyi (Investigation of iron - concrete structures by the method of acoustic emission). Avtomobilni dorohy i dorozhnie budivnytstvo 63:276–282
101. (1984) SNiP 2.03.01—84*. Betonnye i zhelezobetonnye konstrukzii (Building regulations 2.03.01—84*. Concrete and reinforced concrete structures). Strojizda, Moskva
102. Ivanov GM, Veits RI (1968) Statika sooruzheniya (Static of constructions), 2nd edn. Strojizdat, Leningrad
103. Filonenko SF (1999) Akusticheskaya emissiya. Izmereniya, kontrol', diagnostika (Acoustic emission. Measurements, testing, diagnostics). KMGUA, Kiev

Chapter 6
Some Aspects of Applying the Acoustic Emission Method

The direct industrial application of AE as a method of non-destructive testing of the integrity of structures and their elements, under periodic testing or in service conditions, is conditioned by the necessity to overcome a number of serious difficulties that arise due to the presence of significant industrial noise; the considerable loss of information during the passage of acoustic waves through large-scale structures; solving the problem of using multi-channel equipment to locate the defects in large objects; developing effective algorithms of the AE sources' location and data processing for every type of structure of a very complicated geometrical shape and welded and other types of joints, etc. Besides, malfunctions of a device often occur during long-term experiments or NDT. Thus, for example, there is no reliable AET that can be used in aggressive environments, at elevated temperatures and pressure, at vibration, etc. Therefore, the structure of such an AET requires additional technical improvements concerning the operating conditions and service lifespan. It is necessary to develop high-speed and reliable facilities for processing the information recorded by AET and presenting the results, which would enable reducing the extra staff engaged in operating these systems of AE testing.

These difficulties are basically related to the lack of reliable data on the nature of crack propagation under various loading conditions in aggressive environments. Despite the above-mentioned difficulties, the data on the practical application of the AE method accumulated so far, clearly indicate that this type of non-destructive testing permits detecting and evaluating the location, and the stages of sub-critical crack growth in large-scale structures operating in industrial conditions.

Z. Nazarchuk et al., *Acoustic Emission*, Foundations of Engineering Mechanics,
DOI 10.1007/978-3-319-49350-3_6

6.1 Specific Features of Long-Term AE Testing of Industrial Objects

The AE signals caused by crack propagation in metals can be subdivided into two types [1]. The first represents a continuous emission and is conditioned by an avalanche motion of dislocations and the formation of stable micro-cracks in the process of the development of plastic zones. The second is discrete AE caused by the crack front advance in the PZ, and by coalescence of micropores in the crack tip area.

In general, while testing large-scale structures by the AE method, due to considerable damping of signals it is necessary to record the AES caused by the crack front movement. At sub-critical crack growth, these signals usually have the form of short pulses of considerable amplitude. They are emitted for a certain time when the crack front propagates, and then, for several days or sometimes weeks, the AE can be very slight or even absent. The average value of periods of low activity of the AE usually decreases with an increase of the stress state.

On this account, the AE testing of the investigated objects should be carried out over a long period of time; in this way, a great amount of information is gathered. This means that the AE testing system should have a large information capacity. It should also be capable of selecting and quantitatively processing the information related to the relatively short periods of a crack growth that include the AE data obtained during the test time—which can last several weeks or months. In such systems, it is also necessary to foresee the possibility of accounting the parameters of aggressive environments and other features of a technological process that urge the crack initiation and propagation.

6.1.1 Selection of a Frequency Band and AET Placing

The AES spectrum, caused by individual crack jumps at the rate of several thousand meters per second over short intervals, is rather wide and comprises a region of high frequencies. However, while moving away from the source, high-frequency components decay much more than low-frequency ones. A similar type of decaying is observed for different mechanical noises; therefore, the band of working frequencies should be appropriately chosen to maximally increase the distances at which the defects could be detected and, at the same time, to minimize the noise level by increasing the recorded frequency band. The AE system, which consists of a resonance AET and a corresponding preamplifier with a band filter that has a sensitivity maximum at a frequency of 200 kHz with the pass band within the range of 200 kHz [2], is most suitable for practical application, except for the cases in which the noise level is extremely high.

To obtain the necessary sensitivity and the maximum tested area of a structure with a minimum number of AET, it is necessary to carry out the following investigations:

1. Study the statistical distribution of the AES amplitudes near the sources for each material and fracture mechanism. Since a crack often nucleates in the welded joints, it is necessary to perform experiments on specimens with welded joints, choosing the loading conditions and service environments close to real operating conditions.
2. Investigate the character of generation of the mechanical noises caused by the operating equipment, turbulent liquid flows, and friction, as well as other technological reasons by their dependence on the conditions of structure operation.
3. Carry out experiments on the evaluation of sound dispersion in the investigated object, especially during elastic wave propagation along a weld, and at the presence of screening effects conditioned by geometrical features of a structure.
4. Establish the level of electric noises in the measuring system, as well as transitional properties of vibrations transmitted by signals from the AET to the AE testing system, in order to minimize the loss of a signal and provide a reliable screening against electromagnetic industrial noise.

The level of mechanical noises determines the threshold level during AES recording, which together with estimation of the damping value of signals in all important directions, determines the maximum distance from the AE source to the AET, which is capable of detecting the propagating defects. This determines the minimum density of AET allocation on the surface of IO.

During crack growth in welded joints of large-scale steel structures, the AES amplitudes lie mainly in the range of 70… 80 dB with respect to the level of 1 μV, on condition that measurements are done at the standard distance of 20 cm from the source. The AET sensitivity is 67 dB (with respect to the level of 1 V/μBar at a frequency of 200 kHz). In such a case, at the threshold level of 26 dB, the detection distance of a fatigue crack attains 3 m at the initial stages of its growth in various joints that are on an off-shore platform in extreme weather conditions.

Figure 6.1 shows the dependence of the amplitude of AES radiated by a fatigue crack and by a calibrating generator on the distance between the AE source and AET. A generator of pulses with an amplitude of 10 V was used to evaluate the damping of the AES amplitudes. Proceeding from this figure, a maximal possible distance between the AET of AES is determined, using the dependences between sensitivity of the channel of the receiving system, sound attenuation, and the selected threshold level.

In order to unambiguously determine and locate the AE source, only four AETs capable of recording signals from this source are required. In this case, four AET were placed in such a way that the greatest distance from the AE source inside the controlled region of a structure to the receiver did not exceed 5 m.

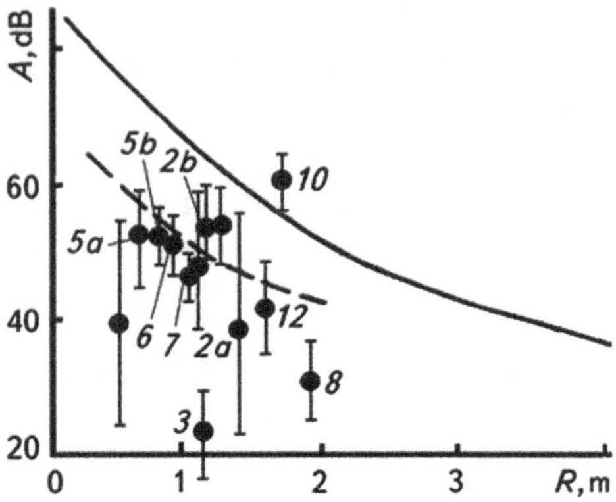

Fig. 6.1 Dependence of the AES amplitudes *A* on the distance *R* between the source of radiation and AET for different types of the AE sources: calibrating generator (*solid line*), fracture of a graphite rod (*dotted line*) and a fatigue crack (*dots with the indicated interval of amplitude variation*)

When the mechanical noise level changes, the threshold level should also change automatically; otherwise, the data recording by the AE testing system will stop at moments of a considerable increase of technological noises.

6.1.2 Calibration of an AE Testing System

The sensitivity of every AET after being mounted on the IO surface could be evaluated using the measurement results of the AES recorded by AET during the progress of an impact-simulated wave. The European working group on AE [3] recommended using the fracture of a pencil lead as an impact wave source located at a distance of 10 cm from the receiving AET center. The parameters of a pencil lead should be as follows: hardness—2H; length—3 mm; diameter—0.5 mm. Typical results of such experiments are shown in Fig. 6.1. Every group of AET located on the structure should make it possible to locate such AE source at any point of the structure region inspected by this group of the AET. At durable AE testings of structures and objects of different types—except for the calibration procedure described above—a generator of ultrasonic vibrations is used that is turned on either automatically or manually, and which is used for periodic verification of service capability of the equipment channels and their sensitivity [4].

6.1.3 Analysis and Presentation of AE Test Results

Specific types and mechanisms of fracture typical of these structures, as well as possible fields of application of the test results, require an appropriate analysis and specific methods of processing and presenting the results of durable AE testing of various structures.

It is well known that sub-critical fatigue crack growth is a slow process that requires durable observations to obtain reliable information on a defect growth. The average time between incremental steps of a crack propagation as well as the causes of the crack appearance, are important characteristics of a fracture. Furthermore, information on the stress state and the parameters of service environment, such as temperature, pressure, vibrations that arise during the operation of a structure, and because of friction, are very important for the evaluation of susceptibility of structures to the initiation and propagation of defects.

High density of the difference of the AES arrival times can be used as one of the most reliable AE parameters of—crack growth. In the case of locating four AET, the values of the difference of the AES arrival times are relative moments of time during which the signals from a source reach each AET, which is evaluated with respect to the time of a signal arrival to the first AET. Since the AET and the source positions are fixed, certain differences of the arrival time will contain a bias error that depends both on the transmission function of the environment through which an elastic wave propagates, and on the direction and distance of propagation, the AET type, and the method of mounting it on the IO.

As described in paper [5], the values of the differences of arrival times for densely localized sources that generate repeated AES usually differed by not more than one micro-second at the distances from the source to the AET of about 6 m in the welded joints of pressure vessels and in other structures. It should be noted that this situation was observed, as a rule, during sub-critical crack growth.

Periods of a strong regular AE, the duration of which, in most cases, does not exceed a few minutes, (as mentioned above), are replaced by long periods (days or even weeks) of the AE of low activity. Since it is impossible to predict the moment of a crack growth start in real structures, the AE testing should be carried out over long periods. Therefore, it is important to summarize the results for a regular period —for example, for a week or a month—using a plot for data presentation similar, for instance, to that in Fig. 6.2. This figure represents fully enough the information necessary for evaluating the degree of a defect danger, namely:

- The sites of the AE sources that can serve as a measure of evaluation of the defect length in projections are shown;
- The density of the AE events that characterizes the depth of a defect location;
- The dependencies of the number of AE events on time that determine the crack growth rate; and
- The value of various technological parameters (for instance, temperature, pressure, and vibration) that enables forming conclusions on the causes of a crack.

6.1.4 Stability of AE Parameters

The AES contain information on the parameters of dynamic processes in a solid that are accompanied by an elastic wave radiation. Usually the following parameters of

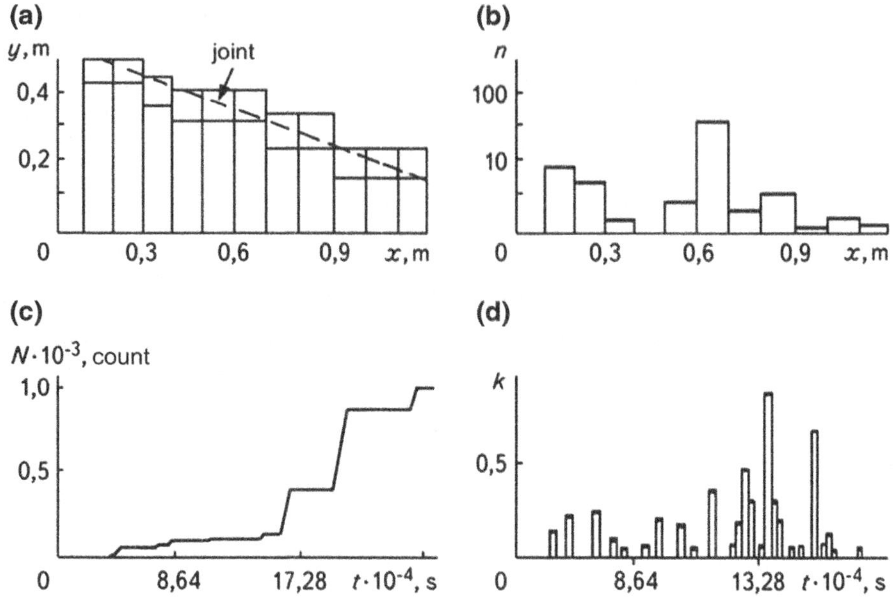

Fig. 6.2 Results of the AE testing of a weld: **a** is the graphic presentation of results of the AE location of a weld region; **b** is the distribution of the number of AE events in different regions of the welded joint; **c** is the time dependence of the cumulative AE count *N*; and **d** is the time dependence of the technological parameter

an AES are used: the number of recorded pulses; the rate of their count; energy and duration of pulses; duration of the forward and back fronts of a pulse; duration of the interval between pulses; frequency spectrum; the AE activity, etc.

When selecting a parameter or a group of AE parameters, one should take into consideration their relationship with the parameters of dynamic processes (e.g., a fracture process) that occur in a solid. Then, it is necessary to give advantage to those AE parameters that bear maximum information and are convenient for selection and processing, and are resistant to various factors that can affect these parameters.

During experiments with the AE application, as a rule, considerable data scattering is observed. To a great extent, this is related to the fact that the values of the recorded parameters depend on such factors as properties of the chosen AET, the distance from AET to the signal source, amplitude-frequency characteristics of the recording equipment, the threshold level chosen, properties and shape of the material, structural features of the investigated object, and so on. The effect of these factors on the recorded AE parameters is certainly unknown, which brings about considerable difficulties when comparing the results obtained in various conditions of the AES recording and interpreting them. Proceeding from the above-mentioned, an important problem to be solved while applying AE is establishing the stability of a particular parameter under the action of different sources of elastic wave generation.

As follows from [6], the effect of the threshold level on the AE parameters can be studied by changing this level within the signal amplitude range and by evaluating the AE parameters investigated. Within the experiment, simultaneously changing a threshold level and any other factor, for instance, the distance between the source and the AET, it is possible to evaluate their mutual effect.

The extent of stability of the AES parameters affected by a definite factor is determined as follows: Measurements of cumulative count, e.g., $N = km$ are carried out, where k is the number of the values of the investigated factor, e.g., the number of threshold levels. For each level of a factor $t = 1, 2, \ldots, k$, m measurements of the AE parameter X_{ti}, $i = 1, 2, \ldots, m$ are done. Thus, there are k groups with m measurements of parameters in each.

For every level of the factor, the average value of a parameter is calculated by the formula

$$X_t = \frac{1}{m}\sum_{i=1}^{m} X_{ti}. \tag{6.1}$$

Having averaged overall measurements, we get

$$X = \frac{1}{N}\sum_{t=1}^{k} mX_t. \tag{6.2}$$

Mean-square divergence of the average values of a parameter in a group from an overall average value is:

$$\sigma_1^2 = \frac{1}{k-1}\sum_{i=1}^{k} m(X_t - X)^2 \tag{6.3}$$

and mean-square divergence of a parameter from its average value in a group is:

$$\sigma_2^2 = \frac{1}{N-k}\sum_{t=1}^{k}\sum_{i=1}^{m}(X_{ti} - X_t)^2. \tag{6.4}$$

If the mean-square divergence between groups exceeds the mean-square of divergence in a group, this indicates that there is a real difference between the groups. The ratio of the mean-square divergences can be used for the quantitative estimation of this difference.

$$R = \sigma_1^2/\sigma_2^2. \tag{6.5}$$

To investigate the mutual effect of several factors, it is possible to introduce a measure similar to (6.5). The mean-square divergence for two factors is calculated

Table 6.1 Mean-square divergence of the parameter values for two factors

Source of changes	Mean-square divergence
First factor	$\frac{mc}{k-1}\sum_{i=1}^{k}(\bar{X}_i-\bar{X})^2$
Second factor	$\frac{mk}{c-1}\sum_{i=1}^{c}(\bar{X}_i-\bar{X})^2$
Interaction of factors	$\frac{m}{(k-1)(c-1)}\sum_{t=1}^{k}\sum_{i=1}^{c}(\bar{X}_{ti}-\bar{X}_t+\bar{X}_i+\bar{X})$

by the formulas presented in Table 6.1, where c is the number of levels of the second factor.

The relationship (6.5), as can be proved under some assumptions, has a non-central F-distribution with $(k-1)$ and $(m-k)$ degrees of freedom tabulated in [7]. As a verification criterion of the hypothesis on the general effect of a corresponding factor in case of accurately calculated level of significance, the ratio of R to the tabular value of F-distribution should be taken

$$R_F = R/F_{k-1;N-k;0.999}, \tag{6.6}$$

and, with a probability of 0.999, we should estimate whether the investigated factor affects the corresponding AE parameter. If $R_F > 1$, the effect of this factor on the respective AE parameter is observed, and the extent of its significance is proportional to R_F. Under the mutual action of several factors, the R_F value describes the variation of the effect of one of them on the AE parameters when the other factor is changed.

In order to evaluate the degree of stability of some AE parameters in the experiment, the distance of the AET from the AE source was changed. The recorded signals were processed at various threshold levels. A piezoelectric transducer, excited by a pulse electric signal, was used as a stable source of acoustic signals that permits receiving reproducible results. The experiment was carried out on a cylindrical pressure vessel 1600 mm high and 600 mm in diameter, with an 80 mm wall thickness made of shell 12X2NMFA steel. Signals were recorded by D9102 AET of a differential type produced by the Dunegan/Endevko Company. The overall amplification of the receiving channel was 80 dB. An amplified signal was recorded by a digital memory oscilloscope and then transmitted to a computer, where the AES parameters were determined. The AES were recorded by the sequential setting of a transducer on the investigated surface every 10 cm, and by conducting 4 measurements at each of its positions. Cumulative count N, maximum amplitude A, energy E, signal duration T, rise time T_r as well as signal propagation velocity v, were calculated from these data. The parameters mentioned were found for the AES recorded at the threshold levels that changed from 0.2 to 1 V with the step of 0.1 V.

Changes in the AE cumulative count depending on the distance to the AET at different threshold levels are shown in Fig. 6.3.

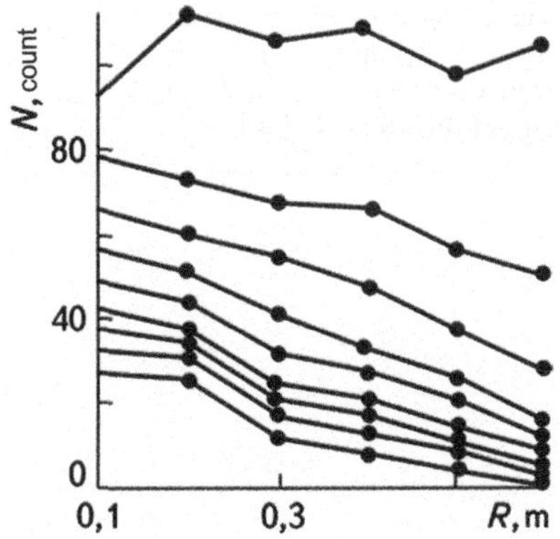

Fig. 6.3 Dependence of the AES cumulative count *N* on the distance *R* between the AE source and the AET at different values of a threshold level, which changed stepwise every 0.1 V starting from 0.2 (*upper curve*) to 1 V (*bottom curve*)

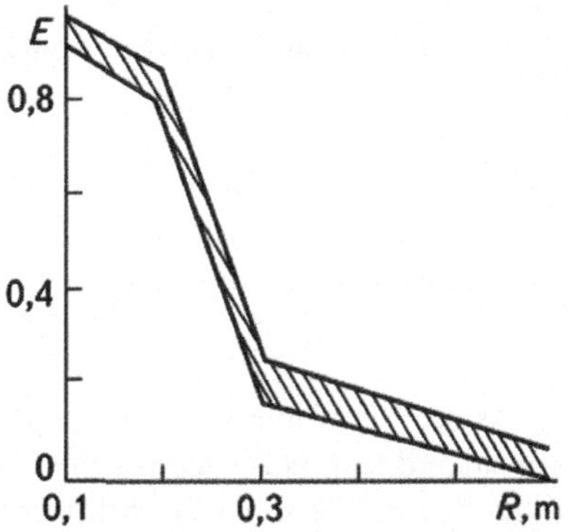

Fig. 6.4 Dependence of the AES *E* energy on distance *R* between the AE source and the AET at various threshold level values, which changed from 0.2 (upper boundary of the *shaded region*) to 1 V (lower boundary)

With an increase of the distance, the value of *N* decreases linearly, and the dependence on the threshold level is close to a logarithmic one. A signal duration changes in a similar way. It was measured as the time interval between the first and the last signal crossing the threshold level.

As seen in Fig. 6.4, a signal energy *E* dropped more sharply during an increase in the distance. However, it only slightly depended on the threshold level. Similarly, the AES amplitude *A* depends on the distance. Signal rise time T_r was determined, starting from the moment of signal crossing the threshold level until the moment of reaching the maximum amplitude.

The T_H versus threshold level dependence at various values of the distance from the AET is shown in Fig. 6.5. The velocity observed of the AES propagation was estimated as a relation of the distance between the source and the AET to the time of a signal arrival, recorded at the moment of its crossing the threshold level or at the moment of reaching the amplitude maximum $A_{\max}$. For all the signal parameters discussed above, the value of stability coefficient R_F, whose values are presented in Table 6.2, was calculated. It was found that the most unstable parameter with respect to the threshold level is a cumulative count, while the most stable is the signal energy.

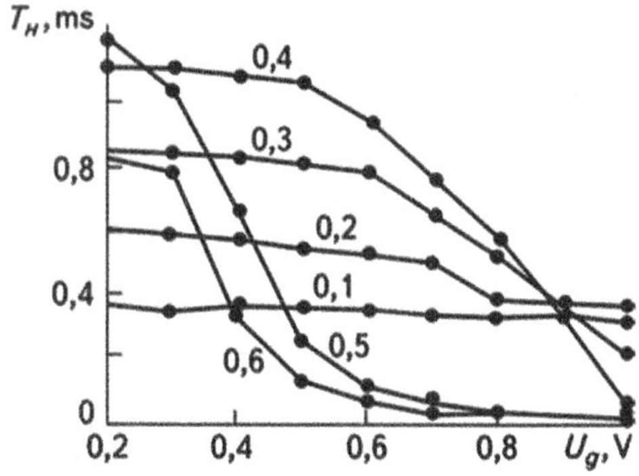

Fig. 6.5 Change in the AES rise time T_H during an increase of the threshold level Ug for distances R = 0.1; 0.2; ...0.5 m from the AE source to the AET

Table 6.2 Stability coefficients R_F for the AES parameters in relation to the change of threshold level and a distance to the source

Factors of the change of stability coefficient	Parameters						
	N	I	E	A	τ	V_p	V_A
Threshold level	53	42	0.1	–	20	16	–
Distance between the AE source and AET	11	2.6	44	55	19	16	79
The interaction of two factors: the threshold level and the distance between the AE source and AET	0.6	1.3	0	–	4.8	2	2

As seen in Table 6.2, a relative stability of a parameter in regard to one of the considered factors is accompanied, as a rule, by a relative instability in regard to the other factor. The sum of stability coefficients tends to preserve to the threshold level and the distance. Stability coefficients for the interaction of factors are lower by about one order than the stability coefficients for each factor separately. Therefore, the order of the values of stability coefficients at interaction is preserved if one of the factors changes. Similarly, it is possible to determine the effect of other factors on the AES parameters, such as the band pass of equipment, the AET type, conditions of its mounting on the specimen surface, its shape, etc.

It was established that various AES parameters show different stability to a threshold level and a distance. Therefore, one should either choose the parameters with a maximum stability and use them as informative parameters of the AES during the AE testing of structures, or provide procedures to reach this stability.

6.1.5 Classification of AE Sources by Their Activity

Difficulties in using the AE as a method of non-destructive testing are mostly related to the absence of reliable criteria that make it possible to determine the risk degree of the defects detected in the structure investigated. The state of an object is most often determined by the dependence of the cumulative count of AE on the SIF [8]. This approach is considered more generally in [9, 10], where, in particular, it is suggested to describe the dependence of the AE cumulative count on any loading parameter, e.g., load value, number of cycles, time, and so on, using a statistical dependence.

According to the classification proposed in [11], the AE sources are subdivided into three groups: C are non-active sources, for which the number of the AE events during observation increases only slightly; B are active sources, where the dependence of cumulative count on the applied load is relatively linear for them; and A are critically active sources, where the number of the AE pulses grows rapidly.

Sources classified as non-active are recorded, but they are not observed later on. Active sources need to be observed, while during the recording of a critically active source, it is recommended that a non-destructive testing be conducted using other methods.

Paper [12] proposes a system of classifying the AE sources used as a basis in a Japanese standard for testing spherical vessels [13]. Each AE source in this system refers to a certain class and type. The class of the source is determined by the total radiated energy and its density emitted from a unit of the source area, and the type of the source is determined by the dependence of total energy on the loading parameter. Four classes and four types of sources are examined in this system. On this basis, they are subdivided according to the degree of danger, into four groups: A, B, C, and D. Group A includes sources, the presence of which requires an immediate interruption of loading; Group B includes the sources whose localization region should be checked using traditional methods of non-destructive testing; and Groups C and D include the AE sources that are not dangerous and do not claim special attention.

The method belonging to the first group of the above-mentioned classification does not have any obvious criteria of structure scrapping, and is based on qualitative approaches that do not take into account a jump-like character of a fracture. The second method differs by a very complicated procedure of estimating the danger of defects, and there are no valid criteria of their classification.

Paper [10] formulated the problem of determining the numerical criteria of estimating the danger of AE sources using the AE test results based on the parameters of AES, which can be easily determined (for instance, the cumulative count of AE, which is related to amplitude and duration of the AE pulses, and, therefore, to the process energy). Since the AE is known to be a discrete process, the growth of cumulative count is also a stepwise function.

If the loading parameter increases by $\Delta\Pi$, then the cumulative count of AE increases by ΔN_i. By approximating two successive values of the Π_i and N_i quantities by power dependence

$$N_i = A_0 \Pi_i^{m_i}, \tag{6.7}$$

where A_0 is the proportionality factor, m_i is the power index. For ΔN_i from (6.7), we get

$$\Delta N_i = m_i A_0 \Pi_i^{m_i - 1} \Delta \Pi_i. \tag{6.8}$$

From dependencies (6.7) and (6.8), we find the power index

$$m_i = \Delta N_i \Pi_i (\Delta \Pi_i N_i)^{-1}. \tag{6.9}$$

Relationship (6.9) permits determining a power index at any stage of an increment of the AE cumulative count. Classification of the AE sources can be performed by m values at each step of the N increase. If m *is* less than one unit, a cumulative count derivative diminishes, and the AE activity of the process for this source drops. Such a source can be classified as non-active for this growth of the parameter. For $m \approx 1$, the AE process develops with a permanent intensity. The corresponding source is active, and on condition that $1 < m < 6$, it is critically active, and when $m > 6$, it is catastrophically active. It is clear that the same source can be classified as non-active at low levels of a load, or as critical and possibly catastrophically active at high values of the applied load. Final classification can be made only after performing the AE research in a given range of the applied loads.

The analysis of an m variation shows that this random value has two areas that are characterized by various average values. This is evident by the dependence of a cumulative count of load in logarithmic coordinates. In the first area, the source can be characterized as non-active or active, while in the second, its activity is critical or even catastrophic.

If we represent the random value of m as a parameter of probability distribution and we take into account the presence of the two above-mentioned areas, it is necessary to find for each of them the distribution by different mean values of m. For this purpose, the following exponential dependencies were chosen:

$$M_1(m) = v_1 \exp(-v_1 m), \quad M_2(m) = v_2 \exp(-v_2 m), \tag{6.10}$$

where $v_1 = 0.99$, and $v_2 = 0.33$. The validity of the empiric distribution in the sample was checked by means of the Pearson fitting criterion. Then the value of k_1^2 for the first approximation is 11.57, and for the second—$k_2^2 = 11.95$. If the level of significance $\alpha = 0.05$ and the degrees of freedom are 6 and 11, the critical values of k_{kp}^2 for approximations (6.10) will be 12.6 and 19.7, respectively. The values of k_1^2 and k_2^2 that were obtained are smaller than the critical ones, so it is possible to consider, with the given probability, that the experimental data are described satisfactorily by the laws of distribution (6.10) with the parameters described above.

To perform an effective acoustic-emission testing, it is necessary to find such criteria whose usage in the dependencies (6.10) permits estimating the degree of the

AE source danger for a structure. The selection of one of the two statistical hypotheses that determines to which type of distribution (6.10) the value of m belongs can be performed by successive procedures [14]. However, for a successive analysis, it is necessary that the distribution in the investigated sample be invariable; this criterion is not satisfied in the given case, because one stage of the cumulative count growth is substituted by the other one at the a priori unknown moment. Therefore, the investigated samples will be mixed up. In this connection, it is necessary to have a criterion that permits distinguishing the second stage of an increase of the AE cumulative count. If the criterion is satisfied, then it is possible to check a hypothesis on the validity of the second distribution (6.10) because it corresponds to the critical and catastrophically active AE sources.

A successive procedure for such estimation is as follows [7]: Some threshold value m_{th} of m is assigned—hardly probable for the first area, and highly probable for the second. In such cases, after a sequence of independent experiments in which realization of the event E_0 in each of them is checked, and which consists of m exceeding the given threshold value, it is possible to get two expressions for a number of experiments, where the event E_0 took place:

$$\begin{aligned} Z_{k1} &= \frac{\ln[\alpha_0/(1-\alpha_1)] + k\ln[(1-p_1)(1-p_0)]}{\ln[p_1(1-p_0)/p_0(1-p_1)]}, \\ Z_{k2} &= \frac{\ln[(1-\alpha_0)/\alpha_1] + k\ln[(1-p_1)(1-p_0)]}{\ln[p_0(1-p_1)/p_1(1-p_0)]}, \end{aligned} \tag{6.11}$$

where α_0 and α_1 are the assigned probability of errors of the hypothesis H_0 on the validity of the second distribution (6.10) and an alternative hypothesis H_1, respectively, p_0 and p_1, are the conditional probability of the event E_0 when the hypothesis H_0 and an alternative hypothesis H_1 are valid, respectively.

$$\begin{aligned} p_0(m \geq m_{\mathrm{th}}/M_2) &= \int_{m_{\mathrm{th}}}^{\infty} M_2(m)\mathrm{d}m, \\ p_1(m \geq m_{\mathrm{th}}/M_1) &= \int_{m_{\mathrm{th}}}^{\infty} M_1(m)\mathrm{d}m. \end{aligned} \tag{6.12}$$

In the coordinates, k is the number of experiments, Z_k is the number of the events E_0, and there are three regions, divided by parallel lines for the successive procedure (6.11).

It follows from the analysis of the m value dependence that a hardly probable value of $m = 4$ for the first stage of fracture appears quite frequently at the second stage. Therefore, putting $m = 4$ (6.12) with the account of (6.10), we get $p_0 = 0.27$, $p_1 = 0.02$. The probability value of the error of the first kind α_0 (probability of an active source missing) should be assigned to be smaller than for the probability of the error of the second kind α_1, i.e., over-rejection. Proceeding from this, we can put

$\alpha_0 = 0.01$, and $\alpha_1 = 0.05$. If for the given p_0, p_1, α_0, α_1, the point (k, Z_k) is located above the region bound by two lines (6.12), it is assumed that the corresponding source can be classified as critical or catastrophically active.

The analysis [10] of the experimental data obtained for specimens made of the pressure vessel material showed that the point (k, Z_k) was as a rule outside the bound area for the third or fourth event E_0. With this conclusion, we can set forth other criteria that follow directly from the dependence of the AE cumulative count on a load.

Such criteria can be formulated as follows:

Putting in the relationship (6.7) as above, $m = m_{\mathrm{th}} = 4$, we get

$$N_i = A_0 \Pi_i^4. \tag{6.13}$$

Then, the increment of the AE cumulative count will be

$$\Delta N_i = 4A_0 \Pi_i^3 \Delta \Pi_i. \tag{6.14}$$

Assuming that

$$\Delta N_{i+1} = 4A_0(\Pi_i + \Delta \Pi_{i+1}), \tag{6.15}$$

and taking into account that

$$N_{i+1} = N_i + \Delta N_{i+1}, \tag{6.16}$$

then it is possible to take the ratio of (6.15)–(6.13) as a criterion

$$\frac{\Delta N_{i+1}}{\Delta N_i} > \left(1 + \frac{\Delta \Pi_{i+1}}{\Pi_i}\right)^3 \frac{\Delta \Pi_{i+1}}{\Pi_i}. \tag{6.17}$$

When this inequality is satisfied, then $m > 4$.

During an accelerated growth of a crack, frequent increments of N occur, and the intervals of the loading parameter in this case are considered to be approximately the same and small ($\Delta \Pi_{i+1} \approx \Delta \Pi$). Then, neglecting in (6.17) the higher order term of smallness, it is possible to get the following criterion:

$$\Delta N_{i+1}/\Delta N_i > 1 + 3\Delta \Pi_{i+1}/\Pi_i. \tag{6.18}$$

However, at the initial stages of loading at large values of $\Delta \Pi$, these criteria do not always correctly characterize the AE source activity. A more exact relationship can be obtained from (6.13) with the account of (6.16)

$$\Delta N_{i+1}/N_i > (1 + \Delta \Pi_{i+1}/\Pi_i)^4 - 1. \tag{6.19}$$

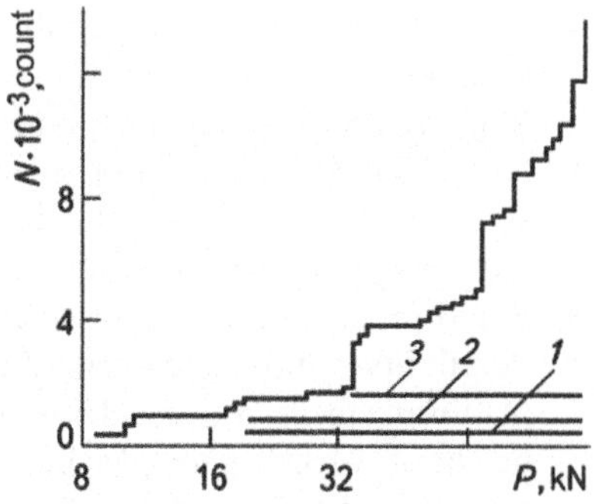

Fig. 6.6 Change in the cumulative count of AES N with load *P* for a 15X1M1F steel specimen with a notch under bending testing. *Straight lines* show the range of fulfillment of criteria (6.17), (6.18), and (6.19), respectively

Taking into account that power index *m* is a random value, the realization of criteria (6.17–6.19) is also a random event similar to the event E_0. The evaluation with a given degree of reliability can be performed by setting additional conditions that determine how many times the criteria should be satisfied. Similar to a successive evaluation, we can require satisfying the (6.19) three times. Hence, a given source should be considered a dangerous one for this structure. As is seen in Fig. 6.6, the criteria (6.17 and 6.18) are satisfied starting from the initial stage of fracture, and the criterion (6.19) is satisfied at the stage of accelerated crack growth. Thus, this criterion more reliably presents the degree of danger of the recorded AE sources. The obtained relationships (6.17–6.19) permit distinguishing the catastrophic stage of fracture, and together with the results of statistical analysis serve as decision-making criteria on the degree of the AE source danger for a structure.

6.2 Using the AE Methods for Testing the Offshore Platforms

For various technical and economic reasons, it is nearly impossible to perform an overall inspection of all kinds of units and joints of the offshore platforms. However, it is possible to check the individual parts that are most sensitive to crack initiation and propagation. A special AE testing system [1, 2] has been developed for this purpose. It consists of 4 × 32 channels and is equipped with 144 AET. The system permits checking various service parameters and all major units, welded joints, and places of fastenings that are on a platform.

The present system of the AE testing has the following characteristics:

- High sensitivity and low level of natural noises;
- Capability of choosing the working frequency range to get an optimum signal/noise ratio under a minimum effect of electromagnetic background (the average value of noises after treatment by the AES processor does not exceed one event per second);

- Distances at which it is possible to reveal the AES of small amplitude that are generated at early stages of crack formation in service conditions are 5…6 m;
- Reliable protection against atmospheric and mechanical damage;
- A specially protected and screened cable for transmission of high-frequency signals of low amplitude at a distance of more than 300 m at a minimum level of induction of electromagnetic hindrances;
- Rapid processing and assessment of the importance of the recorded AES;
- Possibility to take into account the external parameters (in the case of off-shore platforms these are accelerations, rotations, keel swaying, slope angles and height of waves, speed of wind, etc.) in order to predict the danger of the defects more exactly;
- Automatic calibration of the AET and the measuring paths as well as automatic checking of the obtained results; and
- A turn-off system acting in case a defect is located, and as a result the first four AETs of the six that recorded the AES are chosen, ensuring the best calculation of the defect localization.

In this case, for testing the off-shore platforms, the threshold level was selected to be equal to 26 dB (0 dB corresponds to 1 μV, and 120 dB—to 1 V). Sensitivity of the D9203AH AET was 67 dB in the pass band of 100…300 kHz. At this threshold level, the maximum speed of the recorded events was about 100 events per second.

The results of underwater testing of platforms [5] showed that the frequencies of 80 and 240 kHz were most expedient for the reduction of noise. The same frequencies provide a maximal increase of distances at which the AES recording is possible. Above 80 kHz, as the experiments showed, mechanical noise became considerably weaker, but the noise in water was at a very high level that could be reduced further by placing the AET in specially designed protective shrouds. Measurements conducted in various weather conditions showed that the noise of the highest intensity (to 60 dB) was caused by different ship engines, and it spread over considerable distances. To decrease this noise, the required AET sensitivity during the action of a compression wave on the rear wall of the protective shroud was at least 26 dB lower than the sensitivity of the AET front-working surface with which the AET was fixed to the structure surface. The application of such a method made it possible to decrease the number of the recorded AES caused by noises in seawater to only a few events per minute in one channel, and their total number for all channels was close to one event per second.

As the testing of platforms [5] showed, the AES amplitudes during crack propagation in a weld were 50…60 dB at a distance from 0.5 to 3 m from a source. At the chosen threshold level of 26 dB, the range of distances of the recorded AE events at the initial stages of fatigue crack growth in the joints of platforms at extreme weather conditions reached 2.5 m.

To provide the testing of a crack initiation and propagation in the welds of platform attachment fittings, the location of the AET should satisfy the following conditions: For all potential AE sources, the signal damping at a distance of 20 cm for the four nearest AET should not exceed 50 dB, which corresponds to the recording range of approximately 2.5 m. This value depends on the number of welded joints through which a signal passes from a source to AET. Thus, for example, to provide the testing of an attachment fitting of a medium size, six AET are needed.

The systems of AE testing of off-shore platforms can be located both on a platform and outside it. The first one is better in a general case, because there is no need for the permanent presence of an operator. The AE data can be recorded on a hard disk and, if necessary, can be transmitted for on-shore processing. In such a case, an operative supervision over the work of a system implies periodic checking of its serviceability and replacement of disks (on average, two man-days per week).

Test results are analyzed and processed over one or two weeks. To a great extent, this period is determined by the amount of the recorded information. The major problem with AE testing is finding the location of a defect or a group of defects within the region of a platform surface chosen for the testing. The dependences of the cumulative count of AE and of the average state of seaway on time can serve as the measure of danger of defects or their groups. The results of the AE test data processing should include the following information on [5]:

1. Determination of localization of the AE sources in order to identify and group them;
2. Representation of the temporal evolution of each group of sources (cumulative count of AE depending on time);
3. Establishment of correlations with seaway for every group of the AE sources found (accounting the wave height at those moments when each AE event was recorded); and
4. Presentation of the amplitude distribution of the AE events in every group of the Hsu sources [3] (based on the signals recorded by the AET closest to this group).

From the data stored on a disk for every test period (for one or two weeks), only those of interest are preserved. Such sorting of the AE test data is performed according to the results of their analysis outside the platform, in laboratory conditions.

Except for a system of the AE testing of off-shore platforms considered above, in order to inspect the joints of a simple geometrical shape, it is possible to use a simple portable system that has four receiving AET and a calibrating generator of ultrasonic vibrations. The AET can be mounted in the region of a unit having defects and an elevated concentration of stresses, or it can be moved from one place to another where there is a danger of defects forming. Such a system can be also used to check the operation of various mechanisms placed on a platform—for instance, for checking the growth of fatigue cracks in the rings of revolving hoisting cranes [5].

6.3 Using the AE for Testing the Nuclear Reactors

The possibilities of using the AE method in diagnosing nuclear reactors were considered in many papers (see, for instance [15–17]). In particular, in [17] a problem regarding the possibility of the AE testing of the growth of defects in pressure vessels of nuclear reactors was studied both during hydrostatic tests and in operating conditions. Research was conducted in the following three directions:

1. Developing the methods of identification of crack length increment using the AES.
2. Determining relationships between the recorded AE and the crack length increment.
3. Determining a possibility to test the pressure vessels both during hydrostatic tests and in operating conditions.

The realization of this research program started from the laboratory experiments on ATM A533 B steel specimens of Class 1. However, due to the fact that a direct transfer of laboratory results to large-scale structures turned out to be very complicated, the tests were conducted on the pressure vessels with the wall thickness over 100 mm. At the third stage, the AE testing system was mounted directly on the operating reactor.

During the preparation of hydrostatic testing in the central part of a vessel of dimensions of 120 × 700 × 1500 mm, three fatigue cracks were artificially initiated: two on the internal surface and one on the external. The sizes of all three defects were different, which made it possible to obtain different rates of growth for each of them in subsequent testing.

During hydrostatic testing, water temperatures were 65 and 265 °C. Three types of AET were used for AES recording: the AET with metal waveguides twisted into specially drilled holes on the vessel surface; three AET with metal waveguides tightly pressed to the investigated surface; and low-frequency AET directly mounted on the vessel surface. These AETs were used only in hydrostatical testings.

The AE testing equipment consisted of a recording unit, where the AES that were received were processed, and their time dependences, amplitude, difference in the arrival time, and other characteristics were found. Then, this information was translated into a digital code and was stored on a magnetic tape and was simultaneously transferred to a processor, where a defect location was determined, and it was decided whether the given AES was related to the crack length increment. Based on the information gathered, the degree of defect danger was evaluated.

The dependencies of the crack depth on the number of loading cycles at 65 °C were found (Fig. 6.7) using the test results. The crack depth was estimated according to the value of the measured crack opening displacement. The internal defects propagated more quickly than the external ones, due to the action of the water.

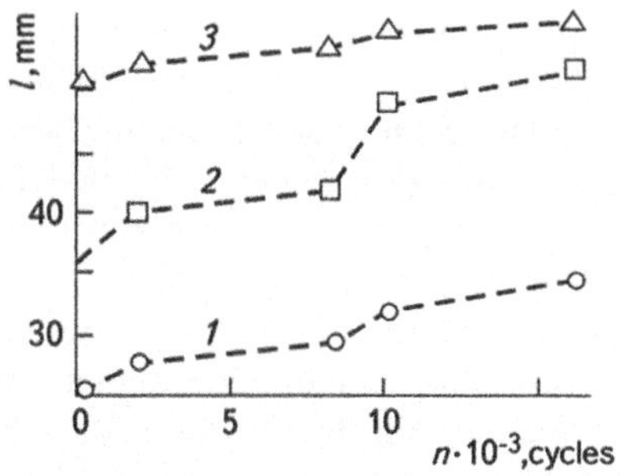

Fig. 6.7 Dependence of the crack depth l on the number of loading cycles n for temperatures 65 °C (*numbers* correspond to enumeration of cracks)

Having analyzed the AES obtained from the investigated section of the vessel, it was established that the emission from this region followed certain regularities. In all tests conducted at 65 °C, the AES had large amplitude values that corresponded to the maximum values within a loading cycle. The amplitude of over 60% of signals was higher than 9 V. Unlike the above-mentioned laboratory testing of specimens, only 8.5% of AES had similar amplitude values. Moreover, it was found that the AES were densely grouped at the maximum load value in a cycle. In laboratory research, only about 2% of the recorded signals corresponded to the maximal load in a cycle, while the majority of the AE events corresponded to the average value of the load. The information on the load value during these tests was transmitted to the AE testing system and stored. For this purpose, the cycle of loading was conventionally divided into 100 parts, and the number of each part that was stored by the system corresponded to each load value.

The dependencies of the AES count rate against time were obtained using the information on the defect location. Together with the results of laboratory tests that enabled finding the rate of the AES count, the possibility of estimating the defect growth velocity was found when no defect growth was observed. It was found that the crack growth rate, determined from AE data, was somewhat higher than that established by measuring the crack tip opening.

Except for the AES caused by the growth of artificially created defects, the AE caused by natural fatigue cracks was recorded under cyclic testing. These sources were concentrated in the area of the welds. Subsequent inspection of these welds—using the methods of non-destructive testing—confirmed the presence of cracks in them. It should be noted that signals from these cracks were recorded at distances of up to 3 m, including the most remote AET. In those tests, the AET were tightly pressed to the surface investigated. They are similarly fastened in the case of AE testing of the reactor.

A special problem of the AES research in hydrostatical tests was to estimate the possibility of detecting the fatigue cracks by means of the signals they emit. However, under periodic loading, when the applied load was 1.1 times higher than the maximum load during cyclic testing, no significant AE was recorded. The results put the efficiency of using the AE for the detection of defects in hydrostatical testings in doubt. It should be noted that the AE activity was considerably higher when a vessel was not preliminarily cyclically loaded up to the level mentioned.

One of the major problems when designing the AE testing system for nuclear reactors is developing a method for separating the AES that are generated by a propagating crack and the signals from the sources that are safe. For this purpose, a numerical processing of signals was carried out, which permitted selecting appropriate features related to spectral composition (for instance, peculiarities of power spectrum of an individual signal, self-correlation function, or statistical moments of power spectrum). This procedure permitted decreasing the volume of information required for a description of the analyzed signal characteristics. Based on the established features of the AE pulses, the regularities were obtained, according to which the recorded signals were classified.

At first, the set of features extracted from the AE pulses did not provide satisfactory results while testing the reactor pressure vessels. Therefore, a new group of the AE signal characteristics was composed, which included the parameters of a self-regression model of the tenth order describing every individual pulse. The advantage of this group of characteristics is that it is not very sensitive to signal amplitudes. The efficiency of these characteristics was demonstrated while classifying seismic signals and while processing the AE data obtained on aluminum specimens. As regards testing reactor pressure vessels, the correctness of such a classification method was 75...80%. The authors [17] felt that despite the improvements introduced, these results still cannot be considered acceptable. The statistical analysis showed that the correct level of accuracy in classifying the defects should provide an error not exceeding 10%. Only in this case is the application of the AE testing system effective.

While testing the reactor power unit, the AE testing system included 16 AET with high-temperature waveguides, which were pressed to the surface of the reactor casing, both in the region of its bottom and near the inlet and the outlet nozzles of the cooler circulation system. The resonance frequencies of the AET were chosen in the range of 450...500 kHz. Tests were carried out at a continuous increase of temperature and pressure up to their working values (292 °C and 15.4 MPa), with subsequent holding at these values. As a result of testing, it was found that the noises were caused by the cooler flow and depended on the service conditions, i.e., values of temperature and pressure, and type of reactor, as well as on the AET type and the construction of the waveguide (Fig. 6.8). It should be noted that the values of signal amplitudes caused by fatigue crack increment presented in the figure were obtained in previous pressure vessel testing. As is clear from the same figure, at a temperature of 65 °C and a pressure of 2.6 MPa, it is almost impossible to select AES that are generated by fatigue crack growth on the background noises, even by using the AET with a resonance frequency of 500 kHz. At higher temperatures and pressure of the cooler, the conditions of AE recordings caused by a crack are much better. Thus, at a temperature of 177 °C and pressure of 2.8 MPa, and at higher values of these parameters, the AET with waveguides of a resonance frequency of 500 kHz, enables selecting the AES effectively enough on the background noises. Besides, it is clear from Fig. 6.9 that when using the high-temperature AET without

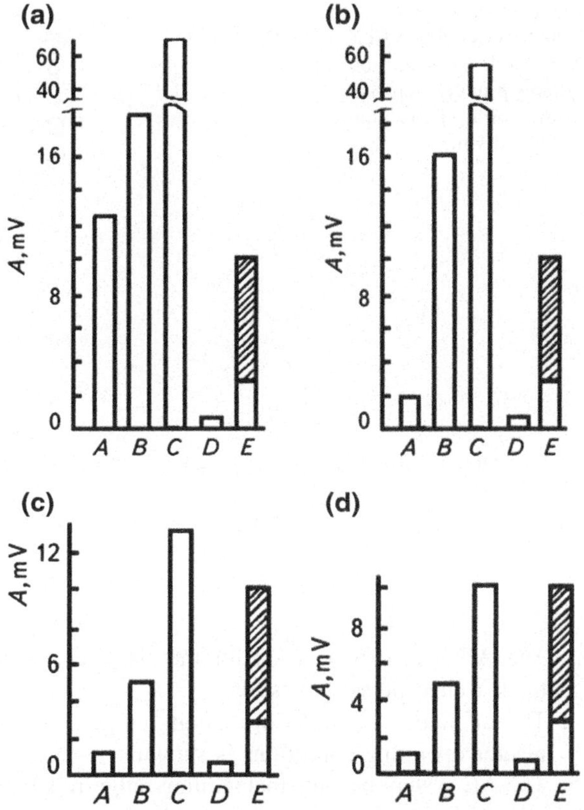

Fig. 6.8 Dependence of the cooler noise level *A* on temperature and pressure: *A*, *B* is recorded by the AET with a waveguide with a resonance frequency of 500 and 375 kHz, respectively; *C* is recorded by high-temperature AET without a waveguide; *D* is a level of electric noises; *E* is the AES level at a temperature and pressure, respectively: **a** 65 °C and 2.8 MPa; **b** 177 °C and 2.8 MPa; **c** 292 °C and 8.4 MPa; d 292 °C and 1.4 MPa

a waveguide, the AES amplitudes slightly differ from the noise level, which is why it is unreasonable to use a waveguide. Such AET efficiency can be improved by providing a waveguide and by decreasing the sensitivity to low-frequency vibrations in this way.

To check whether the AE testing system is capable of recording the AES in service conditions, a series of experiments was carried out. Compact specimens made of instrumental steel were preliminarily loaded mechanically, then mounted on the pipe and tightly fastened to the inlet pipeline. Furthermore, a heated brass bar was pressed to its surface. Due to temperature differences of the coefficients of thermal expansion, the specimen was additionally loaded and ruptured. The AES generated by the specimens that failed in this way were recorded by the AE testing system at various working temperatures, pressures, and conditions of a cooler flow.

As a result of all the tests, the following was revealed:

1. The AE testing permits revealing the increment of both artificially induced and natural fatigue cracks in thick-walled pressure vessels used in nuclear reactors.
2. The possibility of the AES detection in the presence of technological noises of the reactor is shown.

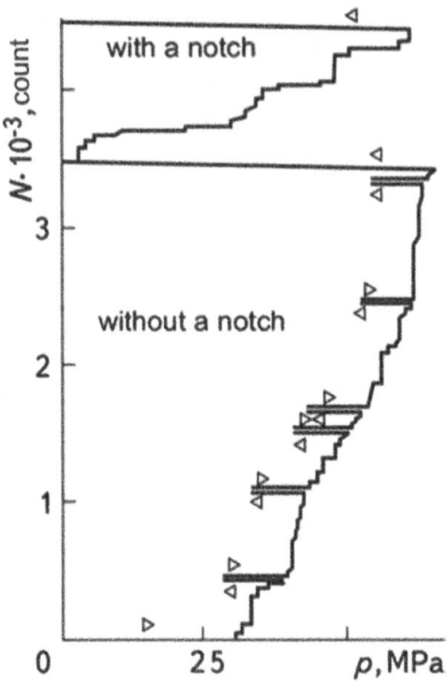

Fig. 6.9 Dependence of the cumulative count of AES N on the level of the applied stresses p under stepwise loading of a pipe without a notch and with a notch

3. The AE data can be used for the estimation of fatigue crack growth rate in thick-walled pressure vessels.
4. The efficiency of applying a high-temperature AET with waveguides in the reactor operating conditions is shown.
5. The AE system permits maintaining rather reliable and continuous testing of the state of an operating reactor for a long period of time (at least one year).
6. The AE testing system permits revealing the defects located in the sections of a reactor that are difficult to access.

6.4 Application of AE Method for Estimation of Strength of Pressure Vessels and Pipelines

Here we present some results obtained while using the AE for checking the initiation and propagation of cracks in the models of a pressure vessel and in large-scale pipelines. For the AES recording, the equipment of the Dunegan-Endevko Company, type 3000, and the multi-channel system 1032 were used [18]. This equipment permitted carrying out a spatial location, time-gating, and selecting the AES pulses according to the parameters of an external loading that made it possible to eliminate the effects of external noise sources and to select information on fracture for subsequent processing and reproduction in a printed form or in a display.

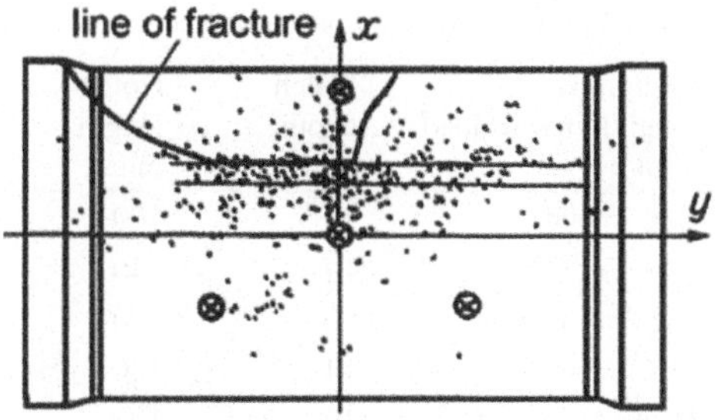

Fig. 6.10 Determination of the coordinates of the AE sources in the region of a weld failure by four AET for the whole observation period

In Fig. 6.10, the results of the AE recording under stepwise loading of the pipe with a 210 mm diameter and 40 mm thick wall made of 12X1MF steel are presented. An external notch located in the axial direction was made on this pipe. Its depth was 0.45 of the wall thickness, and the length was 250 mm. While loading the notched pipe (Fig. 6.10), the AE from the notch region was recorded at a slight internal pressure that confirms the development of this defect. This was also proved by the results of the location of AES due to the fatigue crack increment. The location of AE sources permits estimating the crack sizes and establishing the location of an initiated crack.

Furthermore, a piece of a pipeline of 850 mm in diameter and a 48 mm thick wall made of 22K steel with anticorrosion X18N10T steel coating welded on the internal pipeline was tested. The pipeline contained longitudinal welded joints that provided a considerable heterogeneity of the stress-strain state. The defects that unavoidably arise during welding also reduced the serviceability of the pipeline. The results of the AE sources' location in the area of each longitudinal joint during the loading, up to the moment of failure of the section of the pipeline that was examined, are presented in Fig. 6.11. When the pressure increases, the AE sources that form groups in the area of the welded joints indicate the areas of the location of the defects that develop during testing. At a pressure of 40 MPa, the pipeline fails. It was established that in the area of crack growth, the fracture was caused by the

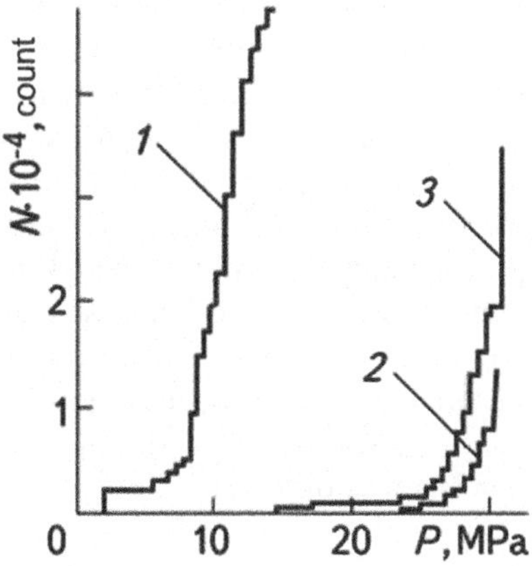

Fig. 6.11 Dependence of the cumulative count of the AES *N* on the level of the applied stresses *P* during hydraulic testing of a section of 22K steel pipeline with cracks (*1*) and welding defects in the area of longitudinal welds Nos.1 (*2*) and.2 (*3*)

Table 6.3 Data on a crack length increment in a pressure vessel

Region of longitudinal welded joint	Notch depth, mm	Notch length, mm	Final crack depth, mm	Number of loading cycles for a visual revealing
Base metal	10	100	13.0	5800
	14	140	24.3	3000
Welding metal	10	100	13.0	6300
	14	140	18.8	3390
Heat-affected zone	10	100	13.0	6000
	14	140	20.8	3200

presence of faulty fusion welding in the joint area of the base metal, surfacing, and the welded metal of a weld. Herein, the AES recording made it possible to find a defect that develops long before a complete failure of the examined pipeline section.

Repeated static testing by internal pressure of real sections of straight pipelines made of 22 k steel with anti-corrosion coating of X18N10T steel of a diameter of 850 mm and wall thickness of 48 mm that contained circular and longitudinal welded joints were carried out as well. In the base metal and in the coating, as well as in the heat-affected zone of longitudinal welds, there were surface notches with a depth of 0.25…0.35 of the wall thickness. Apart from these notches, welding defects of sizes that exceed the admissible ones were left without healing in a welded material. During testing, the loading parameters were recorded, and, using the AE testing system, the AE sources were localized and the AES parameters were recorded. The test results of one of the pipeline sections are presented in Table 6.3 (maximal nominal hoop stresses were 0.8 σ_{02}, while the number of loading cycles of all modes was 7500).

The analysis of recorded AES alongside the visual observation permits proving that the crack initiation conditions are different in different regions of the welded joints; this is explained by the difference of material properties and the presence of typical residual stresses. Therefore, the cracks that started from the notches of identical size but were located in different regions of the welded joints, began to propagate after a different number of loading cycles. The moment of their growth start was revealed much earlier by means of AE than by visual control.

A high concentration of the AE sources in certain areas of a structure and the fact that AES was recorded in the narrow range of the applied load that corresponded to the threshold stress intensity factor values, confirmed the beginning of the crack propagation. It should be noted that this range for the cracks located in different regions of the welded joints was also different, thus proving that the dependence of fatigue crack growth kinetics on heterogeneity of the stress state is related to the material properties in the region of the welded joints. As a result, the depth of fatigue crack growth under the same modes of loading at identical initial sizes of notches was different.

The AES caused by cracks initiated at the notches and at the natural welded defects were recorded separately during the hydraulic testing of pipelines. The results of these tests are presented in Fig. 6.11. It was found that the AE activity caused by the defects that develop from the notches of a depth equal to 0.35 of the wall thickness was higher than by the welded defects. Shallow defects (the notch depth equal to 0.25 of the wall thickness) produced the AE, which was practically the same as the welding defects. Metallographic analysis confirmed the availability of a slight crack growth at small notches and near the welded defects.

Thus, the AE method permits defining the location of the AE sources and their activity, which enables one to conduct the analysis of a structure's imperfectness. The inspection of the AE sources that were found can be further conducted using traditional methods of non-destructive testing, which take into consideration the available standards and provide information on the degree of danger of the defects.

The information obtained using the AE method during hydraulic testing [18] can be used for the operation inspection of pressure vessels and pipelines. In this case, it is enough to check only the areas of the AE activity revealed during hydraulic testing. Based on these observations, it is also possible to establish the specific features of the development of defects, to solve the problem of technical diagnostics, and to predict the lifespan of the equipment.

6.5 AE Inspection of Welded Joints

Most unique structures contain welded joints that should be highly reliable when being used, and special diagnosing systems and complexes have been developed for this purpose. As is known, means of non-destructive testing and diagnostics, which use the AE method, permit testing the required volume zones of a structure or its individual elements regardless of geometrical sizes and the type of material. The defects in the welded joints (gas pores, tungsten inclusions, oxide films, cracks, faulty fusion, etc.) non-uniformly distribute the loads over the volume of the tested object which, in turn, causes some differences in the AE generation due to the availability of the areas with and without defects. Therefore, to effectively use the advantages of the AE method as one of the promising methods of nondestructive testing, one needs to have the preliminary comparative characteristics of the AE activity of structural materials under loading, when the processes of defects initiation and propagation in them are observed, i.e., fracture processes. This enables us to select the necessary conditions of recording the AE signals by a measuring device, to compare the optimum values of their information parameters, such as amplitude, cumulative count of the AE signals, number of events, spectral characteristics, etc., as well as to maximally approach the solution of the problem of signal identification. Such investigations of structural materials and their welded joints are carried out in laboratory conditions and form the basis for the development of new methods of diagnostics and non-destructive testing of products and constructions.

The quality of a welded joint can be inspected using the AE method at different stages. The first stage includes the period of welding from the very start until the end. The second stage includes the period of the weld cooling until it reaches its equilibrium state. The third stage includes the period following the establishment of the equilibrium stage.

The AE testing of extended welds can be carried out simultaneously at the first two stages. Their common feature is that the AES originate under the action of internal local stresses that develop in a near-joint area without additional external loadings (Fig. 6.12). Internal stresses are caused by the material structure heterogeneity, as well as by non-uniformity and non-stationarity of thermal conditions of the process of welding. At the second stage of inspection, a common or local external action on a weld is necessary for the AES generation.

A successful application of the AE for weld testing is related with a possibility to select signals generated by defects from the general amount of signals that are generated against the background noises. A frequency band, in which the AES are selected, is chosen, taking into consideration the spectrum of noises and acoustic waves damping in the material. Usually, this frequency band is set within the interval of 50 kHz…2 MHz.

The AE testing of pipe welds is performed, as a rule, during the cooling of joints, expanding, or hydraulic testing of pipes [19]. During hydraulic tests, the recorded AES were caused mainly by the noise due to the water flow and its friction with the pipe wall, and due to slight leakages that were not recorded by a manometer. However, at hydraulic testing, the defects do not emit AES. As the analysis showed, this is explained by a volume expansion immediately before a hydraulic testing of a pipe, when much stronger forces compared with hydraulic testing are applied to it, and thus the KE is observed.

During expanding the pipe is deformed successively in the sections of length of about 1 m. The degree of deformation attains 0.5…0.7%. In this case, as research

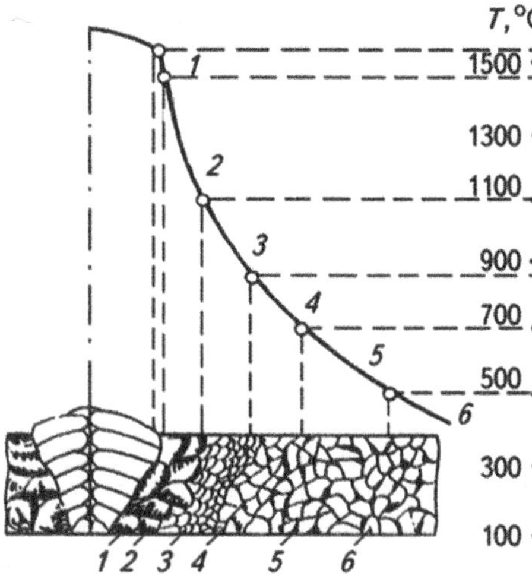

Fig. 6.12 Schematic of temperature limits distribution in the near-weld zones: *1* is the melt area, *2* is the overheat area, *3* is the re-growth area, *4* is the incomplete recrystallization area, *5* is the recrystallization area, *6* is the aging area [9]

has showed, it is very difficult to detect cracks using the AE method, due to the high level of noises caused by plastic deformation of material and by friction of expander elements with the internal pipe surface. During welding, it is difficult to reveal the cracks due to the noise caused by slag film cracking.

More optimistic results were obtained in the AE research at cooling the welds. During their cooling, the AES are emitted with defects for a long time and are characterized by high amplitude and intensity. Specimens with high-quality welds practically do not emit the pulses, and only single rare bursts of the amplitude and intensity are observed (Fig. 6.13) [20]. A slag layer is removed from a weld immediately after its welding. According to the investigation results a good correlation was founded between the longitudinal and transversal cracks and the AES in welds; however, in the case of volume defects (such as pores), the AE was not practically observed.

The welded structures are subdivided into 4 groups according to AES radiation intensity: 0—AES is practically not observed; 1—insignificant intensity of AES radiation (up to 104 pulses); 2—AE of medium intensity (up to 4×10^4 pulses) with the considerable AES amplitudes; and 3—continuous AES of high intensity (up to 12×10^4 pulses).

The initial basic data on the AE character is required for each particular case of the welded joints that would take into account the structure of materials, technology of their welding, and so on. Such data are definitely obtained during laboratory testing. We will briefly discuss some of this data.

6.5.1 *Verification of Selection of Materials, the Type of Specimens, and an Investigation Method [21]*

An improvement of reliability of the welded structures takes place with a simultaneous decrease of steel intensity and is a basic direction of the development of

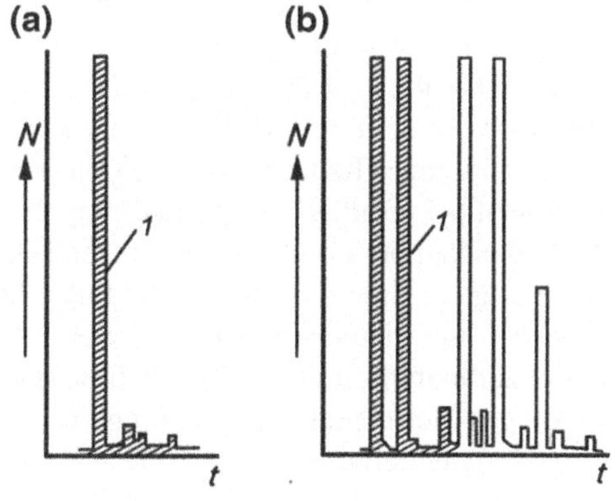

Fig. 6.13 Change in the cumulative count of the AES *N* with time *t* in the case of high-quality (**a**) and low-quality (**b**) welding. *1* are hindrances from a welding arc

science dealing with welding. The tolerance of this or another technological defect of a welded joint can be established today regardless of the structural element location, its stress state, number of cycles of loading alternation, residual stress, etc. The absence of a profoundly considered differentiation of the damages of the materials of the welded joints according to corresponding standards often causes the premature fractures of the welded joints due to defects, while in other cases the defects are quite groundlessly withdrawn without proper consideration. It is well known that the defects being cut off from a component with the following welding does not increase but rather decreases the load-carrying capability of the component. The problem becomes more urgent taking into account the fact that about one-third of all premature fatigue fractures of metal structures, machines, and buildings, reservoirs, and pressure vessels, special technological equipment, etc., are today conditioned by the defects of welded joints [22].

In the world of engineering practice, the production of vitally important welded structures, such as ships, multi-storied buildings, power equipment, aviation and space objects, etc., is accompanied by the allocation of considerable financing for the implementation of facilities and methods of non-destructive testing and diagnostics. A fraction of these finances in some cases reaches 20–25% of the total cost of structures, while labor intensiveness becomes almost the same as for the welding processes [23]. The analysis of the welded structures' failure shows that in most cases it takes place in the near-weld area rather than along the weld. This is explained by the fact that the weld is more durable than the base metal. Thus, the thickness of a strengthening roll under two-sided welding is 1.8…2.2 times higher than the wall thickness. Besides, the weld metal is micro-alloyed with a cast structure. The structure of the near-weld area for arc, plasma, and electro-slag processes is heterogeneous and has considerable residual stresses. Structural transformations in a metal, and considerable plastic strains, appear due to abrupt changes of temperature in this area. The defect at the base metal ends arising due to temperature effect and deformations increase their volume and become dangerous [24]. Herein, a very negative effect of hydrogen should be noted, which in the atomic state is released into micro-cavities during the welding process, and later on having transformed into molecular hydrogen, creates high pressures in them and, as a result, we observe high stresses in the metal. It is the hydrogen mechanism that forms the basis of the phenomenon of the so-called "cold cracks initiation" and propagation in the weld metal and in the heat-affected zone [25, 26]. Thus, taking the above into account, the specimens of structural materials made of the base metal (BM), weld metal (WM), and fusion zone (FZ), became the object of the AE studies.

One of the peculiarities of inspecting the quality of the welded joints using the AE method is that pre-conditions for its realization are available at different stages of the technological process: during welding; during the formation of a joint and after welding; and at transition of the welded joint into the equilibrium state or after termination of this process [9]. All these stages are clearly expressed in the methods of welding, which are related to melting. At the first two stages, the AE is generated under the action of internal stresses, while at the third stage there is an additional local action of an external loading. Exactly then (at temperatures lower than 100 °C), deformation cracks, corrosive cracks, and deformation-corrosive cracks are formed [26]. The first type includes cracks that appear in welds after their cooling and in the process of aging after welding, as well as cracks that appear in a metal during tensile deformation, compression, impact, and other types of mechanical action. The second group includes cracks whose initiation is predetermined by the action of only one corrosive environment without applying an external loading, and the third type of cracks appears in a metal under a simultaneous action of an external loading and a corrosive environment. During exploitation, the cracks that appear due to various strengths of some elements of a structure and phases (for instance, eutectic inter-layers, non-metallic inclusions, pores, etc.), and due to the action of external loading forces appearing in the process of product exploitation, are effectively revealed by the AE method. Thus, in this study we investigate the methods that use the AES testing of the fractured welded joints caused by a quasi-static loading, i.e., the methods for estimation of the static crack growth resistance of a welded joint.

The trends in the development of welding technologies are characterized by the transition to low-alloy steels with a simultaneous reduction of the amount of metal in structures and by an increase of reliable service life. This is confirmed by the fact that today about 20 grades of low-alloy steels are used exclusively for building structures, and, taking into account the quality categories, their number has increased more than threefold [22].

At the same time, in many branches of industry, non-ferrous metals are widely used as structural materials in manufacturing vitally important units and even whole structures. Aluminum alloys are of special interest. Therefore, as representatives of the above-mentioned different grades of structural materials, the steels 10XSND, 09G2S, Steel 3sp, as well as 1201-T aluminum alloy were chosen. Specimens made of these materials were constructed with the account of three-point bending loading or eccentric tension [27], while cylindrical specimens used for the evaluation of mechanical characteristics of 1201-T alloy met the requirements of [28].

According to the test method, the AE signals were recorded under the loading of specimens in a real-time scale. The relationship between the AE (amplitude, cumulative count N of the signals, spectral characteristics) and fracture parameters (stress intensity factor, deformation, crack length increment), and the validity of the KE were established using the test results. Thus, the energy of the AE signals that accompany the processes of the fracture of welded joints was estimated in order to use the obtained results in non-destructive testing of the real products or structures.

6.5.2 Results of the AE Research of the Welded Joints and Their Interpretation

Investigation of the AE indices of fracture of low-alloy steels and their welded joints. Prism-like specimens of 6 × 15 × 170 mm were prepared for testing the 10XSND, 09G2S steels, Steel 3sp., their welded joints, and FZ. The chemical composition is presented in Table 6.4. In these specimens, after mechanical

Table 6.4 The chemical composition

					Elements	Content	%						
Alloy grade	C	Mn	Mg	Si	P	S	Cr	Zn	Ni	Al	Cu	As	Fe
Steel 3sp	0.14…0.2	0.40…0.65	–	0.12…0.30	≤0.045	≤0.055	≤0.3	–	≤0.3	0.013	≤0.3	≤0.08	*
09G2S	≤0.12	1.3…1.7	–	0.5…0.8	≤0,04	≤0.04	≤0.3	–	≤0.3	–	≤0.3	–	*
10XSND	≤0.2	0.5…0.8	–	0.8…1.1	≤0,04	≤0.04	0.6…0.9	–	0.5… 0.8	–	0.4…0.65	–	*
1201-T	–	0.3	0.02	0.03	–	–	–	0.03	0.03	*	6.5	–	0.14

*—Other

treatment by a polishing stress, notches were cut out and then fatigue cracks were induced. Specimens were made in such a way that in the working area, i.e., in the net-section, there was a failure in BM, WM, and FZ. The SVR-5 [29] loading machine that was used to apply a load differed from serial loading devices by a low level of generation of background noises. This permitted selecting the AES at high amplification levels and at low threshold levels in the AE channel [30].

The scheme of testing the specimens was as follows [31]: The loading forces were transmitted from the SVR-5 equipment through an indenter to the investigated specimen, on which the AET was mounted to its lateral surface. The force of pressing the AET with a clamp to the lateral surface of the specimen through a contact layer [3, 32] was 15…20 N. During loading, the diagram "load P – crack opening v" was recorded by an X-Y recorder, while the amplitude A and cumulative count N of the AE signals were simultaneously recorded by a fast-acting recorder H-338/4 in real time. After pre-amplification and processing the AE signals by the AVN-3 device, these parameters were printed on paper.

Before every experiment, a measuring channel was calibrated by the Hsu source. The MBS-9 microscope supplied with a set of the SVR-5 equipment was used to inspect the changes at the lateral surface of a specimen in the crack propagation region. Taking into account the high plasticity of steels (Table 6.5) at room temperatures [32] and service characteristics of a loading machine, we have chosen the following modes of the AE selection and processing: Mode I: amplification factor of a measuring path is equal to 74 dB, threshold level of the AE signals is equal to 0.4 V, working frequency band Δf is equal to 0.12…0.5 MHz with the transmission coefficient of filters of low and high frequencies equal to one. The AET possesses, accordingly, the best working characteristics in the frequency band from 0.2 to 0.5 MHz. The results of testing 10XSND steel specimens, their welded joints, and FZ material obtained in such measurement conditions, are presented in paper [33]. Therefore, we will briefly discuss some of them.

The AE signals were recorded both for the linear area of diagram "P-v and for the non-linear area up to the moment of the final macro-fracture stage. Their amplitudes at the analog output of an AVN-3 device were in the range of dozens of mV to several volts. The AE signals were most actively generated during the testing of the specimens cut out of the welded joint metal, and then with a decrease of amplitude, the number of events, and the cumulative count, there come

Table 6.5 Mechanical characteristics of alloys

Alloy grade	σ_{02}, MPa	σ_b, MPa	δ, %	ψ, %	F_b, %[a]	K_C, MPa$\sqrt{m}$
Steel 3sp	280	430	26	55	92	56.8
09G2S	330	500	28	58	77	69.2
10XSND	400	540	24	50	97	58.3
1201-T	350	440	4.75	7.54	–	44.3 38.04[b]

[a]F_b, is the fraction of a ductile component in a ruptured specimen at t = 20 °C [32]
[b]Is the K_{IC} value

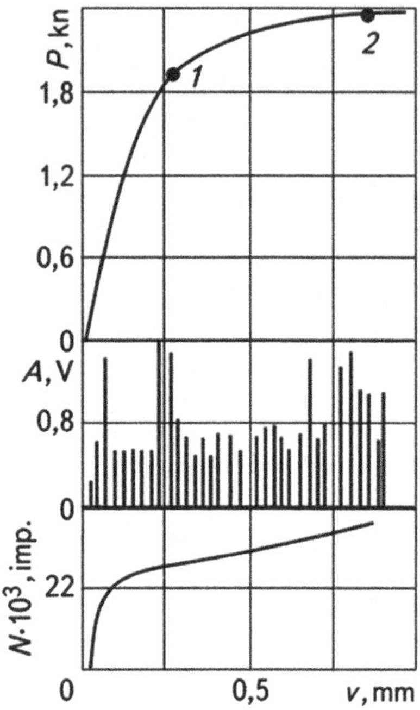

Fig. 6.14 Fracture diagram "*P*-*v*" (three-point bending) for the metal of FZ of a welded joint of 10XSND steel and corresponding distribution of the AES amplitudes *A* and cumulative count *N*

the specimens cut out of BM and WM. Thus, for 10XSND steel and its welded joints, it is shown that the value of amplitudes and intensity of the AE generation at a sub-critical crack propagation in specimens from various welded joint areas are sufficient to be recorded during testing of large-scale structures or the products under static loading (expanding). Then, the industrially produced equipment of the AVN-3, AF-15 types and others can be used. As the experimental data (Fig. 6.14; Table 6.6) argue, there is a margin in the amplification and threshold levels; thus it is possible to reduce the amplification and to increase the threshold level. The AE testing and diagnostics efficiency can be substantially improved using a resonance or narrow-band AET, as well as by using certain methodological approaches to the selection of the AE signals under the effect of sensible background noises [34] in real conditions of diagnostics.

Welded joints of 09G2S steel and Steel 3sp were investigated using a similar scheme. The AES registration regime differed from the previous one by the amplification factor, which was 84 dB, and by the threshold level, which was 0.2 V (Regime II). The other measurement parameters, geometry of the AET location on the specimens, and calibration of measurement path sensitivity were the same. The shading of the area of the crack length increment was done by heating the specimens in an autoclave to a temperature of 300…350 °C and by holding them at this temperature for 600 s with the following impact rupture of the specimens after cooling.

Table 6.6 Results of the AE research

Alloy grade	Area of specimen cutting out	K_I, $MPa\sqrt{m}$	Number of events	Sum of amplitudes A, V	Cumulative count N, 10^3, counts	Maximum amplitude at the AET output A_{max}, mV	Area of crack increment, S, mm^2	Measurement mode
Steel3sp	BM	23.4	23	2.68	5.2	0.027	–	II
	WM[a]	28.3	49	3.4	2.2	0.035	–	
	FZ	34.0	240	15.44	14.5	0.25	–	
09G2S	BM	29.4	96	7.68	9.5	0.05	–	II
	WM	34.4	122	7.24	8.3	0.025	–	
	FZ	32.1	121	8.84	13.0	0.025	–	
10XSND	BM	16.5[b]	246	18.36	19.02	0.104	–	I
	WM	33.15	126	7.72	1.78	0.067	–	
	FZ[1]	32.8	258	65.36	24.2	0.335	–	
	FZ[2]	40.46	376	22.82	38.15	0.335	3.2	
1201-T	BM	440[c]	324	0.14	0.82	0.015	–	III
	WM	42.56	46	0.33	16.8	0.89	173.25	

[1]Point 1 in Fig. 6.14; [2]point 2 in Fig. 6.14
[a]The faulty fusion cavities of the area of up to 4 mm^2 are detected on fracture surfaces
[b]Linear area of the fracture diagram
[c]σ_b value for cylindrical specimen tension

As seen in Fig. 6.14, to develop the criteria approaches in laboratory conditions, it is worth recording the AE signals in the loading ranges of specimens to a certain value in the non-linear area of the fracture diagram (points 1, 2 in Fig. 6.14). This is confirmed by the fact that in real loading conditions under large plastic deformations, the defects in ductile steel structures are easily revealed visually, and in this case the macro-crack opening attains considerable value.

Specimens of 09G2S steel and Steel 3sp were loaded until a plasticity area on the diagram *P-v* appeared; then they were unloaded, and the value of the fatigue macro-crack length increment was calculated, as stated above (Table 6.6). The test results confirm the following: The energy level of signals that accompany the processes of initiation and propagation of micro- and macro-cracks for three grades of steels is the lowest for Steel 3sp. After that, according to an increase of this index and AE activity, come steels 09G2S and 10XSND steels. Concerning the welded joint areas, all grades of steels prove the above regularity of ordering for 10XSND steel: FZ, BM and WM. Besides, for materials of all the welded joint zones, the KE is satisfied [33, 35].

Estimation of peculiarities of generation of the AE signals due to fracture processes in welded joints of 1201-T aluminum alloy. Investigations were carried out in two stages: In the first stage, the BM mechanical characteristics were calculated [28], and simultaneously the AE that accompanies the processes of plastic deformation, macro-crack initiation and propagation was recorded. For this purpose, cylindrical specimens with a diameter of 4 mm were prepared. They had a thick part in front of a threaded part on both sides, where flats were made for mounting the AET [36].

Before each test, specimens were subjected to molding in a special conductor according to the method presented in [37] for removing the non-informing AES emitted from the grips area during loading. Loading forces exceeded the admissible level by 50…80%. After that, the experiments were conducted. A tensile-testing machine of the FPZ-100/1 model with the speed of traverse movement of 0.02… 0.84 mm/min was used as a force-inducing device. The AE chain of selecting, processing and recording the AES was built, based on the AE MISTRAS—2001 system produced by PAC (USA), to which analog signals from the strain gauge force dynamometer from the machine FPZ-100/1 were transmitted, together with the displacement gauge for measuring the value of the specimen elongation or crack faces opening (depending on the type of the specimens tested). Simultaneously, signals from the AET with a working frequency band with linear amplitude-frequency characteristics within the range of 0.1…1.2 MHz were transmitted to the AE chain through a preamplifier. The preamplifier amplified the AES by 40 dB in the path band of 0.02…1.2 MHz, while the threshold level of the AES was 45 dB. The following AES parameters were obtained in this mode of selection and processing (Mode III): time rising to the maximum of the forward front of a signal τ_1, duration of the AE signal above the threshold level τ_2, amplitude, number of pulses N, energy G. Furthermore, the measuring system made it possible to receive spectral characteristics of the AE signals from their waveforms.

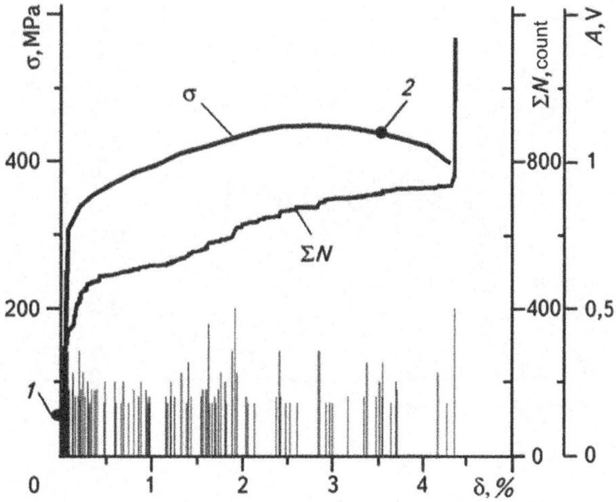

Fig. 6.15 Stress-strain diagram σ-δ of a smooth cylindrical 1201-T alloy specimen and the dependence of a cumulative count N and signal amplitudes A on the strain δ of a specimen

In the second stage, the AE characteristics of fracture processes in materials of 1201-T alloy weld were investigated on 79 × 79 × 7 mm compact specimens, which were manufactured according to standards and typical ratios of geometrical sizes, given in [27]. Welded joints were made by means of electron-beam welding. Figure 6.15 shows a typical diagram of tension and distribution of AES arising in a cylindrical specimen, and Fig. 6.16 presents spectral characteristics for points 1 and 2 of this diagram. From the last figure, it is seen that the AES amplitudes that arise during the process of macro-crack initiation and accompany the appearance of cracks in various inclusions in a material [36] at initial stages of a stress-strain diagram, are of slight magnitude (measured in mV). The main activity of the AES is concentrated in the range of the relative elongation δ change of a specimen within the limits of 0.1…0.4%.

The generation of the AE signals of the BM of 1201-T alloy is lower in comparison with the AE amplitudes of steels during macro-crack growth. With the growth of σ, the spectra of signals (Fig. 6.16) tend to narrow, and dominant frequencies of signals are shifted towards the low frequency values, which concurs with the results of [38]. Figure 6.17 shows the fracture diagram P-v of a WM of a welded joint of 1201-T alloy and the corresponding distribution of AES during sub-critical crack propagation, where AE signals are of a specific character. A slight amount of the AES of large amplitudes is observed, which confirms a jump-like propagation of the macro-crack and the absence of the AES of small amplitudes. This means that the mechanism of sub-critical macro-crack growth is quite brittle, and its diagnostics requires an accurate interpretation of the AES taking into account the initiation and the development of micro- and macro-cracks. The characteristic spectra of the AES in various areas of a fracture diagram confirm the above-described tendency of the change in a spectrum width and in dominant frequencies of AE.

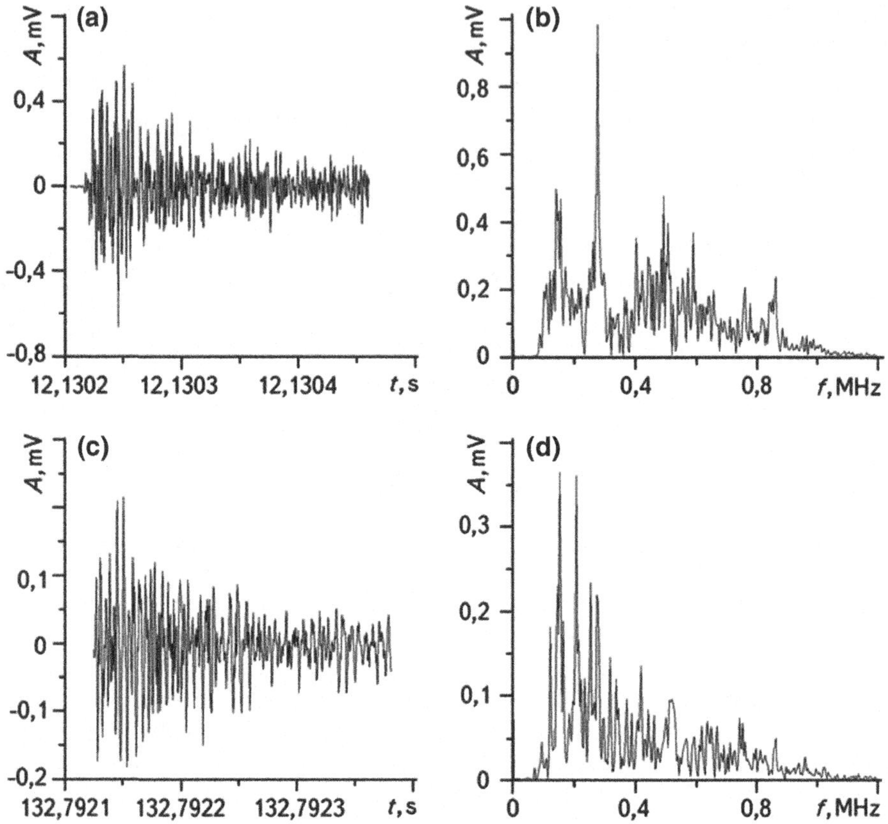

Fig. 6.16 The AES waveforms and their spectral characteristics at point *1* (**a** and **b**, respectively) and at point *2* (**c**, **d**) of stress-strain diagram (Fig. 6.15) of cylindrical 1201-T alloy specimens with the following indices of measuring modes: sampling frequency of the AE signal—4 MHz, sampling number—1024, threshold level—0.178 mV; **a** $\tau_1 = 21$ μs, $\tau_2 = 152$ μs, $N = 15$, $G = 2$ μJ; **c** $\tau_1 = 1$ μs, $\tau_2 = 1$ μs, $N = 1$, $G \approx 0$

Thus, experimental AES investigation of fatigue crack growth resistance of welded joints of low-alloyed steels 10XSND, 09G2S steel and Steel 3sp, as well as of 1201-T aluminum alloy, being the most widespread materials, showed that quantitative indices of the AE events, cumulative count, and the amplitude-frequency characteristics of the AES are specific for every type of material and for particular areas of the welded joints, which enables making a complex evaluation of the mechanisms of generation of the AE signals at the stages of plastic deformation development and the beginning of sub-critical macro-crack propagation. Using the AES spectral analysis, it is necessary to preliminarily achieve the best reproduction of a waveform of every AE event with the account of reflection, damping, and resonance peculiarities of the tested material of a product or of a structural element.

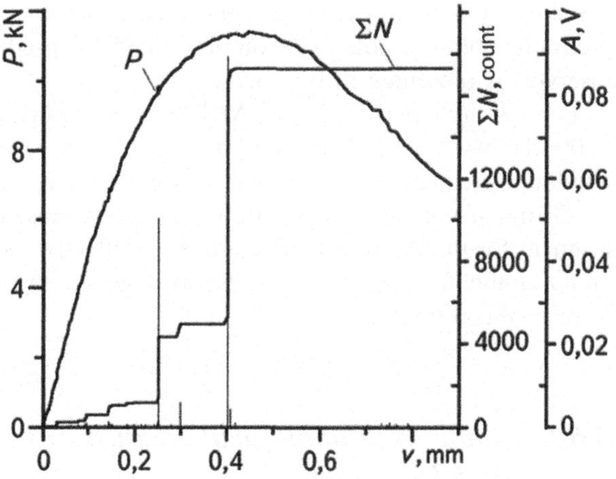

Fig. 6.17 Dependence *P*-*v* for the eccentric tension of compact specimens of a welded joint WM made of 1201-T aluminum alloy and the change in indices of the cumulative count *N* and amplitudes *A* of the AES that accompany the fracture processes

6.6 Selective On-Line AE Hydraulic Testing of an Oil Storage Reservoir

The purpose of the AE testing of a reservoir with a capacity of 75,000 m^3 is to reveal, on-line, the initiation and propagation of the crack-like defects and to evaluate the degree of their danger during hydraulic testing. To record the AES, a two-channel AE system "AKEM" (a programmable-technological complex) was used that permitted estimating the IO state by certain AE criteria. The system is mobile and provides on-line processing of the recorded information and its corresponding representation after termination of AE tests. Its characteristics are:

- Working band of frequencies—100...2000 kHz;
- Non-uniformity of amplitude-frequency characteristics of the amplification path
  - within the frequency band—not larger than ±3 dB;
- Damping of the AE signal outside the working range during the change of threshold
  - frequencies—not less than 30 dB per octave;
- Input resistance of preamplifier—1 MΩ;
- Input sensitivity—not less than 10 μV;
- Dynamic range of amplification—40...72 dB;
- Cut-off band of high-frequency filter—9 kHz;
- Effective voltage of the noise in an amplifying path—not higher than 8 μV;
- Amplification factor of an amplifying path—not less than 60 dB; and
- Amplitude dynamic range of an amplifying path—not less than 50 dB.

A set of devices and equipment that are used for AE testing included the AET with the devices for their installation on IO and materials for providing the acoustic coupling, imitators of the AE signals, and a corresponding software. A chart of a programmable-technological complex “AKEM” used for nondestructive testing of a reservoir is presented in Fig. 6.18.

For research purposes, the AET with the working frequency band of 100… 1200 kHz were used. They were mounted directly on IO by magnetic holders. The amplitude-frequency characteristics of the AET are shown in Fig. 6.19.

Chemical non-aggressive “Ramzay” oil was used as an acoustic contact environment for the AET, which provided an effective acoustic coupling of AET with an IO. In the place where the AET was mounted, the material surface was subjected to mechanical treatment.

6.6.1 Some Methodological Features of AE Testing of a Reservoir

Taking into account the preliminary inspections of a reservoir by means of other non-destructive test methods, a method for on-line exclusive AE testing of the most dangerous parts of an IO liable to fracture was developed. Calculations and the available experience of the AE testing on other objects showed that in this case, these dangerous places are near-bottom joints of a shell and the location of welding of flanges, stripping hatches, etc., which is why the AET were mounted on the IO as close as possible to the indicated places for inspection.

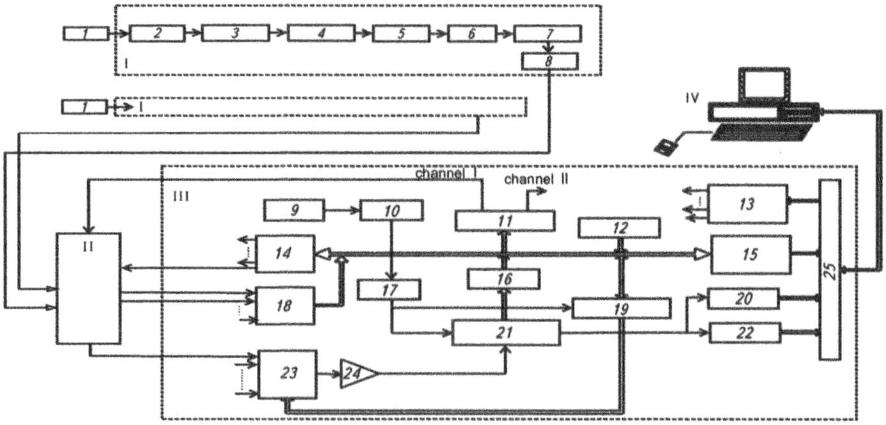

Fig. 6.18 A chart of a programmable-technological AE-complex “AKEM”: *I* is AES amplification path; *II* is a former; *III* is a port of data in-out; *IV* is a computer; *1* is an AET; *2* is an emitter follower, *3* is a preamplifier; *4* is a high-frequency filter; *5* is an amplifier; *6* is a detector; *7* is a low-frequency filter; *8* is an output amplifier; *9* is a generator of 10 MHz; *10* are counters; *11* is a DAC; *12* is a status controller; *13* is an address decoder; *14* is a 16-bit output; *15* is a data buffer; *16, 17, 19, 20, 22, 23* are units of logical control; *18* is a 16-bit input; *21* is an ADC; *24* is an amplifier; and *25* is a personal computer bus

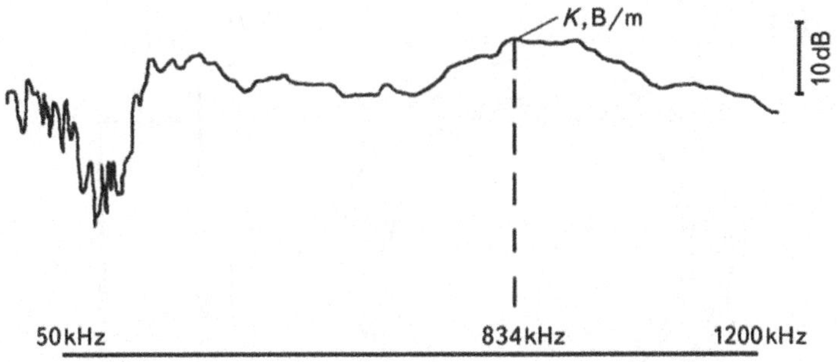

Fig. 6.19 Amplitude-frequency characteristics of the AET used in the AE testing

Taking into account the overall size of a reservoir according to State Standard 4227-2003, it was decided to carry out multi-channel AE testing. During the testing of four channels of the AE system, "AKEM" was used. Before beginning the measurements, according to the requirements of a Standard of the European working group and the requirements of the National Standards of Ukraine, calibration of the sensitivity of the AE path of all channels was performed using the Hsu source. Calibrated in such a way, the channels of AE device were adjusted to the same sensitivity as the elastic vibrations received from the sites of initiation and propagation of crack-like defects (Fig. 6.20).

Proceeding from the above, during hydraulic tests with a simultaneous filling of the external and internal casings of a reservoir, the AET were located on the opposite sides of a structure at the +1.7 m mark on the external casing, as is shown in Fig. 6.21.

In hydraulic testing of the reservoir internal casings, the AET were placed on the internal case of the reservoir according to the chart in Fig. 6.22. In all types of testing, the AE device was located outside the reservoir protective area, in accordance with the accident prevention regularities.

The AES were recorded by four channels during the growth of reservoir loading, and the load was kept at specific levels for an IO for 15 min with technological

Fig. 6.20 Dependence of the change in amplitudes of AE elastic waves on the distance of AE generation by a Hsu source

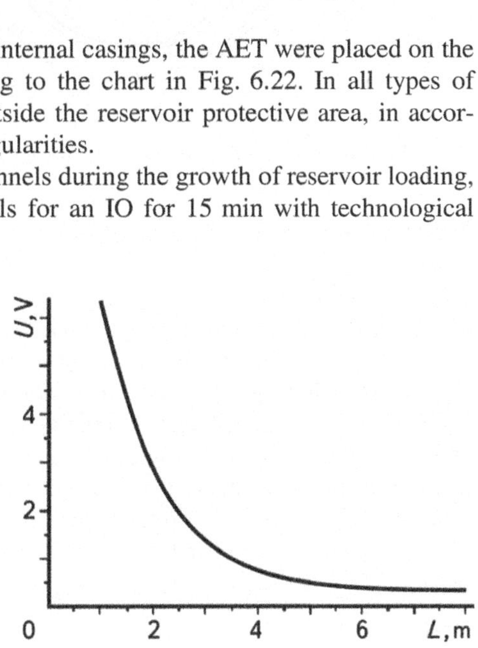

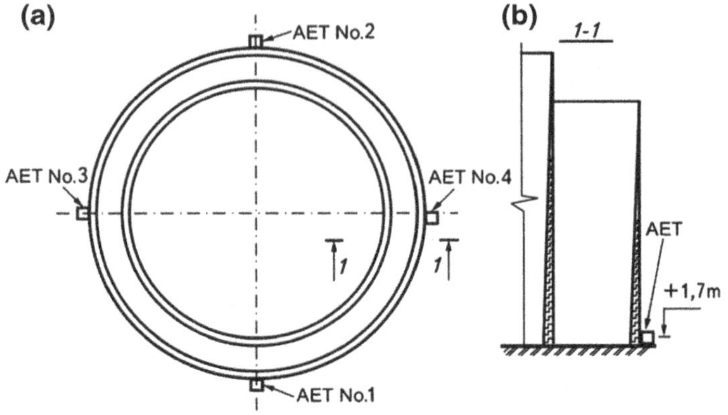

Fig. 6.21 A chart of the AET location during hydraulic testing of the reservoir external casing

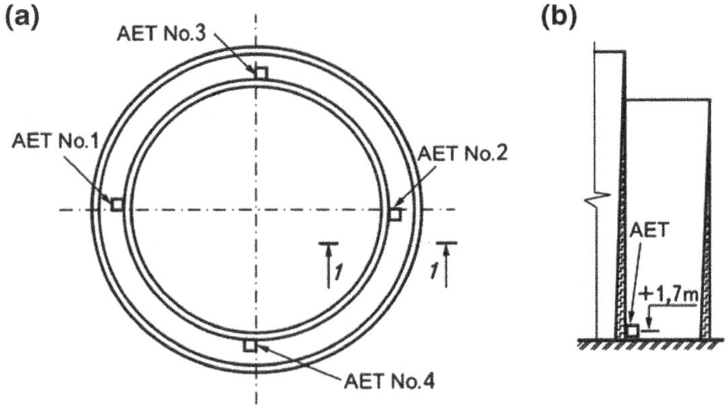

Fig. 6.22 The AET location during hydraulic testing of the reservoir internal casing

interruptions for 4 min. The AE information recording was realized as a continuous monitoring of an object. When the filling of a reservoir with water stopped, the time of the AES recording was 5…6 h.

The external reservoir was filled according to the schedule of testing (Fig. 6.23a), and the internal reservoir—according to the plot in Fig. 6.23b.

The AE criteria taken into consideration during the reservoir testing were as follows: a rapid growth of cumulative count of pulses, their amplitudes, total energy, or the energy that confirms the accelerated growth of defects that caused a fracture. If the parameters of one of the criteria reached a critical magnitude, then a stop of the technological process of reservoir filling was provided in order to investigate the nature of the AE source and estimate the possible danger of further testing.

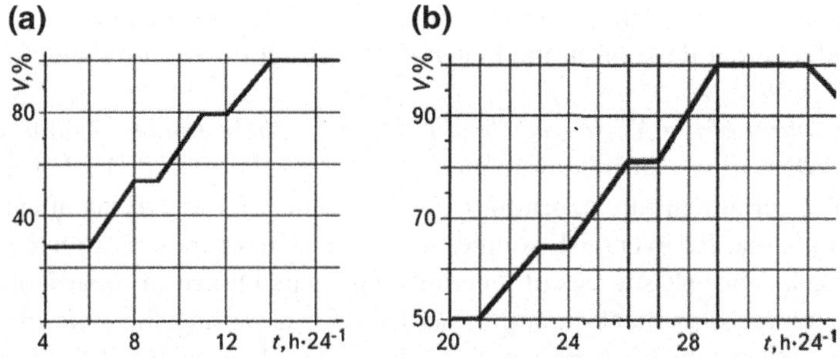

Fig. 6.23 Plot of hydraulic testing of external (**a**) and internal (**b**) reservoirs

6.6.2 *Criteria for Classifying AE Sources*

The coefficient K_p for determining the AE signals caused by cracks characterizes the degree of the energy density change in the recorded AE signal and is used for revealing the crack-induced signals. The following formula is used for evaluation:

$$K_{pj} = lg\left(E_{cj}/\tau_j^2\right) + B + C, \tag{6.20}$$

where: E_{cj} is the energy caused by the j-th recorded AE signal; τ_j is the duration of the recorded AE signal; B is the correction factor for the device sensitivity; and C is the correction factor for a threshold of the amplitude level. Coefficients B and C are determined using the known methods.

The danger of cracking in the material of a structure under loading is evaluated using the analysis of the kinetics of the AE development, including the analysis of common data obtained at the stages of holding the structure under loading. To compare and generalize the results regardless of dimension the analyzed parameters are normalized to unit:

$$\bar{E}H = f(\bar{P}), \tag{6.21}$$

where $\bar{E}H = E_i/E_{imax}$; $\bar{P} = P_i/P_{imax}$; E_i is the energy accumulation of the AE signals while holding a specimen under loading in the selected time interval; P_i is the load of the structure investigated; E_{imax} is the maximal value of the energy accumulation of the AE signals while holding a specimen under loading in the selected time intervals; P_{imax} is the maximal load of a structure during testings. The experimental data are analyzed with approximation as described above, using the equation

$$\bar{E}_H = a\bar{P}^b, \quad \text{at } t_i = \text{const}, \tag{6.22}$$

where a, b are the constants. The absolute value of the power index $b < 3$ proves that the defects that develop in the material of a structure are not dangerous in this case.

Classification of the AE sources by the degree of their danger. Additionally, the results of AE testing can be represented as a list of the recorded AE sources that belong to a certain group depending on the value of the AE parameters. Such evaluation is done for every AE source, while the state of the investigated structure is evaluated by the AE sources of a certain type. The choice of a particular system of the AE source classification and the criteria for assessing the state of an object depends on the mechanical properties of the material of an IO. The classification system and the assessment criteria of the structures' state are chosen using the classification systems, and the criteria for assessing the state of a structure of IO state are given below.

The admissible level of the AE source generation in a fracture is established while preparing a particular structure for AE diagnostics in accordance with existing methods. The AE sources are classified by the following AES parameters: cumulative count, number of pulses, amplitude or amplitude distribution, energy or energy parameter, count rate, activity, and concentration of the AE sources. The system of classification also includes the parameters of loading the tested structure as well as time parameters.

By analogy with approaches [12, 13], the revealed AE sources were divided into four groups: Group I—passive; II—active; III—critically active; IV—catastrophically active. The choice of the classification system of the AE sources and of a possible level (group) of sources is performed each time during AE diagnostics of a particular structure. Every superior group number of the AE source requires a fulfillment of all actions stated for the sources of an inferior group number. A final assessment of admissible AE sources, while applying additional types of non-destructive testing is performed, with the use of defect parameters found by linear fracture mechanics methods, strength calculation methods and other operating standards.

Amplitude criterion. An average amplitude A_{av} of at least three pulses with the individual amplitude A_c is determined for each AE source for the selected observation period. The amplitude is corrected, taking into account the AES damping during their propagation in a material.

The maximum value of an admissible amplitude A_t was determined in the previous experiments:

$$A_t = B_1 \times U_{th} + B_2 \times A_c, \tag{6.23}$$

where U_{th} is the threshold value of amplitude level; A_c is the value of the AE signal exceeding the threshold value that corresponds to the crack growth in a material; and B_1 and B_2 are the coefficients determined within the range of 0…1 from preliminary experiments.

The sources are classified as follows: Group I includes a source for which the average amplitude of pulses was not calculated (less than three pulses are obtained during the observation interval); Group II—the inequality $A_{av} < A_t$ is satisfied; Group III—the inequality $A_{av} > A_t$ is satisfied; and Group IV—a source that includes at least three recorded pulses, for which the inequality $A_{av} \gg A$ is satisfied.

Specific values of A_t, B_1, and B_2 depend on the material of a structure and are determined in the preliminary experiments.

Integral criterion. For every area, the activity F of the sources of AE signals is calculated using the following expression:

$$F = \frac{1}{K}\sum_{k=1}^{k}\frac{N_{k+1}}{N_k}, \tag{6.24}$$

where

$$\frac{N_{k+1}}{N_k} = \begin{cases} 1 & \text{at } N_k = 0 \quad \text{i} \quad N_{k+1} > 0 \\ 0 & \text{at } N_k > 0 \quad \text{i} \quad N_{k+1} = 0 \end{cases}, \quad k = 1, 2, \ldots, K. \tag{6.25}$$

Here, N_k is the number of events throughout the k-th interval of parameter estimation; N_{k+1} is the number of events throughout the k + 1-th interval of parameter estimation; k is the number of intervals of parameter estimation. The observation interval is divided into k intervals of parameter estimation. Estimation of the criterion was done taking into account the dependencies: $F \ll 1$, $F = 1$ and $F > 1$.

Throughout each recording interval, a relative force J_k of the AE source was determined as follows:

$$J_k = A_K \Big/ W\sum\nolimits_{k=1}^{K} A_k, \tag{6.26}$$

where A_k is the average amplitude of the source throughout the interval k; A_K is the average amplitude of all AE sources throughout whole object except for that analyzed for interval k, and W is the coefficient determined in the preliminary experiments.

Local-dynamic criterion. To estimate this criterion in real time, the following AE parameters were used: N_{i+1} is the number of AE pulses in the successive event and N_i is the number of pulses in the preceding event, or: E_{i+1} is the energy of the successive AE event and E_i is the energy of the preceding AE event.

The parameter U^2, which is the square of AES amplitude, can be used instead of energy. For every event, the values W_{i+1} and V_{i+1} are calculated using the following expressions: $W_{i+1} = N_{i+1}/N_i$, or $W_{i+1} = E_{i+1}/E_i$,

$$V_{i+1} = \left(1 + \frac{P_{i+1} - P_i}{P_{i+1}}\right)^4 - 1, \tag{6.27}$$

where P_{i+1} is the external parameter value at the moment of the successive event recording (if time is used as a parameter, then it is an interval of time from the beginning of an observation interval); P_i is the external parameter value at the moment of the preceding event recording (if time is used as a parameter, then it is an interval of time from the beginning of an observation interval). Sources were classified as follows: Group I—$W_{i+1} \leq V_{I+1}$; Group II—$W_{i+1} = V_{i+1}$; Group III—$W_{i+1} > V_{i+1}$; Group IV—$W_{i+1} \gg V_{i+1}$. During testings, an on-line recording of information and processing of the AE data were done. After NDT of an object, data were analyzed and processed by a personal computer that was included in the AE testing system.

6.6.3 Results of the AE Testing and Their Interpretation

During hydraulic testing in the first and the second stages (external and internal reservoirs), a moment of abrupt increase of the AES amplitude was recorded. Using technical means, a criterion was evaluated that indicated the appearance of fracture in the place of a stripping hatch welding. Having adopted certain decisions, agents of corresponding services made a visual observation of the indicated place of the AES generation. As a result, the defect was detected, which after testing by other NDT methods caused the halt of the process of filling the reservoir with water and hatch replacement.

After that, according to the schedule of the reservoir filling with water, the AE testing was continued. At a mark of +13.94 m, the AES were within the limits of the background level. During the subsequent reservoir filling above the mark +13.95 m, the AE intensity increase was recorded (Fig. 6.24a, b). This process was observed up to the mark of +13.97 m of the reservoir filling (Fig. 6.24c, d). The recorded increase of the AE intensity lasted 45 min. After the water supply was stopped, additional inspections and corresponding consultations permitted ascertaining that this increase was caused by an abrupt increase in the water supply rate to the reservoir (to 1000 m^3/h), which affected the background level of the AE signals. When the rate of the water supply decreased to the previous level (500 m^3/h), the recorded AE signals were within the limits of the previous safe background level.

During an internal filling of the reservoir with water, the AES that did not exceed the background level were observed. The processing of the AES using the established criteria showed that they belong to the Group I sources, i.e., passive ones, which under continuous monitoring did not show a tendency towards defect development, although at certain moments, a single, non-dangerous growth of the AES intensity was recorded. At the +17.62 and +18.39 m marks, the AES were also recorded with somewhat higher energy and were classified as Group II sources.

The results of processing the AE signals of higher intensity using the integral and local-dynamic criteria are presented in Table 6.7.

Thus, the results of the AE tests showed that the growth of the intensity of AE signals at certain moments of loading was caused by the following technological

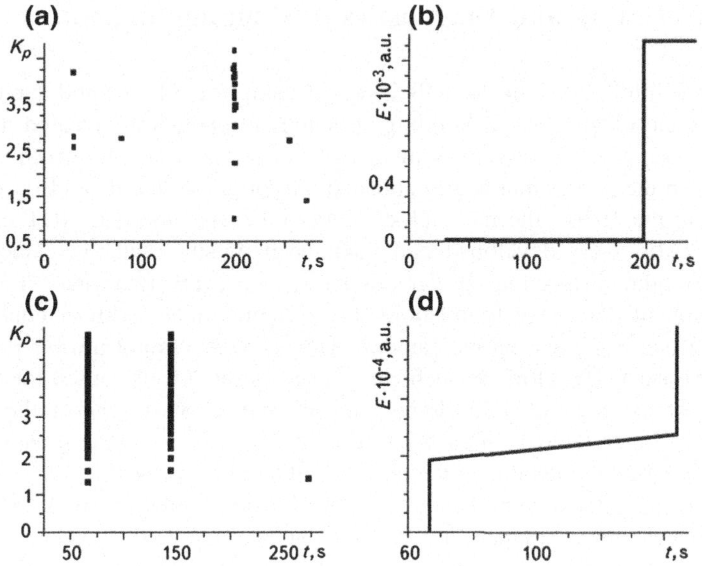

Fig. 6.24 Time variation of the factor K_p (**a**, **b**) and the AES energy (**c**, **d**) during reservoir filling from the +13.94 to +13.95 m mark (**a**, **b**) and from the +13.95 to +13.96 m mark (**c**, **d**)

Table 6.7 Classification of the recorded AE sources

Period of AE signals recording	Number of sources by groups											
	Integral criterion (*W* = 0.1)				Integral criterion (*W* = 0.01)				Local-dynamic criterion			
	I	II	III	IV	I	II	III	IV	I	II	III	IV
External reservoir	0	10	4	0	6	8	0	0	3	2	8	0
Internal reservoir 20–24 days	0	9	0	0	2	7	0	0	0	1	0	0
24–27 days	0	41	0	0	12	29	0	0	–	–	–	–
27–29 days	0	31	0	0	11	20	0	0	–	–	–	–
29–33 days	0	0	0	0	0	0	0	0	0	0	0	0

factors: the presence of cracks propagating under loading, an increase in the rate of water supply, and stress relaxation in the reservoir walls. In the material that was not likely to fail, the AES parameters did not exceed the background level at all stages of loading.

6.7 AE Testing and Diagnostics of Building Structures

The earliest information on the efficiency of using the AE method for industrial inspection and diagnostics of building structures in operation appeared in the late 1960s [39–42]. The method was intensively developed and improved, which is evidenced in the papers that were published. According to the data [43], already in the USA in the 1970s, the national standards for inspection and NDT of bridges using the AES were developed and have been widely used ever since. These standards require inspecting bridges at least once every two years. Earlier, the visualization of defects of foundations and elements of their lower constructions was performed using submarine cameras. However, it was not always possible to reveal the cracks. Therefore, in such cases, even slight defects caused the failure of bridges. For example, a 3 mm-long crack caused a catastrophe at the Pleasant Bridge; the Silver Bridge failed because of a defect of less than 4 mm, and the Mello-Mild Bridge failed due to the presence of fatigue cracks after 12 years of use.

For reinforced concrete bridges of large weight and rigidity, there is only a slight probability of the appearance of irreversible damages due to the presence of fatigue cracks. In such structures, corrosion of a preliminarily stressed reinforcement turns out to be more dangerous. Besides, it is much more difficult to detect defects in reinforced concrete. Proceedings from the above-discussed, now- effective NDT methods are being intensively developed all over the world, and these methods are based on various approaches. However, the methods that use the AE phenomenon are preferable. For example, in Japan in the 1970s, the integrity of old and new railway bridges, local railways, runways, impacting equipment and hoisting cranes began to be checked using these methods [44]. In particular, a four-year AE checking of the runways in airports showed that there is a correlation between the processes of crack propagation, loading, and AES. In [45], the AE and micro-seismic research into the estimation of the methods for testing the growth of damage in the coatings of the same runways is described. The data collected enabled comparing the accuracy of the methods of the AE sources' location, i.e., time-related, zonal, and impact-successive. It has been shown that the last two methods are potentially more accurate than a time-related method. In practice, they provide a satisfactory accuracy of the active areas' location, and it is noted that in the conditions of high damping, an impact-successive method provides the best accuracy. Gradually, the AE method gained more and more recognition, and already in July 1979, information on more than 100 industrial inspections of various structures (bridges, cranes, etc.) with the use of the AE phenomenon was published [46]. About 20 of these structures were at the stage of being assembled, 90 were in the process of current operation, and more than 10 objects were inspected in in-service conditions.

A wide application of reinforced structures in transport building, especially for building bridges, requires corresponding NDT facilities both in the process of work execution and during exploitation. A detailed analysis of the failures of reinforced concrete bridges shows that in the process of construction, a principal cause of failure is the violation of the regularities and technologies of the work performed, while in exploitation there are cracks in structures, primarily due to the action of corrosion of products of a pre-stressed reinforcement. Therefore, alongside traditional methods of NDT of reinforced concrete products, the AE method has also been more widely used lately. Such studies are actual today, which is evidenced by publications in scientific literature. In particular, in [47] the authors showed that in the process of bridge exploitation under dynamic loading, background noise is formed causing a distortion in the AES frequency spectrum. Analytical dependencies are presented therein for obtaining a formula for the calculation of the maximal size of the bridge tested area as well as a method that is described for selecting the optimum working frequency band of AE equipment when distortions of the AES spectrum are minimal.

The paper presents the results of the calculation of normalized AES spectra in three characteristic regions of a metal bridge: 1—with a fatigue crack; 2—unfit riveted joint; 3—non-damaged structural element. The spectra differ both by the level of power and by the shape that can serve as a criterion for characteristics of damages in bridges. In paper [48], the use of the AES spectral analysis is considered to be a reliable method of corrosion cracking and crack propagation detection. Here, the information on the methods of the AES processing based on the application of the fast Fourier transform, as well as the charts of the design of an automated AE diagnostics system are presented. The efficiency of this approach is evidenced by the results obtained in [49], and by the AE testing of reinforced concrete bridge structures. The results of the AE inspection of 36 railway bridges are also described in [50], where some suggestions for the improvement of the AE test procedure and data processing are presented.

The results of experimental studies on the relationship between the AE parameters and cracks in the elements of reinforced concrete bridges under loading and under the action of corrosion products are discussed in [51]. Experimental data on the determination of noises in a reinforced concrete bridge in the conditions of intensive motion of electric vehicles are also discussed. Their classification is given, and the physical processes of initiation and methods of AES extraction from them are considered. The experiments were conducted using the AE analyzers of the AVN-1M and AVN-3 type.

Using a movable multi-channel laboratory for the AE diagnostics, an inspection of the Crimean, Velyko Kamianyi, and Velyko Ustynivskyj bridges, and a metro bridge in Luzhnyky was carried out [52]. The inspection was done in order to assess a bearing strength of structures and their lifespans, to detect potentially dangerous defects, the uniformity of distribution of stresses on bearing units and elements, to predict crack growth resistance of materials, and to study the effect of climatic and other factors on the dynamics of load redistribution. The analysis of the dynamics of the development of AE generation processes in real time and a correlation analysis of distribution permitted, in accordance with the developed criteria, detecting the highly dangerous areas, to localize the areas with defects, and to estimate the local stress state in bearing elements. In paper [53], the authors based on the laboratory research on fracture of a concrete with plasticizer additives elaborated, a method which, they believe, can be used to estimate the quality of a roadbed.

In the 1970s, investigations aimed at using the AE method for assessing the integrity of beam-and-girder constructions began in the U.K. [54]. They were initiated by the Committee for Construction, following the failures of these structures made of a pre-stressed concrete. The NDT of these products by conventional methods did not yield satisfactory results, and the AE method, in the authors' opinion, was most effective for this purpose. This statement was scientifically grounded in paper [55], presenting the results of the experiments on pre-stressed pressure vessels made of concrete and on casings of nuclear reactors in the USA by AE inspection and NDT. The AE phenomenon was used together with radiographic and ultrasonic NDT techniques.

Cracks are formed in large-panel-bearing elements of buildings in the process of their manufacture, transportation, stocking, and mounting. These cracks are mainly from temperature shrink and a force-induced nature. Force-induced cracks are very dangerous for structure operation because they propagate due to the applied loads, considerably violate the conditions of structure exploitation, create emergencies, and lower the reliability and lifespans of buildings. To estimate the effect of cracks on the strength-bearing of a structure, it is necessary to determine not only their origination but also the character of their development under loading. Therefore, in this case, the AE method turns out to be very efficient for determining the degree of danger of force-induced cracking in structures. On the other hand, to arrive at a well-grounded conclusion about the state of a cracking structure, it is necessary to observe the structure for a long period of time, and to collect and analyze statistical data on the crack propagation dynamics and crack parameters during the early years of object exploitation.

In [56], the crack initiation in bearing structures was inspected in the following way: In a structure with cracks, and in a similar one without cracks, both of which were subjected to identical loading (on the same floor of a building), the number of AE pulses was recorded at a discrete increase of a load due to the assembling of the next floor structures. Using the results of AE testing in the structures with defects and in those free of defects during an increase of the load up to the calculated value, empiric dependencies of the number of AE pulses on the level of a load were built, as well as time series of the number of pulses for all the values of the second. The results of testing the state of bearing wall panels having force-induced cracks, testified to an available discrepancy between the designed operation conditions of structures and their real service conditions. It was shown that an increase of the rigidity of buildings with the growth of the number of stories, obviously, resulted in a considerable and more uniform redistribution of forces and slowed down the propagation of cracks. Therefore, structure-strengthening measures were not always justified. This was proved by the results of durable AE observations of cracks during the first two years of exploitation that showed the invariability of crack parameters. As a result of the AE research, the method of estimation of crack formation in reinforced concrete structures using the AES was developed and put into practice.

The AES in mountain rocks during cement solution pressurization into a dam was analyzed in paper [57], where highly promising economic capabilities of the AE method for diagnosing the state of such buildings were confirmed. A similar result was obtained by the authors [58], where a possibility of applying the AE for inspection of the bearing strength of frozen foundations during their defrosting was considered. The method for laboratory and field testing was described, and the test results were presented. The possibilities of integrating various diagnostic systems were considered as well.

The problem of utilizing the AE phenomenon for estimating the displacement of soils and for the development of displacement prevention measures was comprehensively discussed in [59]. The results of both laboratory and field studies were presented. The pertinent data of such experimental testing published by scientists from many countries were analyzed, and the positive results showed good prospects for using the AE method, its high resolution and information capacity, and its advantages as a technique for predicting of the state of the objects tested, such as buildings and structures.

6.8 The AE Inspection of Bridges in Ukraine

Bridges form only a minor part of the total length of transport communications, but their technical condition affects the transportation safety to a great extent. This is conditioned by the fact that bridges concentrate traffic and it is more difficult to provide their reliable and durable operation compared to roads, because they are more complicated engineering structures and are subjected to the action of various loadings and effects [60]. Apart from the action of transport loading and their own weight, span bridge structures are subjected to dynamic wind effects and longitudinal loadings during the braking or acceleration of vehicles, etc. Bridge footings are also subjected to an additional pressure of ice, the impacts of floating materials, and in earthquake-prone regions.

There is a wide network of motor roads and railways in Ukraine; it includes 16,300 road-transport national and 358 km long local bridges, 4082 communal bridges that are 184.8 km long, and 8050 railway bridges that are 210.4 km in length [61]. Many bridges that have damage and defects are in use in Ukraine. Data of large-scale inspections of the bridges carried out by various organizations in the 1980s stated that [62, 63] bearing structures mainly contain the following defects: fracture of protective coatings and corrosion of steel bridges—41% of the inspected bridges; splitting off, cavities, and cracks in a concrete—65%; corrosion of reinforcement—40%; and carbonization of a concrete protective layer—60%. The actual lifespan of reinforced concrete bridge structures is 25…30 years, at which point the necessary expensive repair work does not promote the potential properties of reinforced concrete as a material [64]. The principal causes of this are the corrosion of reinforcement and concrete.

Bridge constructions have a tendency to brittle fracture, i.e., to fracture by propagation of crack-like defects. The fracture process in such materials is not immediate because some time passes from the moment of crack initiation to the beginning of reaching its critical value. Therefore, the early detection of such defects and estimation of sub-critical crack growth stages is an important scientific and technological problem. It is difficult to detect a considerable part of defects in bridges because their metallic elements are covered with paint, the reinforcement is inside the concrete in reinforced concrete structures, and crack initiation and propagation often occur inside the material. Thus the problem of providing a reliable and long-term exploitation of bridges becomes more and more urgent because the ages of the bridges increase, the weight, intensity of motion and dynamic effects on the bridge structures grow, and the aggressiveness of the environment increases. The growing complications in providing a reliable and durable exploitation of bridges require new approaches to their technical diagnosing [65].

To estimate a macro-crack initiation in bridge structures by means of the AE method, an approach proposed in [66] is used. It consists of using the estimation criterion K_{pj} which is based on the rate of the recorded variation of the AES energy density.

$$K_{pj} = \lg\left(E_j/\tau_j^2\right), \tag{6.28}$$

where

E_j is the $j-th$ AES energy; τ_j is its duration.

In order to digitally process the AES, expression (6.28) was transformed into:

$$K_{pj} = \lg\left[\sum_{i=1}^{n} (A_{ji})^2/\mathrm{LE}_j^2\right] + \lg(\Delta U^2/\Delta t), \tag{6.29}$$

where Δt is the time interval of the AES sampling (Δt = const); A_{ji} is the number of bits of an analog-digital converter for the i-th sampling of the j-th AES amplitude; n is the number of amplitude counts for the j-th AES; and LE_j is the j-th AES duration ($\mathrm{LE}_j = n$).

If the sensitivity of the AE equipment ΔU and sampling frequencies of the input signal are set, the second part of equation (6.29) becomes constant B, that is

$$K_{pj} = \lg\left[\sum_{i=1}^{n} (A_{ji})^2/LE_j^2\right] + B. \tag{6.30}$$

The AE *inspection of the bridge across the Western Bug River.* An object under investigation was the bridge roadway near the village of Yahodyn in the Volyn Oblast. The bridge was built in 1953–1954 from metallic structures taken from the demountable bridge built in the 1930s. The bridge is of a split-type with three spans, and its cross-section consists of two girders according to a 3 × 62 m scheme. The size of the bridge roadway is G-7.0 + 2 × 1.55 m. Metallic girders are riveted with parallel bridge booms with a carriageway on the bottom and triangular grillwork. The distance between the axes of girders is 8.65 m, and top and bottom booms are of an H-like cross-section. Diagonal webs are of an assembled, double-T, through and grated type. The assembled buckstays are also of a box-like cross-section, and there are horizontal crossing connections in the plane of the top and bottom bridge booms. The bridge roadway is made of monolithic reinforced concrete arranged on metallic sections. It lies on a beam cage made of cross-beams of variable cross-sections that are fastened to the girder joints, and longitudinal double—T beams fastened to the supports of cross beams. The coating of the bridge roadway is asphalt-concrete, without coating on the sidewalks. Bridge supports made of concrete are massive and are located on reinforced concrete hanging piles. Unmovable supports are made of metal with top and bottom equalizers and a cylinder hinge between them, while movable supports are metallic roll bearings.

The span structures of the bridge were subjected to static and dynamic loadings. All three spans were loaded alternately by trailer trucks weighing from 200 to 380 kN each. The "AKEM" programming complex recorded the AES.

Static loading was applied as follows: Trailer trucks drove successively to the roadway of every bridge span (Fig. 6.25) and then stopped. The trucks weighed 200…380 kN, and there were a maximum of eight trucks on the span roadway. After the arrival and stopping of every truck, the AES were recorded and processed. At the same time, the bending of girders in the middle of a span was measured with a deflectometer, while strains of the most stressed girder elements were measured with electromechanical strain gauges.

The AE data were recorded for 45…120 s, and the AET was mounted in the region of maximum tensile stresses (a middle part of the span; see Fig. 6.26). Bridge spans were numbered, beginning in Ukranian territory. Before mounting the AETs on a bearing beam and a stiffening rib, they were cleaned of paint and rust; the diameter of the cleaned surface area was about 20 mm. After applying an acoustic transparent contact layer, the AETs were pressed to the metal surface by clamps.

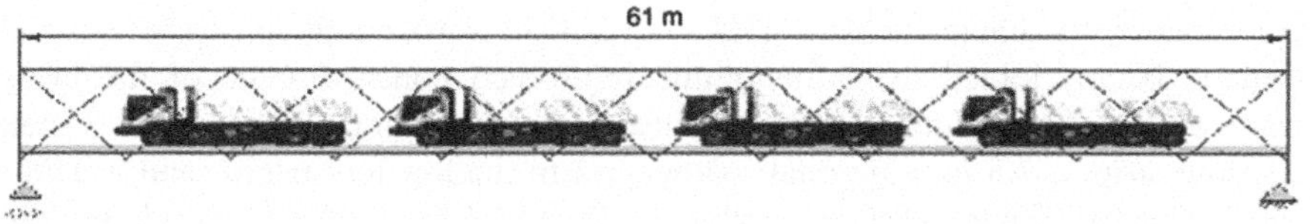

Fig. 6.25 Static loading on the bridge span

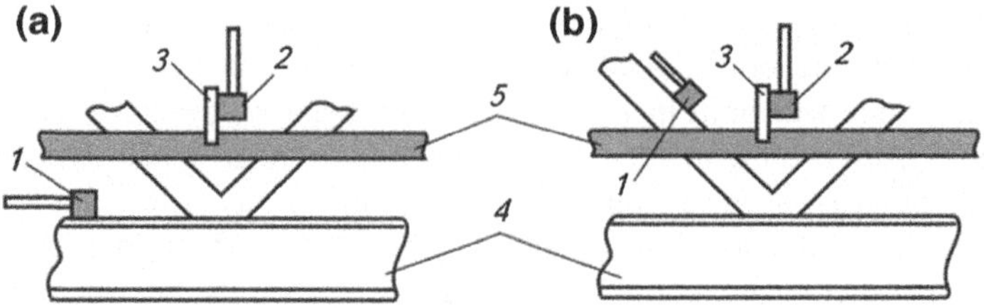

Fig. 6.26 Location of the AET under the loading of span 0–1 (**a**) and spans 1–2 and 2–3 (**b**): *1* is AET No. 1; *2* is AET No. 2; *3* is a waveguide; *4* is a lower girder boom; *5* is a reinforced concrete plate of the bridge roadway

While testing all bridge spans, AET No. 2 was located on the waveguide placed directly in the concrete. The waveguide was made from a reinforced rod of 8 mm in diameter and 150 mm long. On one of the edges of the waveguide, parallel to its axis, the flat of a depth of 4…3 mm was made in order to mount the AET. The waveguide was inserted into a hole (7 mm in diameter, 10–15 mm deep), which was drilled into the concrete. The AET was mounted on the waveguide with tension, and two channels were simultaneously recording and processing the AES.

During the first kind of dynamic test, one truck with a fixed weight was driven at a constant speed of 5, 10, 20, and 40 km/h along each bridge span. The recording and processing of AES began at the moment the trailer truck arrived at a span, and ended when it drove away from the bridge span. During the second kind of dynamic test, the truck crossed a barrier placed in the central part of the bridge span. The weight of the truck was fixed, and the speed was the same as for the first type. A 10 cm. thick wooden square beam was used as a barrier.

The criteria estimation of the detected AE signals caused by cracks was performed in accordance with a crack evaluation by the K_p factor. For instance, at the first, second, and fourth loading degrees of span 0–1 with one, two, and three trucks, respectively, crack propagation was not detected. The AES caused by cracks were recorded only at the fourth stage of loading (four trucks). The analysis of the recorded AES parameters (amplitude, energy, duration) shows that they are of low magnitude, which testifies to the stabilization of the development of micro-processes in the material of a structure. Further testing of the bridge using the AE method showed that under service loadings, no defects whose growth would endanger the span structures appeared [65].

The AE inspection of the road-transport bridge across the Pivdennyi Buh River [67]. A bridge is located near the village of Lupolove on the Kiev-Odessa highway. The span structure of the bridge is made of non-cut steel iron concrete, made according to the scheme (32.9 + 43.3 + 43.3 + 32.9 m), and it consists of six main metallic beams joined by a monolithic reinforced concrete plate of the bridge roadway and constraints. The width clearance of the bridge includes a 3-meter dividing strip, a 7.5 m wide dual roadway, 2 m thick safety bands, and a 0.78 m wide sidewalk. The total bridge width is 14.14–153.3 m long, and a facade-building

Fig. 6.27 General view of the bridge during testing

height is 2.1 m. There are tower-shaped intermediate supports made of reinforced concrete (Fig. 6.27).

The bridge span structure was subjected to static and dynamic loadings. Trucks weighing 250 kN each were used as a test load. AE signals were recorded and processed by means of an "AKEM" programming complex [66]. The aim of the research was to detect the AE sources related to the defects that are active under static and dynamic loadings of bridge structures, as well as to reveal the regularities of the AE process.

A maximal bending moment was formed in a span 3–4, and the AET was placed on the rib of the upright dual-T metallic beam (Fig. 6.28b) in accordance with the loading Mode 1. The following two modes (Nos. 2 and 3) enabled creating maximal forces in the reinforced concrete cantilever of a roadway plate, and the AET was mounted on the concrete (Fig. 6.28c). Before placing the AET on the metal, the paint was cleaned from its surface; after this, a layer of "Ramzay" acoustic-transparent filler was applied and the AET was pressed to the surface by clamps. The AET was pressed to the concrete by a magnet that was fixed by its one end to the AET, and the other end was fixed to the metal of the dual-T beam top shelf (Fig. 6.29).

During static testing, the AES recording began at the moment the trucks stopped in the places determined by the testing modes, and the data were recorded for 40… 60 s. During dynamic tests, the AE was recorded from the moment of the arrival of the trucks at the bridge up to the moment of their leaving it. The AET location was similar to that for static testing according to Modes 2 and 3.

The criterial assessment of detection of the AE signals caused by cracks was processed, as shown above, by means of the K_p factor. When the AET was placed on the metal surface, the value of the criterion for extraction of the AES caused by cracks (the K_p factor) was assumed to be 3, and for concrete it was 6. The AE sources that were revealed, as described above, were divided into four classes:

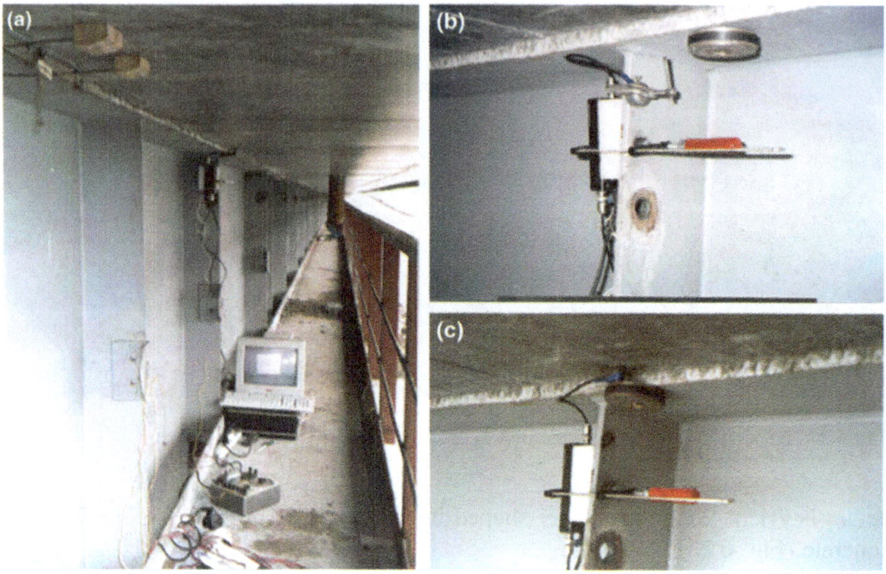

Fig. 6.28 The AE "AKEM" equipment (**a**) and methods of mounting the AET on metal (**b**) and concrete (**c**)

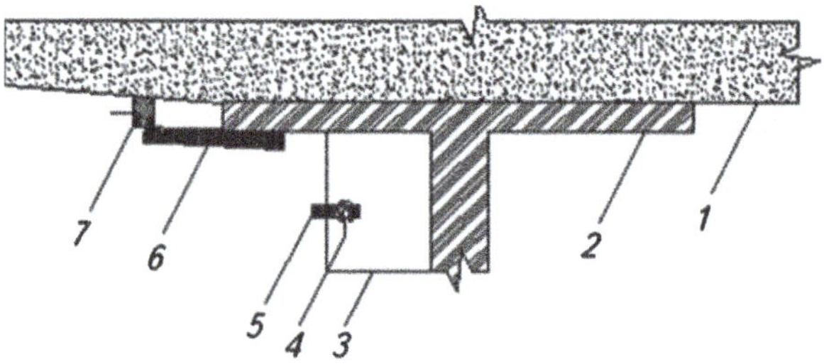

Fig. 6.29 The AET location during bridge testing: *1* is a reinforced concrete roadway plate; *2* is a metallic beam; *3* is a metallic beam edge; *4* is the AET placed on metal; *5* is a pressing clamp; *6* is a magnet; *7* is the AET placed on concrete (Modes *2* and *3*)

Class I is a passive source; II, an active source; III, a critically active source; and IV, a catastrophically active source.

Using the integral criterion, which was also described above, the AE sources' activity was calculated from the expression:

$$\tilde{A} = a\tilde{\Pi}^{b}, \quad \text{at } t_i = \text{const}, \tag{6.31}$$

where $\tilde{A} = A_{\mathrm{T}}/A_{\max}$; $\tilde{\Pi} = \Pi_{\mathrm{T}}/\Pi_{\max}$; A_T, $\Pi_{\max}$ are the current and maximal values of parameters. The absolute value of the power index $b < 3$ proves that the defects that develop in the material of a structure are not dangerous.

Under the static testing of the bridge span structure by Mode 1, the AES from the defects in the metallic beam material were recorded. This is confirmed by the corresponding values of the K_p parameter. The analysis concerning the development of dangerous defects in the material structure under the loading of the span showed the following: A relationship between the load level and the accumulated AE energy is observed beginning at the 12th second. In this case, the values of coefficients in the approximating expression are as follows: $a = 0.0878$; $b = -1.76897$; $\sigma^2 = 0.02824$. The absolute value of the power index is $b < 3$, and while loading the bridge span structure by Mode 1, the defects that develop in the structure of a material are not dangerous.

Under the static testing of the bridge span structure by Modes 2 and 3, the AES caused by micro-cracks in the reinforced concrete plate of the bridge roadway are recorded, which is proved by the corresponding values of the K_p parameter that do not exceed 6. The following was shown by analyzing the development of dangerous defects: Under the loading Mode 2, beginning at the 20th second, a relationship between the accumulated AES energy, and the load of a span was observed. The analysis of the relationship using data formalization shows that it can be described by the expression with the parameters of approximation: $a = 0.26273$; $b = 0.05307$; $\sigma^2 = 0.36816$. Under loading by Mode 3, the dependence between the accumulated AES energy and the loading of a span appeared after 7 s. After 11 s: $a = 1.08086$; $b = 1.61458$; $\sigma^2 = 0.04489$. The absolute value of power index is $b < 3$, i.e., during the bridge span structure testing by loading Modes 2 and 3, the defects that develop in the material structure turned out not to be dangerous.

Thus, bridge testing using the AE method showed that there are no dangerous defects in the bridge structure that could hinder a reliable and safe use of span structures. Moreover, these defects did not show any tendency to develop. It is recommended that such tests be repeatedly performed after a certain period of time in order to estimate the state of the bridge span structures and to create a system of their monitoring. This would permit developing a database for establishing the dependencies used to estimate the bridge lifespan according to the AES parameters.

The AE inspection of the bridge across the Prut River in Chernivtsi. This bridge is located on the main highway of the third category M20 Zhytomyr–Chernivtsi–Terebleche (Fig. 6.30 [68]). The bridge is made of metal, has 6 spans, and was built according to a chart: 38.60 + (39.14 + 2 × 38.90 + 39.14) + 38.60. The bridge is 247.18 m long.

A permanent bridge passage with metallic span structures on rubble concrete footings was built in 1927–1931 and crossed the Prut River at a right angle. During the World War II the bridge was destroyed twice and rebuilt twice. The foundation of the footing 2 was not destroyed and a 5-m high primary rubble concrete footing was built on it. A temporal frame bricked with its basis in rubble concrete was arranged over it; the bridge girders were supported by a metallic frame through a grillage foundation. In 1961 a wooden roadway of the bridge was replaced with reinforced concrete.

The span structures consist of two through metallic girders with parallel booms and a carriageway on the bottom boom. Girders are of an open type and do not have wind constraints on the upper boom. The filling is constructed as a triangular web with additional vertical tower bodies.

Fig. 6.30 A general view of the bridge across the Prut River in Chernivtsi

Coastal spans 0–1 and 5–6 are covered with cut girders. The river bed spans 1–2…4–5 are covered with a non-cut, four-span girder. Every span girder is divided into 10 panels (3.8–3.87 m long). In a transversal cross-section, the distance between the axes of girders is 6.51 m, and their height is 4.04 m.

The sections of girder elements are made of the rolled elements of channels and angle bars that are reinforced on the top and on the bottom booms by metallic sheets in the middle panels and in the supporting areas of non-cut girders. In a transversal direction, the rigidity of span structures is provided by horizontal diagonal connections of angle bars (2*L* 90 × 90 × 10 mm) located at the bridge boom bottom level.

Abutment spans are made of rubble concrete and are monolithic with undersides. Supports consist of two parts and end with cornices. The lower part is widened, and its height is about 5.0 m, with the top part about 1.5 m high. A reinforced concrete underside is made as a monolith with sidewalk cantilevers about 7 m long. Two concrete diaphragms are mounted on the bottom of sidewalk cantilevers at the up-river bridge side and at the down-river side.

Intermediate supports are rubble concrete, hollow (except for the restored support 2), slush plastered, and made of two parts: The lateral sides of the bottom parts of intermediate supports have inclinations with a ratio of 15:1. The up-river bridge side is executed as an elongated ellipse and acts as a breaker; the down-river bridge is semi-circular. Foundations of all supports are caissons laid 20…35 m deep. Girders are set on the supporting mobile (bowled, rolled) and immobile (balance) parts.

The reinforced concrete plate of a bridge roadway is an assembled structure 11 cm thick; its cross-section is a trapezoid that consists of two Γ-shaped units. A longitudinal weld is located along the axis of the bridge roadway. The length of a unit is equal to the plate length, and transversal welds are located over the crossbars. The plate is placed on a beam cage that consists of longitudinal and transversal beams.

Transversal beams are assembled, riveted from angle bars and double-T sheets 0.74 cm high, have a calculated length of 6.15 m, and are installed in every joint of a bottom boom. They are tightly joined with the support of the main girders and form a semi-frame. Two stiffeners made of 90 × 90 × 10 mm angle bars are riveted to the vertical ribs on both sides to provide local stability.

Longitudinal beams are made of an I-bar 30 and are tightly connected at one level (by a supporting table) with crossbars. Six longitudinal bars are placed in the transversal section with an axes distance of 1.0 m and an estimated length of 3.7…3.75 m. A plate is set on the bar cage using reinforced concrete square beams of a 13.5 × 12.5 cm cross-section, made as a concrete monolith with the top boom of the longitudinal beam. The length of the squared beams is equal to the length of the panel.

Sidewalks are placed on the external bearing-out cantilevers made of a rolled metal (angular bar and a channel) that are riveted to the girder parts. The channels are inlaid in a cantilever, on which corrugated iron flooring is placed. The space between the girder elements is covered by a metallic sheet. Communications facilities are laid under the flooring of sidewalks.

A coating is applied over the bridge roadway that consists of a flashing, draining triangle, and a 7 cm. thick layer of asphalt concrete. The roadway is protected by a 0.75 m high metallic barrier.

The bridge railing is metallic, riveted, and specially designed. On the underside walls of a support 10, the railing is reinforced with concrete parapets. On support 6, the railing is metallic, and mounted on sidewalk blocks. On the up-river side of the bridge, the light towers made of metallic cone-shaped pipes are mounted. Over supports 0, 1, 5, and 6, deformation welds are covered with sliding metallic sheets at the level of a coating.

Under a static loading of the bridge 13, the modes of sequential loading of all girders were realized. As a test loading, two tipper "KrAZ" trucks were used, loaded with ballast weighing 270 kN each (Fig. 6.31). The deflections (vertical displacement) were measured in the middle part of girders 0–1, 1–2, and 5–6 by drum-gear deflectometer PAO-6 of the Aistov system; stresses (relative deformations)—in the middle of girders 0–1, 1–2, 5–6 at the top and bottom shelves of the main bars—were measured by Aistov electromechanical strain gauges, and micro-indicators with a 200 mm base.

The bridge was dynamically loaded by the "KrAZ" tipper trucks with a ballast that was moving at various speeds on the bridge roadway, having passed the barrier made of a wooden squared beam of the 10 × 10 cm cross-section, stopped and were unmovable across the middle of a girder. The following loads were realized: one "KrAZ" tipper truck was driven at speeds of 10, 20, 30 km/h in the direction from support 6; another "KrAZ" truck moved across the 10 × 10 cm barrier at 20 km/h.

Fig. 6.31 Loading the bridge span 1–2 by two “KrAZ” tipper trucks

Under all modes of dynamic and statistic loadings, the AES were recorded from bearing structures of the bridge, as in previous cases, using the “AKEM” program complex that was designed for a personal computer that used “PCLabCard” technology. To evaluate the dynamic characteristics, there a VIBROPORT 30 device equipped with a computer produced by the “SCHENCK” company (Germany) was used. A vibration sensor was mounted on the top boom of the girder in the middle of the investigated bridge span (0–1, 1–2 and 5–6) [68].

The recording and analysis of AES during static testing began at the moment the truck stopped in the places determined by testing charts. Information was recorded for 40…60 s. The processing of the criterial estimates of the AES caused by cracks was also done using the K_p factor as in the aforementioned research.

As a result of static and dynamic testing of the bridge across the Prut River on the Zhytomyr–Chernivtsi highway, it was found that bridge structures do not contain defects and damage that could reduce their bearing capacity. To lengthen the lifespan of a bridge, it is necessary to finalize the studies on corrosion protection of metallic structures of a bridge using modern technologies and materials and do the repairs of bridge supports by arranging protective iron concrete shirts. It is recommended to set the limit of the load carrying capacity of the bridge with a maximal admissible temporary load of up to 300 kN per one span.

6.9 Prospects for Further AE Application

Thus, as numerous published data show, the second half of the twentieth century gave an incentive to the development and application of the AE phenomenon for NDT and technical diagnostics of various products [35, 69, 70]. For instance, already in 1965 the Aerojet-General Company (USA) under NASA contract performed hydrostatic testing of the casing of a SL-1 rocket engine with a diameter of 6.6 m produced by the Tiocol Company [35, 70]. Using the AE method, the moment of sub-critical crack initiation and propagation under pressure that constituted 56% of admissible value was detected. The method of triangulation made it possible to determine the fracture location with an accuracy of 305 mm. This promoted the application of the AE method in research on the strength characteristics of structural materials, their fracture and metal science problems, and consequently for the determination of the coordinates of developing defects [71]. Further research effectively used the methods of AE diagnosing for testing the state of wing crossbars in aircraft [72] and in other elements of aviation structures in which initiation and propagation of fatigue cracks is possible [73].

Progress in microelectronics and, especially, in the development of a theoretical basis of the AE method, urged further studies. Estimating the state of rocket engines by 52-channel AE equipment [74] showed that using the criterion approaches, the defects can be divided into three groups: (a) safe; (b) those that need inspection and study; and (c) obviously dangerous. Publications appeared where the AE method was effectively used for the quality control of rods [75, 76] under the action of temperature and mechanical loading [77–79], and in gas flows [80]. The AE methods for determining the characteristics of structural materials, especially their crack growth resistance [81, 82], for locating the defects [83, 84], for determining their orientation [85], especially during exploitation of the objects [86], are still urgent.

Paper [87] describes the laboratory AE research of crack initiation from the holes for rivets under fatigue loading. The corrosion of rivets was accompanied by the AE that was recorded in the frequency band below 150 kHz. Therefore, the AE caused by fatigue fracture was recorded by the AET at the natural resonance frequency of 450 kHz. Basic information on the mechanisms of the fatigue fracture of rivets was obtained by amplitude distribution of the AES. Signals differ by a short time of amplitude growth, which is typical of crack growth. The distribution of the number of AE pulses vs. the number of loading cycles was calculated.

It is also shown there that there are other methods of utilizing the AE in diagnosing the state of structures and products. The most widespread are the methods of the AE inspection of leakages in vessels that operate under pressure [69–71, 87]. The Kaiser effect is most often used for this purpose, because the AES arise from the very start of a crack growth. They are recorded either during primary application of pressure in a vessel or under the repeated loading above the previously attained maximum level. During such tests, noises arising due to the leakage of liquids and gases from vessels give a substantial contribution to the recorded AES. Even at a slight leakage, the AE can be higher than the AE recorded during a crack growth. Therefore, in conditions of leakage detection by means of the AE, it is necessary to eliminate the leakage and then perform a subsequent AE testing of the object. The state of deep-diving into a submersible chamber was examined [87] using this method.

Pipelines and pipeline systems can be diagnosed using the AE method during hydrostatic testing by applying overpressure [88, 89]. The AET location depends on the thickness and diameter of a pipe, the type of protective layer, the composition of the soil (for an underground pipeline), the working environment in a pipeline, etc.

The wide use of the AE test methods for industrial equipment is also recognized [87], apart from those described in this monograph. Among them are the methods for assessing the state of the equipment for oil catalytic cracking, hydro-cracking, blast furnace, autoclave of pulp and paper production, mine fastening, cryogenic vessels for the storage of ammonia, spherical vessels, products of reinforced glass-fiber material, hydrolysis equipment, devices for hydrothermal treatment, absorption tower, materials of clock springs, surface coatings, quality of mechanical and heat treatment of components, and food, as well as honeycomb elements of airplane structures, buildings, geological processes, and others.

When analyzing the state and prospects of the development of studies using the AE phenomenon to estimate damaged products, it is necessary to distinguish between the following basic directions:

- Development of a theoretical basis in order to establish correlations between the AES parameters and the defects, and the elaboration of effective experimental methods for determining the crack growth resistance of structural materials;
- Creation and verification of new methods of detection of defects in structures during semi- or full-scale testing, which, in turn, use the results of theoretical and experimental research; and
- Development and production of new facilities for loading and for extracting the AE signals, as well as units for their processing in order to perform the NDT and technical diagnosing of the IO.

References

1. Webborn TJC, Stewens PG (1983) On-line monitoring by acoustic emission. In: 4th national conference on condition monitoring, London, 15–16 Mar 1983
2. Rogers LM, Monk RG (1983) In service monitoring of structural integrity by acoustic emission analysis. In: 2nd national conference on condition monitoring in the process industries, London, 10–11 May 1983
3. (1981) The EWGAE AE code for acoustic emission examination of sources of discrete acoustic events. NDT Int 14(8):181–183
4. Rogers LM (1978) Application of acoustic emission source location to on-line condition monitoring of fabrication, machinery and process plant. In: International physics conference on machine aided image analysis, Series No. 44, cr. 4., London, 5–8 Sept 1978
5. Rogers LM (1985) Measurement and interpretation of the acoustic emission from propagation cracks in steel structures. In: Proceeding of British Society for Strain Measurement: annual conference "structurual integrity." Lancaster, 10–13 Sept 1984. Newcastle upon Tyne
6. Tuikin OP, Ivanov VI (1985) Faktornyy analiz ustoychivosti parametrov akusticheskoy emissii (Factor analysis of acoustic emission parameters stability). Defektoskopia 8:39–44

7. Johnson N, Lion F (1981) Statistika i planirovanie eksperimentov v technike i nauke. Metody planirovaniya eksperimenta (Statistics and planning of experiments in engineering and science. Methods of experiments planning). Moskva, Mir
8. Dunegan HL, Harris DO, Tatro CA (1968) Fracture analysis by use of acoustic emission. Eng Fract Mech 1(1):105–122
9. Ivanov VI, Belov VM (1981) Akustiko-emissionnyy kontrol' svarki i svarnych soedineniy (Acoustic emission testing of welding and welded joints). Mashinostroyeniye, Moskva
10. Ivanov VI, Bykov SP (1985) Klassifikaziya istochnikov akusticheskoy emissii (Classification of acoustic emission sources). Diagnostika i prognozirovanie razrusheniy svarnych konstrukziy (Diagn Predict Fail Welded Struct) 1:67–74
11. (1976) Standard recommended practice for acoustic emission monitoring of structures during controlled stimulation. ASTM E, pp 569–576
12. Watanabe T, Huchirizaki S, Arita HA (1978) Method of evolution the harmfulness of flaws in structures using acoustic emission techniques. The fourth acoustic emission symposium, Tokyo
13. (1980) Acoustic emission testing of spherical pressure vessel made of high tensile strength steel and classification of test results. NDIS 2412:6–8
14. Vald A (1960) Posledovatel'nyy analiz (Sequential analysis). Fizmatgiz, Moskva
15. Vakar KB (ed) Akusticheskaya emissiya i ee primenenie dlya nerazrushayuschego kontrolya v atomnoy energetike (Acoustic emission and its application for NDT in nuclear power engineering). Atomizdat, Moskva
16. Hutton PH, Kurtz RJ, Pappas RA (1984) AE/Flaw characterization for nuclear pressure vessels. In: Review of progress in quantitative nondestructive evaluation: proceedings of 10th annual review, Santa Crus, California, 7–12 Aug 1983
17. Hutton PH, Kurtz RJ (1985) Acoustic emission for on-line reactor monitoring. Results of intermediate vessel test monitoring and reactor hot functional. In: Review of progress in quantitative nondestructive evaluation: proceedings of 11th annual review, San Diego, California, 8–13 July 1984
18. Kuranov VN et al (1981) Ispol'zovanie metoda akusticheskoy emissii pri ozenke konstrukzionnoy prochnosti sosudov vysokogo davleniya i truboprovodov (Application of acoustic emission method in evaluation of structural strength of pressure vessels and pipelines). In: Trudy ZNIITMash (Proceedings of Central Research Institute of Engineering Technology (CNIITMash)), vol 165, pp 3–12
19. Troitskyi VA et al (1986) Nerazrushayuschiy kontrol' kachestva svarnych konstrukziy (Nondestructive testing of welded structures quality). Tekhnika, Kiev
20. Paton BY et al (1982) Perspektivy razvitiya kontrolya kachestva s ispol'zovaniem EVM pri proizvodstve svarnych trub bol'shogo diametra (Prospects of development of quality testing with the use of computer in production of welded pipes of large size). Automaticheskaya svarka 10:40–47
21. Skalskiy VR, Serghiyenko OM, Holaski L (1999) Generuvannya akustichnoyi emisiyi trischinami, scho rozvivayut'sya u zvarnich z'yednannyach (Generation of acoustic emission by cracks which develop in the welded joints). Tekhnichskaia diagnostika i nerazrushajuschii kontrol (Technical diagnostics and nondestructive testing) 4:23–31
22. Trufiakov VI (1991) Puti povysheniya nadezhnosti pri odnovremennom snizhenii metalloemkosti svarnych konstrukziy (Ways of improvement of reliability with simultaneous decrease of specific content of metals of welded structures). Ibid 3:3–8
23. Troitskiy VA, Rad'ko VP, Demidko BG (1983) Defekty svarnych soedineniy i sredstva ich obnaruzheniya (Defects in welded joints and tools for their detection). Vyshcha shkola, Kiev
24. Nedosieka AYa (2001) Osnovy rascheta i diagnostiki svarnych konstrukziy (Basis for calculation and diagnostics of welded structures). Indprom, Kiev
25. Makarov EL (1981) Cholodnye treschiny pri svarke legirovannych staley (Cold cracks in of alloyed steels welding). Mashinostroyenie, Moskva
26. Smiyan OD (1985) Prognozirovanie obrazovaniya i razvitiya cholodnych treschin v konstrukzionnych materialach s pomosch'yu segregazionnych kart primesey vnedreniya

(Prediction of cold cracks initiation and propagation in structural materials with the use of segregation maps of interstitial impurities). Diagnostika i prognozirovanie razrusheniya svarnych konstrukziy (Diagnostika i prognozirovanie razrushenia svarnykh konstruktsii) 1:59–67

27. (1985) GOST 25.506–85. Raschety i ispytaniya na prochnost'. Metody mecha-nicheskich ispytaniy metallov. Opredelenie charakteristik treschino-stoykosti (vyazkosti razrusheniya) pri staticheskom nagruzhenii. Vved. v deystvie 27.03.1985g. (State Standard 25.506–85. Calculation and testing for strength. Methods of materials mechanical testing. Determination of crack growth resistance characteristics (fracture toughness). Implemented 27.03.1985). Izdatel'stvo standartov, Moskva
28. (1985) GOST 1497-84 Metally. Metody ispytaniy na rastyazhenie (Standard 1497-84. Metals. Methods of tensile testing). Izdatel'stvo standartov, Moskva
29. Skalskiy VR et al (1998) Pristroyi i ustanovki dlya vyznachennya trischynostiykosti konstrukziynych materializ metodom akustychnoyi emisiyi (Equipment and devices for evaluation of crack growth resistance of structural materials by the method of the acoustic emission). Preprint, NAN Ukrayini, Fizyko-mechanichnyi institut, 1(1998), L'viv
30. Andreykiv AYe et al (1990) Metodicheskie aspekty primeneniya metoda akusticheskoy emissii pri opredelenii staticheskoy treschinostoykosti materialov (Methodical aspects of application of the acoustic emission method for evaluation of static crack growth resistance of materials). Preprint, NAN Ukrayini, Fizyko-mechanichnyi institut, 165(1990), L'viv
31. Lysak MV, Skalskiy VR (1997) Metodychniy pidkhid dlya eksperymental'noyi akustyko-emisiynoyi ozinky trischynotryvkosti konstrukziynych materialiv (Methodical approach for experimental acoustic emission evaluation of the crack growth resistance of structural materials). Fizyko-chimichna mechanika materialiv (Physicochem Mech Materi) 5:17–29
32. Pustovoi VM et al (1990) Vplyv nyz'kych temperatur na statychnu ta dynamichnu trischynostiykist' materialiv pidiymal'no-transportnych mechanizmiv (The effect of low temperatures on static and dynamic crack growth resistance of materials of lifting-transport mechanisms). Ibid 6:80–84
33. Skalsky VR (1996) Investigation of AE properties of the hoists. In: Panasyuk VV (ed) Advances in fracture resistance in materials. Tata McGraw-Hill Publishing Co. Ltd, New Delhi, pp 639–644
34. Andreikiv OY, Skalskyi VR, Lysak MV (1994) Sposib kontrolu rostu trischyn u zrazkakh materialiv (A method of checking the growth of cracks in the material specimens). Patent of Ukraine N2914, MPK: G01N29/14, Bul. 5-1, 26 Dec 1994
35. Greshnikov VA, Drobot YuB (1976) Akusticheskaya emissiya (Acoustic emission). Izdatel'stvo standartov, Moskva
36. Skalskiy VR et al (1998) Doslidzhennya za sygnalamy AE prozesiv ruynuvannya u konstrukziynych stalyach ASO (Researches of fracture processes in artillery infantry armament structural materials using the AE signals). In: Zbirnyk praz' II Mizhnar. konf. "Artyleryis'ki stvol'ni systemy, bojeprypasy, zasoby artyleryis'koyi rozvidky ta keruvannya vognem" (Proceedings of II international conference on "Artillery trunk systems, ammunitions, facilities of artillery reconnaissance and fire-control", Kiev, 27–29 Oct 1998)
37. Mykytyshyn SI, Hrytsyshyn PM (1981) Sposob mekhanicheskikh ispytaniy obraztsov na prochnost' (A method of mechanical strength testing of specimens). USSR Inventor's certificate, 879373, G01N3/00, Bul. No. 41, 07 Nov 1981
38. Andreykiv AY et al (1993) Spektral'nyy analiz signalov akusticheskoy emissii rastuschey treschiny (Spectral analysis of acoustic emission signals of a growing crack). Tekhnichskaia diagnostika i nerazrushajuschii kontrol (Tech Diagn Nondestr Test) 3:75–84
39. Climent FJ, Cribbs R, Eckert T (1968) Nondestructive testing and performance monitoring techniques (Updated instrumentation techniques for monitoring embankmens and concrete structures for civil works). In: Final report on contract SA N.o 157161 with the State of California, Department of Water Resources. Aerojet-General Corporation. Sacramento, California, Nov 1968

40. Li ST, Ramakrishnan V, Russell JE (1970) Where stands nondestructive testing of concrete and whither? Int J Nondest Testing 2:281–300
41. Muenow RA (1973) Large scale applications of acoustic emission. IEEE Trans Sonics and Ultrasonics 20(48):85–96
42. Sevall GW Jr (1973) Nondestructive testing of construction materials and operations. Champaign, Illinois
43. Galambos CF, McGorney CH Opportunities for NDT of highway structures. Mater Eval 33 (7):168–175
44. Hardy HR Jr, Belesky RM, Ge Maochen (1988) Acoustic emission/microseismic source location in geotechnical applications. In: Proceedings of 4th European conference, London, 13–17 Sept 1987, vol 2. Oxford, 1988, pp 3066–3075
45. Muenov RA (1974) Uses of acoustic emission in the construction industry. In: Proceedings of the second acoustic emission symposium, Tokyo, Japan, 2–4 Sept 1974
46. Eitzen DG, Wadley HNG (1984) Acoustic emission: establishing the fundamentals. J Res Nat Bureau Stand 89(1):75–100
47. Otsu M (1988) Acoustic emission control of concrete constructions. Konkurito kogaku 25 (12):5–11
48. Braginskii AP et al (1984) Akustikoemissionnoe obsledovanie zheleznodorozhnych mostov (Acoustic emission investigation of railroad bridges). In: Sbornik dokladov I Vsesoyuzn. konf. "Akusticheskaya emissiya materialov i konstrukziy" (Proceedings of the first all-union conference. " Acoustic emission of materials and constructions", Rostov-upon the-Don, 11–13 Sept 1984), vol 1. Rostov-na-Donu
49. Sakuda T et al (1988) Acoustic emission control of bridge reinforced concrete constructions. Khikhai kensa 37(9):866–965
50. Gong Z, Nybor EO, Oommen G (1992) Acoustic emission monitoring of steel railroad bridges. Mater Eval 50(7):883–887
51. Muravin GB, Yerminson AL (1984) Ispol'zovanie akusticheskoy emissii dlya kontrolya sostoyaniya zhelezobetonnych mostov (Usage of acoustic emission for testing the state of railroad bridges). In: Sbornik dokladov I Vsesoyuzn. konf. "Akusticheskaya emissiya materialov i konstrukziy" (Proceedings of first all-union conference. "Acoustic emission of materials and constructions", Rostov-upon the-Don, 1984), vol 2. Rostov-na-Donu
52. Chausov NG et al (1998) Ispol'zovanie metoda akusticheskoy emissii dlya ekspress–kontrolya razrusheniya betonov s dobavkami plastifikatorov (Usage of the method of acoustic emission for express-evaluation of fracture of concretes with plasticizers additives). Tekhnichskaia diagnostika i nerazrushajuschii kontrol (Tech Diagn Nondestr Test) 3:12–16
53. Makhonin AI et al (1984) Akustiko-emissionnaja diagnostika mostovykh i stroitel'nykh konstruktsij (Acoustic emission diagnostics of bridges and buildings). In: Ispol'zovanie metoda akusticheskoy emissii dlya ekspress—kontrolya razrusheniya betonov s dobavkami plastifikatorov (Proceedings of first all-union conference. "Acoustic emission of materials and constructions", Rostov-upon the-Don, 1984), vol 2. Rostov-na-Donu
54. Arrington M, Evans BM (1977) Acoustic emission testing of high alumina cement concrete. NDT Int 7:81–87
55. McClung RW (1972) On nondestructive testing in the USA of prestressed concrete pressure vessels for nuclear reactors. In: Nondestructive testing for reactor core components and pressure vessels, pp 527–540
56. Gorbunov IA et al (1984) Razrabotka i vnedrenie ekspress-metodov ozenki sostoyaniya nesuschich elementov krupnopanel'nych zdaniy na osnove statisticheskogo analiza i akusticheskoy emissii (Development and implementation of express-methods of evaluation of the state of large-panel structures using statistic analysis and acoustic emission). In: Ispol'zovanie metoda akusticheskoy emissii dlya ekspress–kontrolya razrusheniya betonov s dobavkami plastifikatorov (Proceedings of first all-union conference. "Acoustic emission of materials and constructions", Rostov-upon the-Don, 1984), vol 2. Rostov-na-Donu
57. Takao U et al (1992) AE waveform analysis for rock mass in grout injection of dam. Takenaka gijutsu kenkyu hokoku 47:61–71

58. Potapov AI et al (1993) Akustikoemissionnyy kontrol' nesuschey sposobnosti merzlych osnovaniy sooruzheniy i prinzipy postroeniya avtomatizirovannych diagnosticheskich sistem (The acoustic emission control of bearing capacity of frozen foundations of buildings and principles of construction of automated diagnostic systems). Defektoskopia 3:22–26
59. Muravin GB et al (1991) Metod akusticheskoy emissii v issledovaniyach podvizhki gruntov. (Obzor) (Method of acoustic emission in investigations of soils moving (a review)). Ibid 11:3–17
60. Lantukh-Liashchenko AI (1998) Problem of creation of the national system of bridges exploitation. In: Proceedings of Ukrainian scientific-practical interbranch seminar "modern problems of planning, building and exploitation of constructions on the roads". Kiev
61. Bondar' NG. (1986) Kak rabotayut mosty? (How do bridges work). Naukova dumka, Kiev
62. Vinogradskii DYu, Prudenko YuD, Shruratovskii AA (1985) Ekspluataziya i dolgovechnost' mostov (Operation and life time of bridges). Budivelnyk, Kiev
63. Strakhova NS (1992) Ekspluataziya ta rekonstrukziya mostiv (Exploitation and reconstruction of bridges). NMK VO, Kiev
64. Vasyliev AI, Polevko VP (1995) Dolgovechnost' zhelezobetonnych mostov i mery po uvelicheniyu sroka ich sluzhby (Life time of iron-concrete bridges and measures for improvement of their service life). Avtomobilnye dorogi 9:30–32
65. Koval PM (2003) Vikoristannya metodu akustychnoyi emisiyi pry doslidzhenni mostiv (Use of the acoustic emission method in investigation of bridges). Avtoshliakhovyk Ukrainy 1:34–37
66. Filonenko SF (1999) Akusticheskaya emissiya. Izmereniya, kontrol', diagnostika (Acoustic emission. Measuring, testing, diagnostics). KNUTCA, Kiev
67. Koval PM, Stashchuk PM, Fal AY (2003) Doslidzhennya prohonovoyi budovy novoho stalezalizobetonnoho avtodorozhn'ogo mosta z vykorystannyam metodu akustychnoi emisii (Research of a span structure of new steel iron-concrete motor bridge with the use of the acoustic emission method). Diagnostyka, dovgovichnist' ta rekonstrukziya mostiv i budivel'nych konstrukziy (Diagn Durab Reconstr Bridges Build Constr) 5:85–93
68. Koval PM et al (2003) Vivedennya z avariynoho stanu metalevoho mosta cherez riku Prut v m. Chernivzi (Getting out of metallic bridge across the Prut river in Chernivtsi from the emergency state). Ibid 5:72–84
69. Green A, Lockman C, Steek R (1964) Acoustic verification of structural integrity of polaris Chambers. Mod Plast, 137–139, 178, 180
70. Drouillard T (1979) Acoustic emission: a bibliography with abstracts. IFU/PLENUM, New York
71. Tatro CA, Liptai RG, Harris DO (1971) Acoustic emission technique in materials research. Int J Nondestr Testing 3(3):215–275
72. Shnaiderman IA (1974) Novye nerazrushayuschie metody kontrolya kachestva i technicheskoy diagnostiki. Obzor (New nondestructive methods for quality control and technical diagnostics. A review). Zarubezhnauya radioelectronika 5:74–82
73. Green JS, Toney BW (1972) Acoustic monitoring of airframe structural proof testing. J Environ Sci 15(1):20–23
74. Kishi T, Mori Y (1979) Evaluating the severity of rocket motor case during burst test using acoustic emission. In: Acoustic Emission Monitoring Pressurized Systems Ft Lauderdal, Fla, pp 131–148
75. Mashechkov VV, Muravin GB (1982) Akustiko emissionnyj sposob kontrolia izdelij sterzhnevoj i trubchtoj formy (Acoustic emission control method of products in the form of a rod and a pipe). USSR Inventor's certificate 905781, G01N29/04, Bul. No. 6
76. Green EJ, Rogers LM (1987) Location of discrete sources of acoustic emission in complex tubular joints. In: Non-destructive testing: proceedings of 4th European conference, London, 13–17 Sept 1987
77. Rapoport YM (1982) Sposob kontrolia kachestva izdelij (A test method for product quality). USSR Inventor's certificate 905777, G01N 29/04, Bull. No. 6
78. Zaretskii-Feoktistov GG, Rapoport YM (1982) Sposob opredelenija parametrov naibol'shego termocheskogo vozdejstvija perenesennogo izdeliem (A method for determination of

parameters of the largest thermal effect endured by a product). USSR Inventor's certificate 954875, G01 N29/04, Bul. No. 32
79. Andronov VM, Korshak VF, Kuznetsova RI, Nerubenko VV (1990) Sposob kontrolia narushenija sploshnosti pri termicheskom vozdejstvii na splavy (A method for integrity disturbance testing under thermal effect on alloys). USSR Inventor's certificate 1599758, G01N29/04, Bul. No. 38
80. Nechaiev YuA. (1982) Sposob nerazrushauschego kontrolia poverkhnostnykh defektov izdelij (A method for nondestructive testing of surface defects in products). USSR Inventor's certificate 911324, G01 N29/04, Bul. No. 9
81. Nakasa H (1980) Application of acoustic emission techniques to structural integrity assessment. In: Acoustic emission paper meet, Bad Nauheim, Apr 1979. Oberursel
82. Yamaguchi K, Oyaizu H (1988) Distributed fracture monitoring system by high speed processing of acoustic emission microdata. In: Non-destructive testing: Proceedings of 4th European conference, London, 13–17 Sept 1987, vol 4. Oxford
83. Tirbonod B, Hanacek L (1988) Some properties of acoustic emission signals measured in the vicinity of and crack during the cyclic pressure loading of and vessel. Ibid
84. Andreykiv OYe, Lysak NV (1989) Metod akusticheskoy emissii v issledovanii prozessov razrusheniya (A method of acoustic emission in investigation of fracture processes). Naukova Dumka, Kiev
85. Ohtsu M (1987) Determination of crack orientation by acoustic emission. Mater Eval 35 (9):1070–1075, 1082
86. McBride SL et al (1988) Acoustic emission detection of crack presence and crack advance during flight. Rep Progr Qnant Nondestract Eval 8B:1819–1825
87. Collacott RA (1989) Diagnostika povrezhdeniy (Diagnostics of Damages). Mir, Moskva
88. (2001) Rekomendazii schodo akustyko-emisiynoho kontrolyu ob'yektiv pidvyschenoyi nebezpeki. R 50.01-01. Vved.27. 01. 2001 r. (Recommendations on acoustic-emission testing of the highly dangerous objects. P 50.01-01. Introduced 27. 01. 2001)
89. (2003) Nazional'nyi standart Ukrayini DSTU 4227–2003. Nastanovy schodo provedennya akustyko-emisiynoho diagnostuvannya obyektiv pidvyschenoyi nebezpeki. Chynnnyi vid 2003 – 12 – 01 (National standard of Ukraine 4227–2003. Instructions on conducting acoustic emission diagnostics of the highly dangerous objects. Valid since 01 Dec 2003)

The manufacturer's authorised representative in the EU is Springer Nature Customer Service Centre GmbH, Europaplatz 3, 69115 Heidelberg, Germany. If you have any concerns regarding our products, please contact ProductSafety@springernature.com

Printed and bound by CPI Group (UK) Ltd, Croydon, CR0 4YY
30/09/2026
02225230-0005